The historical development of the important discoveries concerning the plasmasphere in the 1960s are presented in this monograph by the pioneering discoverers themselves. The plasmasphere is the vast 'doughnut-shaped' region of the magnetosphere, filled with ions and electrons of ionospheric origin trapped along geomagnetic field lines, forming a cold thermal plasma cloud encircling the Earth to an equatorial distance of 30 000 km. At this radial distance the plasmasphere terminates abruptly over a sharp surface of discontinuity or 'knee' in the magnetospheric plasma density distribution known as the plasmapause.

The volume commences with an as yet unpublished historical account of the difficulties met in the USSR by Gringauz to release his early discoveries from Soviet rocket measurements, and the contemporaneous breakthroughs in the USA by Carpenter from ground-based whistler measurements. The monograph goes on to present experimental results of the past three decades obtained from spacecraft and whistler observations, to discuss plasma physical processes, and to outline an up-to-date picture of this wide magnetospheric region. Mathematical and physical theories proposed to explain the formation of the plasmasphere and plasmapause are reviewed and discussed.

This monograph is the first to describe the historical development of ideas, experiments, observations and models of the plasmasphere, bringing plasmaspheric research up-to-date, and as such will be invaluable for researchers in space physics. It will also appeal to those interested in the history of science.

The Earth's Plasmasphere

Cambridge Atmospheric and Space Science Series

EDITORS

Alexander J. Dessler
John T. Houghton
Michael J. Rycroft

OTHER TITLES IN PRINT IN THE SERIES

M. H. Rees
Physics and chemistry of the
upper atmosphere

Roger Daley
Atmospheric data analysis

Ya. L. Al'pert
Space plasma, Volumes 1 and 2

J. R. Garratt
The atmospheric boundary layer

J. K. Hargreaves
The solar-terrestrial environment

Sergei Sazhin
Whistler-mode waves in a hot plasma

S. Peter Gary
Theory of space plasma
microinstabilities

Martin Walt
Introduction to geomagnetically
trapped radiation

Tamas I. Gombosi
Gaskinetic theory

Boris A. Kagan
Ocean-atmospheric interaction and
climate modelling

Ian N. James
Introduction to circulating atmospheres

J. C. King and J. Turner
Antarctic Meteorology and Climatology

The Earth's Plasmasphere

J. F. Lemaire
*Institut d'Aéronomie Spatiale de Belgique
and Université Catholique de Louvain*

K. I. Gringauz
Formerly of Space Research Institute, Moscow

With contributions from
D. L. Carpenter
*Space, Telecommunications and Radioscience Laboratory,
Stanford University*
V. Bassolo
Space Research Institute, Moscow

CAMBRIDGE
UNIVERSITY PRESS

CAMBRIDGE UNIVERSITY PRESS
Cambridge, New York, Melbourne, Madrid, Cape Town, Singapore,
São Paulo, Delhi, Dubai, Tokyo, Mexico City

Cambridge University Press
The Edinburgh Building, Cambridge CB2 8RU, UK

Published in the United States of America by Cambridge University Press, New York

www.cambridge.org
Information on this title: www.cambridge.org/9780521430913

First published 1998

A catalogue record for this publication is available from the British Library

Library of Congress Cataloguing in Publication Data

Lemaire, J. F.
    The earth's plasmasphere / J. F. Lemaire and K. I. Gringauz with
contributions from D. L. Carpenter and V. Bassolo.
    p.   cm. – (Cambridge atmospheric and space science series)
    Includes bibliographical references and index.
    ISBN 0 521 43091 7
    1. Plasmasphere.   I. Gringauz, K. I. II. Title.   III. Series.
QC809.P5L46   1997
538'.766–DC20   96-38481 CIP

ISBN 978-0-521-43091-3 Hardback
ISBN 978-0-521-67555-0 Paperback

Contents

Preface

This book is much more than a monograph about a scientific topic; it also provides a historical account of the growth of a new field of research by some of its pioneers. As such it is a case study of the long road from observations to phenomenological description culminating in true physical understanding typical for the geophysical sciences. It also illustrates the strong dependence on international collaboration, in this particular case the stimulus provided by the International Geophysical Year (IGY). This field of research grew out of ground-based whistler observations conducted during the IGY on the one hand and the first *in situ* measurements in the space environment surrounding the Earth made possible with the concurrent advent of the space age, on the other.

The theme of this book, the plasmasphere of the Earth, had its beginning in the study of the Earth's ionosphere, the thermal (cold) plasma originating from the interaction of solar ionizing radiation with the Earth's neutral atmosphere. As a long-time practicioner in this discipline, I often was frustrated in my early years when asked where the upper boundary of the Earth's ionosphere was, or when reading in publications some completely arbitrary altitudes assigned to such a boundary. With the new concept of the Earth's magnetosphere it first appeared probable that cold plasma originating in the ionosphere, because of the magnetic control of charged particles, might extend throughout the closed geomagnetic field lines up to the magnetopause. When whistler observations made by one of the contributors to this book (Don Carpenter) indicated that the electron densitiy distribution in the equatorial plane exhibited a precipitous drop well within the magnetosphere, it appeared at first difficult to accept this 'whistler knee' as a physical reality. Independently, the *in situ* observations of thermal ions on Russian satellites and space probes by one of the co-authors (the late Konstantin Gringauz) showed the same behaviour, although there was a time when *in situ* measurements of thermal electrons on American spacecraft appeared to contradict (erroneously) the Russian data. From the early whistler observations a direct correlation between the location of the 'knee' in the thermal plasma distribution (today called the plasmapause) with magnetic activity was found. With the then known connection between magnetic activity and the solar-wind interaction with the geomagnetic field it became obvious that the boundary of the thermal plasma distribution must be a direct consequence of the solar-wind interaction responsible for convective motion within the magnetosphere. From the empirical relationship between the L-shell of the plasmapause and the planetary magnetic index K_p it appeared that the average quiet time plasmapause was located at about $L = 5.8$. I remember the excitement I felt when reading a paper by the first author of this book (Joseph Lemaire) explaining this fact as the corotation limit of the thermal plasma constrained by the Earth's dipole field lines.

This limit is the result of the fact that the balance between gravitational attraction and centrifugal force (still appreciable for the cold plasma) which in the unconstrained case corresponds to 6.6 Earth radii (the so-called geostationary orbit) is partly compensated by the electromagnetic force along the field line. Thus, the field-constrained corotation reduces this equatorial distance by the factor $(2/3)^{1/3}$ leading to the observed value of the quiet time plasmapause. This fact, caused by the balance between the field aligned projection of the gravitational force, to 2/3 by the centrifugal and to 1/3 by the electromagnetic force has been called the 'Two-Thirds Law' by H. Alfvén who has used this concept in his theory of the evolution of the solar system. Although it was first published by him in ICARUS, a journal of planetary science, already in its early years of existence, it then certainly was not generally read by plasma physicists concerned with the Earth's ionosphere/magnetosphere, and thus unknown to Lemaire who in his interpretation of the quiet time plasmapause independently arrived at the same conclusion. This is but one example commonly occurring in new fields of research, that earlier contributions in another area are not always in the realm of general knowledge. Another example that comes to my mind is the altitude distribution of a minor ion in an ion mixture first observed in the Earth's ionosphere which, because of the presence of the polarization field, is quite different from that of the equivalent neutral species. This law had to be 're-invented' by scientists working on the physics of the Earth's upper atmosphere, although a detailed academic discussion of this problem relevant to stellar atmospheres can be found decades before in Eddington's famous book, *The Internal Constitution of the Stars*. These two examples serve but as a reminder that, particularly in present days, broad general knowledge is the exception rather than the rule, and that in a new area of research, practising scientists often have to find their own road to truth.

This book can be highly recommended not only to anyone who wants an up-to-date account of the Earth's plasmasphere, but also to those who want to relive the ups and downs, the wrong starts and controversies that finally lead to a satisfactory understanding of the physical environment of our planet Earth.

S. J. Bauer

Foreword

The origin of this monograph is a thesis entitled 'Frontiers of the Plasmasphere' that I submitted in 1985 at the Université catholique de Louvain in fulfillment of the 'Agrégation de l'Enseignement Supérieur'. This D.Sc. thesis described a new physical theory for the formation of the plasmapause.

As a result of this work, Professor M. J. Rycroft, Editor of the Cambridge Atmospheric and Space Science Series, asked me to prepare a monograph on the Plasmasphere. I was honoured by this proposal, but I wanted to decline it in view of the formidable effort that this project would involve, the time that it would take to review the large body of observations as well as the set of controversial theories put forward over twenty years, and then the time needed to compile a comprehensive synthesis all that material. But both the Editor and the Publishing Director of Cambridge University Press (CUP) argued that there was no topical monograph on the Earth's plasmasphere currently on the market and that such a book would be useful in Space Science Laboratories and their Libraries. Since the referees consulted by CUP were also very positive about such a project, I finally accepted.

Of course, a comprehensive monograph on the Earth's plasmasphere should not only describe theoretical aspects as in my thesis. It needed to contain a comprehensive review of the observations collected in the plasmasphere and at the plasmapause. These observations come mainly from whistler as well as from *in situ* satellite measurements. Only experimentalists who themselves had contributed to observations of the plasmasphere could be responsible for this important part of the monograph.

Fortunate circumstances led me in 1985 to meet in Moscow the late Professor

K. I. Gringauz from the Institute for Space Research (IKI). He is the renowned scientist who adjusted the antennae of SPUTNIK 1 just before its launch in 1957. He discovered, among many other outer space plasma frontiers, a sharp drop in the magnetospheric plasma density distribution which, later on, received the name 'plasmapause'. This historical discovery by Gringauz was made with particle detectors flown on board LUNIK 1 and 2, Soviet lunar probes traversing the distant Earth's environment on their journey to the Moon. Therefore, provided he would agree, Konstantin Gringauz would be the right co-author of a monograph on the plasmasphere.

During a meeting in Brussels, November 1990, K. I. Gringauz announced to me his agreement to co-author this publication. At the same time he suggested that Valya Bassolo, one of his co-workers at IKI, be associated with writing Chapter 3 on all relevant *in situ* satellite observations of the plasmasphere from the very early Soviet space probes up to the much more complex American DYNAMICS EXPLORER or Japanese AKEBONO missions.

For the ground-based whistler observations of the plasmasphere, we felt that D. L. Carpenter, from the STAR Laboratory at Stanford University, CA, would be fundamental. D. L. Carpenter could recall accurately the important and pioneering work based on whistler measurements. Indeed, he discovered the 'knee' in the equatorial plasma density distribution from early whistler observations, and in 1966 named this new magnetospheric frontier the 'plasmapause'.

On 22 November 1990 we called Don Carpenter from Brussels. He welcomed the idea of a monograph on the plasmasphere. He agreed to help us in the project and to contribute the chapter on radioelectric sounding observations and plasma wave observations. However, his agreement was subject to the condition of not being co-author. The Editor agreed that the words: 'with contributions from D. L. Carpenter and V. Bassolo' should be included in the jacket. These became the terms of reference which were agreed by all of us.

I wish to address here my deepest gratitude to Don Carpenter not only for agreeing to write the important Chapter 2 on radio sounding observations, but also for his judicious advice during the preparation of this book and for his help in editing the manuscript. I enjoyed very much Don's visits to the Institut d'Aéronomie Spatiale de Belgique (IASB) where he came several times to work on this monograph.

We decided to report in Chapter 1 the stories behind the discovery of the plasmapause, in the former Soviet Union and in the United States. They are not well known and have never been published before. We thought it interesting to report, for instance, the difficulties that Gringauz met when publishing his discovery. It is also relevant for history to recall where and when this discovery was first presented to the public at the 1963 URSI Assembly in Tokyo where Carpenter and Gringauz first met. The first chapter of this book recalls also the controversies and objections that this discovery immediately provoked. Who better than K. I. Gringauz and D. L. Carpenter to write these invaluable contributions to the history of science?

Unfortunately, Konstantin Gringauz passed away on 10 June 1993. When I called him for the last time, two days before he died, his voice was feeble and poignant in tone. I reassured him that the main parts that he had already provided were almost in final form and that his contribution would be published soon. So, with Konstantin Gringauz's demise, the first part of Chapter 1 constitutes, in some manner, part of his scientific testament and his view of events as he lived them in Moscow in the early days of the space race between the Soviet and the US Space Agencies.

The way that events developed independently in the Western World, culminating in Carpenter's historical discovery of the plasmapause from whistler observations, is revealed in the second part of Chapter 1.

Those who are interested in the history of modern science will find in this chapter an invaluable source of information and references. Many of these stories are unknown to most space scientists, even to most of those who have been working in the field since the early 1960s. The younger generation of scientists may also learn in reading how, in those days, investigation of the frontiers of space physics was challenging. I wish to thank Siegfried Bauer, Don Carpenter, Marius Echim, Anatoly Remizov, Michael Rycroft and Bernie Shizgal for reading this text, or part of it, and improving its comprehension.

It took much more time than originally expected to finalize the manuscript for hundreds of 'good and not-so-good' reasons. The long sickness and death, in 1993, of Konstantin Gringauz, was the most painful episode thwarting our project. The geographical distances between the authors and the difficulties of communication were other excuses, in addition to the normal workloads that we all endure.

Without the encouragement of friends and the worthy cooperation of colleagues from Laboratories and Institutes in all parts of the World, our task would probably not yet be finished. It is difficult to cite them all without forgetting one, perhaps the most neighborly of them. In particular, I would like to thank P. and Y. Corcuff, P. M. E. Décréau, Yu. Galperin, J. L. Horwitz, T. E. Moore and R. W. Schunk for useful discussions. Many thanks also to Carl McIlwain who provided his software subroutines to compute the electric field models E3H and E5D quoted at several places in this monograph, and to C. R. Chappell and R. H. Comfort who kindly invited me to Marshall Space Flight Center (MSFC) and University of Alabama, Huntsville, where I was offered access to the unique DE/RIMS data set and where I developed some of the software used in the numerical simulations discussed in Chapter 5 of this monograph. I wish to thank the members of the Center for Atmospheric and Space Sciences (CASS), at Utah State University, and especially Bob Schunk for inviting me to work in collaboration with him on his theory project at one of the *hauts lieux* in space plasma physics and atmospheric sciences. V. Afonin, R. R. Anderson, D. L. Carpenter, B. L. Giles, Yu. Galperin and K. Torkar kindly provided original figures to illustrate this monograph and are gratefully acknowledged for their cooperation.

Without the influence exercised by professors L. Ledoux, M. Nicolet and L. Bossy on me, and their encouragement, I would never have had the opportunity to work in the field of space physics or to contribute any monograph on the plasmasphere. I wish to testify my deepest gratitude to these great mentors.

I acknowledge also the continuous support of Baron M. Ackerman, Director of the Institute for Space Aeronomy in Brussels, and of his most efficient administrative and technical staff headed by Jean Palange, whom I warmly thank for his cooperation. Their continuous logistic support has facilitated this enterprise considerably. Let me especially point out the good work of M. De Clercq, L. Fedullo, A. Simon, J. Schmitz and E. Rigo, who did the typing, computing, drawing, and checking of references for this book, respectively.

Finally, our wives, Betty Carpenter, Irina Gringauz and Cécile Lemaire, are warmly acknowledged for bearing this book patiently.

Despite the pain and many difficulties which had to be overcome, I enjoyed working on this project with Konstantin, Don and Valya. We would all be rewarded if this compilation proves to be useful to those around the world who will come to our field and study the plasmasphere in the next decade.

Joseph Lemaire

Dedications

Ce livre est dédié à nos petits enfants Masha, Jérome et Lou. Il l'est également aux étudiants et chercheurs passionnés par l'exploration de notre planète, et, soucieux d'en étudier et préserver son environnement.

K.I.G. & J.F.L.

Introduction

The plasmasphere is the vast 'doughnut' shaped region of the magnetosphere that is filled with trapped ions and electrons of ionospheric origin; their energy is less than 1–2 eV. These charged particles are trapped on geomagnetic field lines, forming a cold thermal plasma cloud around the Earth out to geocentric equatorial distances of 4–5 Earth radii (R_E).

The outer boundary of the plasmasphere forms a rather characteristic 'knee' in the equatorial plasma density profile. This field-aligned surface is called the 'plasmapause surface or region', or more simply the 'plasmapause'. The plasmapause was discovered in the 1960s independently from *in situ* space probe measurements and from ground-based whistler observations. A first-hand account of the history of this discovery of the boundary is presented in Chapter 1. In the first part of this chapter K. I. Gringauz reports the prevailing situation in the former USSR, his design of the first ion traps flown in outer space and how he had to fight to get his experimental findings published and accepted in his country as well as in the Western World. In the second part of Chapter 1, D. L. Carpenter describes the situation in the US and the history of his discovery of the plasmapause from dynamic spectrograms of whistlers.

Whistler waves are audio frequency radio waves produced by lightning in the atmosphere. They propagate back and forth along field-aligned plasma density irregularities in the magnetosphere. Their travel time from one hemisphere to the magnetically conjugate point in the opposite hemisphere is mainly determined by the electron concentration in the distant magnetosphere where the VLF waves cross the geomagnetic equatorial plane.

In Chapter 2, D. L. Carpenter explains how this powerful observation technique, introduced by Owen Storey, was used to track the equatorial plasmapause and to determine the overall density distribution of plasma in the plasmasphere, its local time variations, its cross-L motions during quiet and disturbed geomagnetic conditions, the time constant for the refilling of magnetic flux tubes after substorm erosion events, etc. This chapter also contains additional experimental results deduced from radio and plasma wave observations with electric antennae onboard magnetospheric satellites. It emphasizes the timeliness of a mission devoted to High Altitude Radio and EUV Imaging of the Magnetosphere that would substantially increase our understanding of the plasmasphere.

In Chapter 3, K. I. Gringauz and V. Bassolo review *in situ* satellite observations relevant to the study of the plasmasphere and plasmapause. It is divided into three parts, corresponding to three steps in the design of space-borne plasma instruments. In the early days, before 1970, rather simple detectors were used as on LUNIK 1 and 2, ELECTRON-2 and 4, OGO-1, 3 and 5, and IMP-2. During the second epoch, 1970–80, a new generation of instruments was developed and used on PROGNOZ-5 and 6, GEOS-1 and 2, and ISEE-1 to collect observations in the plasmasphere. The third epoch is 1980–92, when most sophisticated experiments were flown on DYNAMICS EXPLORER, AKEBONO (EXOS-D), and more recently CRRES. When this chapter was completed in January, 1993, the new results obtained with this latter spacecraft were not yet all published or available to K. I. Gringauz, who was then already ill and hospitalized. This chapter therefore presents in historical order, the major findings obtained from *in situ* particle measurements and published before 1993.

Chapter 4 is an attempt to offer a unified phenomenological picture of what is currently known concerning the plasmasphere, the plasmapause, and the underlying upper ionosphere that is the main source of particles. It contains additional sets of observations which have not been included in the previous chapter, but which are most relevant to the study of plasmaphere–ionosphere coupling and the erosion of the plasmasphere during a geomagnetic storm.

The refilling of magnetic flux tubes by upward flow of ionization from the ionosphere is discussed in this chapter. Included are comments on the variety of hydrodynamic refilling scenarios that have been proposed since 1971, when the well-known 'top to bottom refilling' model was first proposed. These fluid transport models are compared with kinetic approaches describing the evaporation of thermal ions and electrons out of the topside ionosphere. The 'velocity filtration effect' due to the velocity-dependent Coulomb collision cross-section and the non-Maxwellian velocity distribution function are discussed. Recent Monte Carlo simulations describing these evaporation processes – with and without ion heating by wave–particle interaction mechanisms – are also discussed. All these physical processes must be taken into account to explain the large temperatures observed in the outermost part of the plasmasphere as well as the anisotropic pitch angle distributions often present in low-density flux tubes of the plasmatrough.

The plasmasphere sometimes contains unexpectedly large concentrations of helium and heavier ions. The origin of these abnormal ionic compositions is not yet understood, although some suggestions have been made in the literature.

Empirical models of the plasma density distribution in the plasmasphere and plasmatrough have been proposed by various authors. These models are based on large numbers of whistler and *in situ* satellite measurements. These models, which are useful tools to test theoretical models, are reviewed in the second part of Chapter 4. The equatorial distance of the plasmapause, its local time and K_p dependence are reviewed. The relationship between the plasmapause and other plasma boundaries in the inner magnetosphere and ionosphere are discussed. The existence of plasma density irregularities outside and inside the plasmasphere are reported.

The distribution of convection electric fields in the magnetosphere is of major importance for the plasmasphere and plasmapause. Indeed, the electric drift is most important for the low-energy charged particles forming the plasmasphere. Furthermore, the topology of electric equipotential surfaces has been assumed to determine the position of the plasmapause within the magnetosphere. The time dependence of the convection electric field is expected to determine the deformation of the plasmapause during episodes of enhanced geomagnetic activity. The radial and local time distributions of the convection electric field are usually assumed to depend on Universal Time through the changing value of the planetary geomagnetic index K_p. Although this assumption has been popular for several decades, it is shown that this might not be the best assumption. Indeed, the magnetospheric electric field intensity is known to change over time scales much shorter than that of the 3-hourly K_p-index. Furthermore, these changes in the electric field intensity are not concomitant (i.e. in phase) at midnight, noon and dusk local times. The existence of discontinuities and the formation of Alfvén layers in the nightside magnetosphere is generally not considered in most popular large-scale convection electric field models. Therefore, new types of dynamical electric field models as well as magnetic field models must be developed, based on coordinated, simultaneous and multi-point observations from the ground and from space, as well as from computer simulations of magnetospheric processes. Modelling the electric fields induced in the magnetosphere by the changing interplanetary magnetic field intensity is another challenge facing future generations of modellers.

A phenomenological description of the plasmasphere has been sketched in Chapter 4 by putting together the available pieces of the 'puzzle'. Each of these pieces is obtained from the observations presented in Chapters 2 and 3, as well as from theoretical studies. It must be admitted that many pieces are still missing and that we are not even sure that all of them have been put in the right place.

The next generations of magnetospheric missions, supported by interdisciplinary groups of physicists and engineers, will be challenged to complete the unachieved blueprint proposed in this Chapter 4.

The next chapter (Chapter 5) is devoted exclusively to theoretical aspects and modelling. Many theoretical models have been offered in the literature to represent the cold plasma density distribution in field-aligned whistler ducts, or in the equatorial plane of the magnetosphere, including the sharp 'knee' at 4–5 R_E. They are discussed in this chapter.

Unlike the case of the energetic Van Allen trapped particles, the drift velocities of plasmaspheric electrons and ions are very sensitive to the electrostatic and gravitational potential distributions. Therefore, not only the magnetic forces – determined by the gradient of the magnetic field intensity and curvature of the magnetic field lines – but also the electric, gravitational and inertial forces acting on these particles of very low energy must be taken into account to calculate their motion and density distribution in the magnetosphere.

In the ideal-MHD approximation of plasma physics, it is assumed that magnetospheric electric fields only have a component perpendicular to the magnetic field direction. But as a consequence of gravitational and thermoelectric charge separation between the heavy ions and more mobile electrons, magnetospheric electric fields always acquire a non-zero parallel component. Although neglected in the past, since they are much smaller then the perpendicular convection electric field intensity, the parallel electric field, the gravitational field and the centrifugal forces must be taken into account in any realistic model for the density distribution of the cold plasma in the Earth plasmasphere. In the first part of Chapter 5, J. F. Lemaire outlines how the parallel component of the Pannekoek–Rosseland electric field can be derived from the partial hydrostatic equations for the ions and electrons. It is pointed out that when plasma is flowing out of the ionosphere, as in the polar wind, the parallel component of the ambipolar electric field is larger then the value corresponding to the Pannekoek–Rosseland field. It is this larger upward directed electric field that accelerates the polar wind ions to supersonic speed. In empty plasmatrough flux tubes and even in the outer plasmasphere, a similar upward directed ambipolar electric field accelerates ions out of the ionosphere.

The interplay of these gravitational, centrifugal, electric and magnetic forces determines the equilibrium distribution of plasma along magnetic field lines as well as across the equatorial plane. There are two extreme cases of basic (hydrostatic) equilibrium: diffusive equilibrium and exospheric collisionless equilibrium. Both have been examined in detail to set the stage for the discussion of time-dependent models simulating refilling of empty flux tubes.

Interchange motion of plasma elements in the gravitational field is another important physical mechanism that has not always been properly treated in earlier convection models of the plasmasphere. This is why Lemaire's kinetic description of plasma interchange motion applied to the plasmasphere has been included explicitly in Chapter 5. Indeed, it completes the more formal mathematical descriptions which were given later. Thermally driven interchange motion is also relevant in the plasmasphere. It is shown to be most important when kinetic pressure is not balanced across plasma density irregularities in the plasmasphere.

But there are other issues as well which have been addressed here, for instance, the effects of induction electric fields and of weak double layers which decouple high-altitude plasma clouds from the ionospheric plasma. This latter effect should be included anyway in future theoretical studies of magnetospheric convection and also of the detachment of plasma elements from the main body of the plasmasphere.

In the remaining part of Chapter 5 we recall in historical order all theories proposed since 1967 to explain the formation of a sharp knee in the cold plasma density distribution. The early popular theory, which still remains in textbooks, was based on the assumption that the plasmapause – with its characteristic bulge in the duskside sector – coincides with the Last Closed Equipotential (LCE) of the large-scale convection electric field. However, this definition of the plasmapause must be abandoned, as already pointed out in the 1970s when alternative MHD and non-MHD theories were proposed by various theoreticians. The role of plasma interchange motion for the formation of the plasmapause was first pointed out in 1974 and recalled in this chapter.

The smoothing of the plasmapause density gradient by different physical mechanisms, including plasma interchange motion, has been discussed. The formation of multiple plasmapauses and 'staircase density profiles' during a series of substorms with decreasing intensity is described. The key importance of Subauroral Ion Drifts (SAID) and Alfvén layers forming at the front of plasmasheet clouds, moving earthward in the subauroral region, and eventually penetrating into the nightside plasmasphere, has also been pointed out here.

Several topics have not been discussed in this monograph, not only because of the lack of space, but because of lack of expertise in these specialized subject matters. Among these topics which warrant more attention, and perhaps full monographs, one could mention wave–particle interactions and their effects on the plasmasphere, as well as the wide variety of hydrodynamical codes and models published in the literature since 1971 to describe the refilling of the plasmasphere. These topics have been touched only briefly in Chapters 4 and 5. A more exhaustive review would have required much more effort and was beyond of the scope of this monograph.

There are also problems and issues that have never been addressed anywhere else so far. One of the most interesting and pressing of these is to model the depletion of plasmaspheric flux tubes during erosion events. Indeed, the subsequent refilling phases of the many alternative models that have been proposed so far critically depend on the boundary conditions at the end of the depletion or erosion phase. This issue needs more attention from experimentalists as well as from theoreticians.

This indicates that the topic of this monograph is not a closed subject. On the contrary, beyond all existing observations of the plasmasphere and current theoretical and empirical models for the plasmasphere, there is space for novel accomplishments based on novel experimental methods as well as on innovative models of the plasmasphere and plasmapause.

Chapter 1

Discovery of the plasmasphere and initial studies of its properties

1.1   Introduction

Until the early 1950s no one suspected that the ionosphere of Earth extends far into the geomagnetic field and forms what is now called the protonosphere. The discovery of this region, filled with low-energy charged particles of ionospheric origin, is an interesting one from a historical perspective. It started with the pioneering whistler wave investigations by Storey, and continued with the theoretical work of Dungey, who developed the idea of a Chapman–Ferraro cavity. Then, towards the end of the decade, the first *in situ* measurements of the high-altitude plasma and of a steep density dropoff at $L \sim 4$ were made by Gringauz and his colleagues in the USSR. Meanwhile, Carpenter was beginning the series of whistler studies that led to the identification of what is called the 'knee' in the equatorial electron density profile, and eventually to the description of the worldwide structure and first-order dynamics of what he called the plasmasphere. The historical account of these early discoveries is not generally well known by the younger generation of space physicists; some of its aspects are unknown even to the older. Since it is interesting to record for future studies of the history of science the paths followed by the pioneers, we devote this first chapter to events that happened in the early days of the space age.

In reading what follows, one should realize that in the middle and late 1950s the world space physics community was small and widely scattered. New experimental methods were being developed and applied by groups with varying experience, and it is not surprising that many scientists were unaware that a growing body of knowledge about the Earth's plasma envelope was being obtained from whistlers. Nor were they necessarily aware of the suggestions made by Dungey (1954, 1955a,b) about a boundary of the Earth's magnetic field at $\sim 8R_E$, inside which there existed the dense plasma that Storey had reported earlier from his whistler studies (1953). And in particular there was no prediction of the existence of the rather sharp cold plasma density boundary which is now called the plasmapause. Meanwhile, there were some who regarded Storey's remarkable work with skepticism, among them his supervisor at the Cavendish Laboratory in Cambridge, H. J. A. Ratcliffe. At the General Assembly of the International Radioscience Union (URSI) in Sydney in 1952, Ratcliffe presented the essentials of Storey's work and stressed the need for certain additional experiments. He also asserted, however, that Storey's theoretical interpretation of whistlers was probably not correct (URSI, 1952). A measure of Ratcliffe's later acceptance of the whistler method of magnetospheric probing was provided in 1972 when he published his introductory book on the ionosphere and magneto-sphere. On the paper cover of the book was a diagram showing whistler components propagating on both sides of the knee, or plasmapause.

In the following two sections, Konstantin Gringauz and Don Carpenter recall the circumstances of their early discoveries. Necessarily, each emphasizes the particular point of view held by himself and his near associates at the time.

1.2   First plasmaspheric research in the USSR

1.2.1   Background to the discovery of the plasmasphere

To place in perspective the initial *in situ* detection of the plasmasphere and of its outer boundary, the plasmapause, by the experimental equipment flown on board the Soviet space vehicle LUNIK in 1959, it is necessary to recall our understanding of the terrestrial plasma environment and of interplanetary space in the middle of the 1950s.

Until the late 1950s, information from traditional methods of probing the upper atmosphere of the Earth was limited to an altitude of about 1000 km, which is the upper boundary where polar aurorae are formed. The observation of such high-altitude aurorae produced by energetic electrons exciting atoms or molecules in the upper atmosphere was definite evidence that atmospheric gases are still present in a significant amount at that altitude. The ionized regions of the upper atmosphere were rather well explored by means of ionospheric radio sounding up to the height of 300–350 km. This corresponds to the altitude of the ionospheric F2-layer, where the electron and ion densities reach their maximum values. Unfortunately, ground-based radio sounding techniques at frequencies above ~ 30 MHz were unable to determine the ionization density above this altitude. The only *in situ* electron density profile in the upper atmosphere prior to 1958 was obtained from a US rocket flight and reported by Berning (1951). This ionospheric density profile was reproduced in a large number of publications at that time. According to this profile, the ionospheric plasma density decreased with altitude above the F2 maximum much faster than it increased below this altitude. At the level of 400 km the ionization density became very small.

In 1958 it was found that Berning's result was incorrect, and that the plasma density decreases more slowly above the F2 maximum than it increases with altitude below that level. These new results were obtained in February 1958 by radiophysical measurements from a Soviet geophysical rocket launched vertically to an altitude of 470 km (Gringauz, 1958). The electron density measured at 470 km was $n_e = 2 \times 10^6 \, \text{cm}^{-3}$, corresponding to a year of maximum solar activity. In May 1958 satellite measurements of the ion number density with spherical ion traps using the retarding potential method showed that the ion density was larger than 1000 cm^{-3} at a height of 1000 km (Gringauz, Bezrukikh and Ozerov, 1961).

The existence of the 'magnetosphere' was not yet known at that time – the word magnetosphere was only introduced in 1959 by Gold (1959). Therefore, in 1958 it was still assumed that the upper atmosphere was in direct contact with the interplanetary medium.

At the end of 1958, when the instruments for LUNIK were developed, several models of the interplanetary medium were available in the literature. According to Bartels (1932), there were some stable separated active regions on the Sun, emitting corpuscular streams with velocities of the order of hundreds of km/s. These charged particle streams were affecting the Earth in a recurrent manner with the period of the solar rotation, i.e. 27 days.

According to Chapman (1957), interplanetary space was filled with hot ionized gas extending in hydrostatic equilibrium from the base of the solar corona out into interplanetary space beyond the orbit of Earth. He assumed also that the temperature distribution in this extended solar coronal plasma was determined by heat conduction. The density corresponding to Chapman's hydrostatic and conductive coronal model was 600 cm^{-3} at 1 Astronomical Unit (AU).

The third model for the interplanetary medium had been proposed by Bierman (1957), who considered that the Sun constantly emitted corpuscular fluxes with velocities of the order of hundreds of km/s. He came to this conclusion from his observations of cometary ion tails that were deflected by this continous flow of solar particles.

In 1958, when the instruments for the first LUNIK were produced, the theory of the continous dynamic solar wind expansion was first published by Parker (1958). In Parker's initial model the value of the solar wind radial expansion velocity was estimated to be 500 km/s, and its number density was of the order of $500 \, \text{cm}^{-3}$. (Subsequent space measurements have shown that this value is too large by two orders of magnitude.) In this early model of the solar wind, the flux at 1 AU was assumed to be $1.5 \times 10^{10} \, \text{cm}^{-2} \text{s}^{-1}$. At that time Parker also predicted the spiral structure of the interplanetary magnetic field.

All these early theoretical estimates of the interplanetary plasma densities near the orbit of the Earth were based on observations available at that time. Behr and Siedentopf (1953) had interpreted zodiacal light observations as a consequence of the scattering of solar photons by free electrons present in the interplanetary medium. From the intensity and polarization of the zodiacal light these authors estimated the value of the electron density at 1 AU to be equal to $600 \, \text{cm}^{-3}$.

On the other hand Storey (1953) found from whistler observations that the electron concentration at a distance of 12 000 km from the Earth was approximately $400 \, \text{cm}^{-3}$. Following Dungey's (1954, 1955) suggestions, Storey (1958) postulated that the plasma observed by whistlers was an upward extension of the ionosphere consisting of protons.

The value of the velocity of the solar corpuscular stream was determined from the delay time between phenomena observed on the surface of the Sun and their effects on the Earth. This velocity was estimated to be of the order of hundreds of km/s. Combining this value of the plasma bulk velocity and the data on the interplanetary plasma density, values of the corpuscular fluxes were estimated to range between 10^{10} and $10^{11} \, \text{cm}^{-2} \text{s}^{-1}$. The LUNIK data showed that the solar wind flux at 1 AU is almost two orders of magnitude smaller than these values. To place the discovery of the plasmasphere in the right perspective it is important to recall these premises. It allows us to appreciate the many achievements that have been made during the first 35 years of the space age.

1.2.2 Description of the experiments on LUNIKs

The Soviet LUNIK 1 and 2 spacecraft were equipped by K. I. Gringauz with various instruments to detect and study the plasma in interplanetary space. These lunar rockets, launched on 2 January 1959 and 12 September 1959, respectively, were the first vehicles to penetrate the interplanetary medium after their acceleration to the escape velocity of 11 km/s. As a consequence of stringent limitations on instrument weight and on telemetry rate, and also because of the restricted time for the development and testing of the scientific instruments (about one year), these instruments had to be of the simplest design.

By the end of 1957 the group headed by K. I. Gringauz had finalized the design and production of the spherical ion traps to be used in Earth orbit for experimental studies of the ionospheric plasma. The description of this measurement technique had already been published prior to the launch of the first

SPUTNIK in 1957 (Gringauz and Zelikman, 1957; Gringauz, Bezrukikh, Ozerov and Rybchinsky, 1960a). The equipment was then installed on a Soviet satellite that was launched in May 1958; the instrument operated sucessfully on board the third SPUTNIK (Gringauz, Bezrukikh and Ozerov, 1961). Since the group had thus acquired a certain experience in building plasma instruments for rockets and satellites, it was now instructed to develop an instrument for the exploration of the interplanetary plasma.

After studying the publications on the interplanetary medium available at that time, Gringauz pointed out that there was no generally accepted model for the interplanetary medium, as noted in the previous section. For this reason it was decided to limit the scientific objectives to investigations of the following simple questions: is the plasma in the interplanetary medium in stationary equilibrium, or is it formed of separated and sporadic corpuscular streams? And what is the approximate temperature of the interplanetary plasma ions?

To determine the ion temperature the method of retarding potential was chosen. Unfortunately, the telemetry rate of the LUNIK spacecraft was too low to transmit the whole ion retardation curves (i.e. the characteristic current-voltage curves) within a reasonable time interval. For this reason it was decided to mount four identical hemispherical charged particle traps, with four different but constant potentials applied to the external grids. These external grid potentials ranged between 25 V and -10 V. For comparatively 'cold' (very low energy) stationary plasma, the collector currents in these traps should be different. On the other hand, for corpuscular streams consisting of ions with velocities of several hundreds of km/s (i.e. with energies of the order of hundreds or thousands of electron volts (eV)), the collector currents should be independent of the value of the external grid potential, i.e. equal in all four ion traps. LUNIK spacecraft were not spin stabilized, but tumbled during the flight.

A schematic diagram of the ion traps is shown in Fig. 1.1, and its design is illustrated in Fig. 1.2. The hemispherical shape of the external grids was chosen to provide a larger effective cross-section of the probe in the case in which the ambient ions have an isotropic velocity distribution and the velocities are substantially higher than that of the spacecraft. The sensitivities and dynamic ranges of the current amplifiers were identical for all four ion traps. Furthermore, besides the positive currents, the amplifiers could also register negative currents carried by electrons with energies >200 eV when the electron fluxes exceeded the ion fluxes.

The collector currents in interplanetary space were of course expected to be far smaller than those normally measured in the ionosphere, in which case the photocurrents from the collectors could exceed those produced by the particles from the ambient plasma. Thus an inner grid 3, at a fixed potential of -200 V with respect to the spacecraft, was added to suppress the photoelectrons from the collector. This design has since become widely used in such plasma probes as a means of reducing photoemission currents. When this suppressor grid 3 is illuminated by the Sun, it becomes a source of photoelectrons that produce a negative collector current. However, this secondary current is weaker by two orders of magnitude than the positive current produced by photoemission from the collector in the absence of the suppressor grid.

1.2.3 Results of the first measurements

In the first article published in 1960 by Gringauz, Bezrukikh, Ozerov and Rybchinsky only the results of the second LUNIK flight were presented, although both rockets were equipped with identical ion traps and the measurements of both instruments were not inconsistent with each other. The reason

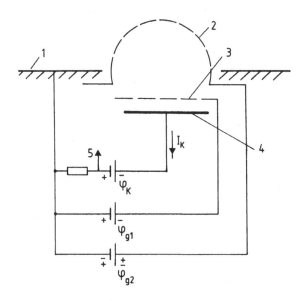

Figure 1.1 Schematic diagram of ion traps with three electrodes. 1- Spacecraft main body; 2- External grid; 3- inner grid; 4- Collector; 5- to the amplifier of the collector current (after Gringauz *et al.*, 1960a; c).

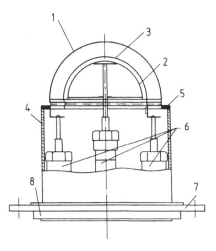

Figure 1.2 Gringauz's design for the ion trap developed for LUNIK 2; 1 is the external grid, and 3 is the internal grid (after Gringauz, 1961a).

Figure 1.3 Drawing of LUNIK 2, launched in September 1959. Lunik 2 was the first spacecraft to reach another body in the solar system; it was also the first spacecraft to record measurements of the solar-wind plasma flow. Three of its four ion traps are visible, one on the lower hemisphere and two on the upper one (after Gringauz, 1961a).

was that for the first flight the ion traps were located along a great circle of the spherical spacecraft, so that all four traps could be illuminated simultaneously by the Sun. As a consequence, photoemissions could take place in all four of the traps. Therefore the authors had some doubts about the relevance of the first set of data.

LUNIK 2 was launched on 12 September 1959. For this second flight the positions of the ion traps were changed and located at the corners of a tetrahedron embedded below the surface of the spherical spacecraft (see Fig. 1.3). With this new geometrical arrangement, all four ion traps could not be illuminated at the same time. The results of the measurements appeared to be unexpected:

(a)   At distances less than 25 000 km from the surface of the Earth, positive collector currents were recorded in the ion traps with external grid potentials fixed at 0, − 5, and − 10 V. The current intensity oscillated at the spin rate of the spacecraft. The upper envelope of these current oscillations was highest for the lowest value of the external grid potential; i.e. − 10 V. At altitudes greater than 15 000 km a sharp drop was observed in the intensities of all collector currents.

(b)   No effect was observed when the spacecraft traversed the outer radiation belt where the energetic particle fluxes measured at that time by Van Allen and the group of Vernov and Chudakov were estimated to be as high as 10^{11} cm^{-2} s^{-1} for electrons of energy greater than 20 keV in the heart of the outer zone (Van Allen, 1959, 1960). Such intense fluxes of energetic electrons should have produced a considerable negative collector current in all the ion traps, but no such current was observed by the LUNIKs.

(c)   Beyond the outer radiation belt the spacecraft penetrated a region where only negative currents were recorded by all four ion traps, regardless of orientation. Evidently, these currents were due to high fluxes of electrons whose energies exceeded the negative potential of the inner grid 3: i.e. > 200 eV.

(d)   Finally, when radio communication with LUNIK was resumed after it had been interrupted due to the rotation of the Earth, all four ions traps recorded identical periodic positive collector currents. Between 245 000 km and the surface of the Moon these currents were independent of the values of the external grid potentials. These positive currents were undoubtedly produced by corpuscular streams.

All these experimental results were of fundamental importance and were widely discussed later on. In particular, the result (b) forced us to reduce estimates of the maximum flux of energetic electrons in the outer radiation belt by at least three orders of magnitude below the value accepted at that time; i.e. from 10^{11} cm^{-2} s^{-1} to less than 10^{7}–10^{8} cm^{-2} s^{-1}. The result (c) represented the first observation of a population of hot magnetospheric electrons outside the Van Allen zones. The result (d) represented the first record of the solar wind plasma outside the geomagnetic field, and enabled us to decrease the estimated value of the solar particle flux and density by 2 to 3 orders of magnitude below the values assumed in 1959 by the scientific community. But the result (a) is the most relevant one in the context of this monograph, since it was the first *in situ* evidence concerning the presence of an abrupt plasma boundary, the existence of which was being independently investigated by Carpenter (1962a–c, 1963a–b) at that time through a series of whistler observations. Since it is directly related to the subject of this monograph, the result (a) is the only one that will be discussed in more detail in the rest of this section.

Figure 1.4, illustrating the result (a), is taken from the first publication of plasma measurements obtained along the LUNIK 2 trajectory by Gringauz, Bezrukikh, Ozerov & Rybchinsky (1960). From this figure it can be seen that the collector current intensity was strongly influenced by a negative grid potential of −5 V at altitudes below 25 000 km. Furthermore, a grid potential of +15 V completely retarded the ambient plasma ions and prevented them from entering the trap at all altitudes above 3 000 km. Therefore, it is clear that on this part of

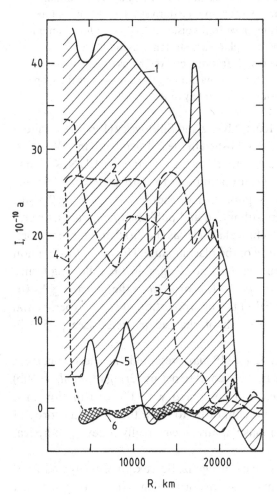

Figure 1.4 Upper and lower envelopes of the collector currents measured by the four different ion traps of LUNIK 2 at altitudes less than 25 000 km. The upper bounds represent traps with external grid potentials respectively equal to − 10 V (curve 1); − 5 V (curve 2); 0 V (curve 3) asnd + 15 V (curve 4). The lower bounds correspond to traps with external grid potentials of − 10 V, − 5 V and 0 V (curve 5 for both) and + 15 V (curve 6). Note the steep drops at altitudes of about 20 000 km (after Gringauz *et al.*, 1960a).

the trajectory LUNIK passed through a region where the plasma was composed of ions with energies less than a few eV, i.e. with a kinetic temperature less than several tens of thousands Kelvin.

This conclusion was formulated in 1960 in the first published article concerning these measurements (Gringauz, Bezrukikh, Ozerov and Rybchinsky, 1960a). This paper was originally submitted on 5 February 1960 by Gringauz. It was published in the April 1960 issue of *Doklady Akademii Nauk SSR* (Vol. 131). An English translation appeared in 1960, Vol. 5 of *Soviet Physics Doklady*. A new translation was published in 1962, Vol. 9 of *Planetary and Space Science* where on p. 107 the discovery of the 'plasmasphere' and of its outer boundry at $4R_E$ are reported in the following terms:

> At distances R from the earth's surface up to four earth radii a plasma was observed with temperature not greater than tens of thousands of degrees . . . The existence of plasma at these distances from earth is confirmed by results obtained on the first space rocket in January 1959 and on the third space rocket in October 1959 (in the latter case up to 7000 km, since at this distance the first radiocommunication session with the automatic interplanetary station ceased). Questions connected with evaluations of plasma concentration observed by us together with the possible concentration of interplanetary plasma (for large R), are outside the scope of the present report and are considered separately in Ref. 5.

Ref. 5 is a paper by Gringauz, Kurt, Moroz and Shklovsky (1960b) submitted nine weeks later (12 April 1960) and published in the following volume (Vol. 132) of *Doklady Akademii Nauk SSSR*. A detailed technical paper on LUNIK results was also submitted by this same group of authors to *Astonomicheskii Zhurnal SSSR* on 12 April 1960, and was published in the July–August 1960 issue. Although these latter papers were submitted and published afterwards, several soviet colleagues consider these technical articles as the 'first' and more important ones in relation to the discovery of the plasmapause. However, this opinion was not shared by K. I. Gringauz who acknowledges, in the following terms, 'Prof. I. S. Shklovsky, V. G. Kurt and V. I. Moroz from the Sternberg Astronomical Institute, who participated in the interpretation of some experimental results pertaining to the outermost belt of charged particles' (see p. 553 in Gringauz, 1961a).

The first paper by Gringauz *et al.* (1960a) presented the electric current measured by the ion traps, while in the second paper by Gringauz *et al.* (1960b) ion number densities were given and compared with the distribution in a hydrostatic equilibrium density model in order to show that the sharp drop of density at four Earth radii – the plasmapause – was really a new geophysical phenomenon.

The first publication quoted above (Gringauz, Bezrukikh, Ozerov and Rybchinsky, 1960a) had a rather dramatic pre-history which is interesting to report in this historical section because it illustrates how difficult it is sometimes for young researchers to publish unexpected new results. The history of sciences is full of similar stories; we wish to add this unpublished one, for the benefit of future historians in space geophysics.

1.2.4   A lesson from history

Each of the results (a)–(d) obtained with the ion traps of the LUNIK automatic stations was unexpected. In particular, we were unaware of the existence of a plasma envelope surrounding the Earth, extending up to a distance of four Earth radii, and in particular one having a rather abrupt break at its outer boundary.

On 11 February, 1960 the article by Gringauz, Bezrukikh, Ozerov and Rybchinsky (1960a) was submitted for publication in the magazine *Doklady Akademii Nauk SSSR* by the Academician A. L. Mintz, who was the Director of the Radiotechnical Institute of the Academy of Sciences where the authors were employed at that time. This journal normally provided rapid publication, without review by referees, of short articles written by or submitted by members of the Academy of Sciences of the USSR. Among the authors K. I. Gringauz was the only one who at that time had a scientific degree, and he had only three publications in the open literature; his earlier papers were classified, since they were related to rocket studies of the ionosphere and flights of the Soviet rockets were not announced in the USSR before 1957.

In those early days of the space era the Soviet Government considered that results from space research had political significance as well as scientific interest. Therefore, errors or misinterpretations in scientific publications had to be avoided at all costs. This is why there was a policy of publishing scientific results only after permission had been granted by the person ultimately responsible for the space research programme at the Academy of Sciences of the USSR, i.e. the President of the Academy, at that time M. Y. Keldysh, a specialist in the fields of mechanics and mathematics. The existence of this policy was never announced in public.

Since the President of the Academy of Sciences was not himself involved in geophysics, plasma physics or astrophysics, he decided to consult with specialists in these fields, and invited P. L. Kapitza – member of the Academy of Sciences of the USSR and of the Royal Society – and two correspondent members of the Academy (and now Academicians); S. N. Vernov and V. L. Ginzburg. He invited also Ya. L. Alpert, the author of well-known books on ionospheric physics, and A. I. Liebedinsky and I. S. Shklovsky, two distinguished specialists in geophysics and astrophysics, respectively, to consider the manuscript.

The result of their review and discussions was negative with regard to publication; the most vehement objections came from Ya. L. Alpert, who argued in particular that an abrupt plasma density decrease like that reported to occur over such a small altitude range was impossible . . . 'indeed in nature everything varies according to exponential laws'. . . S. N. Vernov and A. I. Liebedinsky also considered the results to be erroneous, while P. L. Kapitza and V. L. Ginzburg did not express their opinions. K. I. Gringauz was convinced that the experimental results were reliable, but could not propose any likely geophysical explanation of the observed plasma boundary at about 25 000 km. The only participant in this peer review meeting who supported the publication of the article was I. S. Shklovsky from the Sternberg Astronomical Institute. After this consultation, the president of the Academy did not authorize publication of the article.

However, A. L. Mintz, Director of the Radiotechnical Institute, insisted on publishing. He considered the prohibition to publish in Doklady Akademii Nauk the article that he had presented to be a deprivation of his rights as an Academician. Thereafter, M. Y. Keldysh sent the article to the Department of Physical and Mathematical Sciences of the Academy asking for an official referee report and definitive conclusions. A commission consisting of specialists of the Kurchatov Institute of atomic energy, chaired by Academician M. A. Leontovitch was then formed , and charged with the task of analyzing the observational data set and the manuscript submitted to the official Journal of the Academy. After a careful study of all the data, including the telemetry films, this commission concluded that the manuscript should be published. However, these conclusions did not dissipate all the doubts of M. Y. Keldysh, who was strongly influenced by his first consultation with the specialists from the Academy. Then A. L. Mintz visited M. Y. Keldysh and said:

> if the article contains errors, the errors will be those of the four authors, but if these results are found acceptable, the credit will belong to the Soviet Union.

The president of the Academy did not find objections to this argument, and he authorized the printing of the manuscript; it appeared in Doklady Akademii Nauk in April 1960. All of these consultations and discussions had delayed the publication by several weeks.

Subsequently, K. I. Gringauz invited I. S. Shklovsky to take part in the interpretation of the additional data obtained at distances up to 100 000 km altitude; two additional papers were published this same year with the collaboration of I.S. Shklovsky (Gringauz, Kurt, Moroz and Shklovsky, 1960b, 1960c). In these articles the authors discussed the various factors and physical phenomena which could have affected the measurements of charged particle traps along different parts of the LUNIK trajectory; i.e. photoemission, interaction with the trapped radiation belt particles, spacecraft charging, etc.

Figure 1.5, which was presented in these articles, gives the ion density distribution deduced from the observations corresponding to the ion trap with 0 V external grid potential; the different curves shown were calculated under the

assumption that the ions were mainly protons with different temperatures :
1800 K, 10 000 K and 50 000 K, respectively. The dotted curve shows the theoretical proton density distribution corresponding to isothermal hydrostatic equilibrium (barometric formula) for plasma temperature of 1800 K. From this
figure it can be seen that, in the region where the gradients were small, the
plasma density distribution could easily be explained, but that it was more
difficult to understand the nature of the plasma measured in the higher altitude
region, where the ion density decreased sharply with altitude. The minimum ion
density that could be measured was $70 \, \text{cm}^{-3}$, as indicated in Fig. 1.5.

1.2.5 Presentation of the results to an international conference

At the COSPAR meeting in Florence, in May 1961, Gringauz (1961b) presented
a report entitled 'The structure of the Earth's ionized gas envelope based on
local charged particle concentrations measured in the USSR'. In this talk he
presented the data that had been obtained in 1958–59 from Soviet rockets
launched up to an altitude of 470 km, from the third Soviet satellite, and from the
LUNIK space probes. In this paper the region above 1000 km was described as
an extension or continuation of the ionosphere.

Neither the global shape of the outer boundary of this ionized envelope nor
the dependence of its radius as a function of local time could be determined from
these early observations. But it was noted in these papers that the electron
densities deduced by Storey for an altitude of 12 000 km and all those deduced
from other whistler observations should not be related to the interplanetary

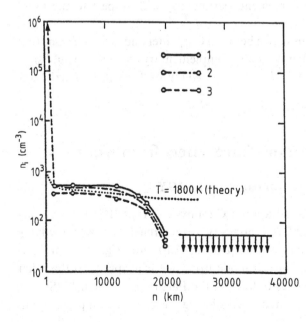

Figure 1.5 Ion density distribution as a function of altitude deduced from the
LUNIK 2 ion trap collector
current measurements; the
different curves correspond
to density deduced under
the assumption that the
plasma was formed of protons and electrons with temperatures of 1800 K (curve
1), 10 000 K (curve 2) and
50 000 K (curve 3). The dotted line corresponds to the
density distribution of a proton plasma in hydrostatic
equilibrium at a constant
temperature of 1800 K (after
Gringauz et al., 1960c).

medium, but rather to the ionized gas envelope of the Earth. It was also mentioned in these papers that the zodiacal light is not due to scattering of solar radiation by free electrons in interplanetary space (as had been postulated until then) but is caused by diffraction by microscopic dust grains orbiting around the Sun (Gringauz *et al.*, 1960c).

The papers by Gringauz *et al.* (1960a and 1960c) were translated into English and published in various journals. However, there was almost no response to these papers between 1960 and 1962 in the international scientific literature.

1.2.6   Meeting of Carpenter and Gringauz in Tokyo

In the autumn of 1963, K. I. Gringauz took part in the XIVth URSI Assembly in Tokyo. At one point the desk personnel of the Imperial Hotel where the Soviet delegates stayed called Gringauz's room. They informed Gringauz that an American would like to meet him. The young man in the lobby of the hotel introduced himself as Don Carpenter from Stanford University. He explained that he was studying whistlers, and had confirmed the results from the LUNIK observations on the plasma in the Earth's environment and on its upper boundary. He presented Gringauz with a preprint of his talk, in which an equatorial electron density profile was shown with an abrupt drop at an altitude of $\sim 2.5R_E$ (approximately 16 000 km), which he called a 'knee' in the plasma density distribution (Carpenter, 1963b). In one of the figures of Carpenter's preprint the LUNIK 2 ion density profile was shown, together with the whistler data obtained by Carpenter (1965) (see Fig. 1.6). The data sets shown in this figure appear to be in reasonable agreement, particularly with respect to the rapid density decrease.

A friendly relationship thus began between Carpenter and Gringauz, one that continued over the years. Their mutually complementary papers were published in the *Proceedings of the XIVth URSI Assembly* of 1963 (Gringauz, 1965; Carpenter, 1965).

1.3   The discovery of the 'knee' effect from whistlers

1.3.1   A puzzling observation early in the IGY

In early 1958, during the International Geophysical Year (IGY), a puzzling observation was made at Stanford University by researchers who were studying whistlers, highly dispersed very low frequency (VLF) radio signals from lightning that propagate in a slow-wave mode from one hemisphere to the other along paths that follow the lines of force of the Earth's magnetic field (Storey, 1953; Helliwell, 1965). On 13 January whistler signals traveling on paths terminating in the vicinity of Seattle, Washington (SE), appeared to have arrived at the same time that signals from the same lightning sources arrived at Stanford, California (ST), ~1500 km to the south (see map of Fig. 1.7a). Because the geomagnetic field line paths ending near Seattle were substantially longer than those terminating near Stanford, the Seattle signals had arrived roughly 0.5 s earlier than expected. This puzzle was not solved immediately, but it set in motion a process of discovery that culminated within two years in the recognition of the plasmapause phenomenon.

1.3.2 Storey's pioneering work on whistlers

The time was ripe for discovery; seven years earlier, in his brilliant PhD work, L. R. O. Storey (1951; 1953) had unlocked the secret of the whistler path, and in so doing had found evidence of something quite unexpected, a region of dense plasma lying well above the known ionosphere. At that time, whistlers were known to originate in lightning (Barkhausen, 1919; Burton and Boardman, 1933), and a form of their dispersion law had been derived by Eckersley (1935), who speculated that the propagation took place in the ionosphere. However, the actual form of the propagation path was not known, Eckersley having suggested that the waves 'traverse the spherical channel formed by the ionosphere and are reflected at the polar regions'. Using both experiment and theory, Storey deduced that the dispersion of the whistler arose from propagation along the

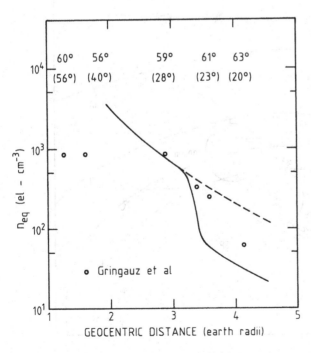

Figure 1.6 Early measured ionization density profiles. The dots correspond to ion densities measured by the ion traps on LUNIK 2, while the solid curve is an idealized equatorial electron density profile based on whistler measurements. Both profiles show a region of steep gradient near 3.5 R_E. The numbers with and without parentheses represent, respectively, the approximate invariant latitudes and the latitudes associated with the ion measurements (after Carpenter, 1965).

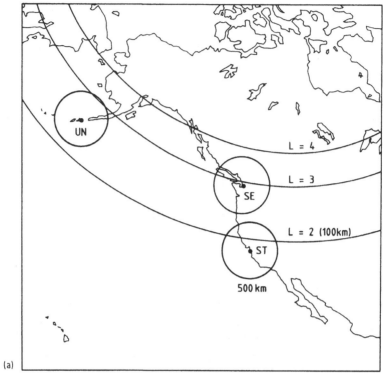

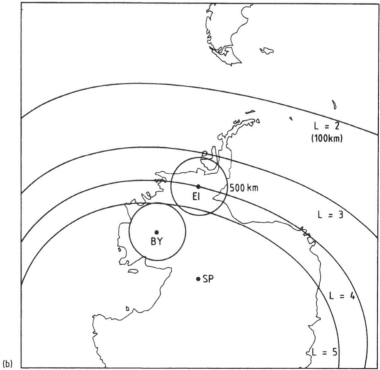

geomagnetic field lines between conjugate hemispheres. Furthermore, he launched the era of whistler probing of what later came to be called the magnetosphere (Gold, 1959), concluding from the observed properties of whistlers that
there was a dense, ionized plasma lying well above the regular ionosphere in a
region extending to at least two Earth radii altitude. Conventional ionospheric
theory, based upon extrapolating upward from the F layer peak using the scale
height of atomic oxygen or nitrogen ions, predicted essentially no plasma of
ionospheric origin in that region.

Storey (1953) initially speculated that the dense high altitude plasma consisted of protons and electrons whose origin was in the solar corona. Later (in
1958), influenced by Dungey's (1955b) prediction of a magnetized region extending to ∼ 8 Earth radii that is terminated by an abrupt transition to the region of
interplanetary wind, he suggested that the proton plasma inside this boundary
was controlled by the Earth's magnetic field and, although still of unknown
origin, could be considered part of the ionosphere. Within another year, Allcock
(1959) published an estimate of the form of the equatorial electron density profile
in the altitude range ∼ 400–12 000 km. The concentrations were at levels consistent with those previously estimated by Storey (1953; 1956), and were based
upon the average frequency-versus-time, or dispersion, properties of whistlers
received at several ground stations in the period January–June 1957. The crude
but practical assumption was made that the ionospheric endpoints of the paths
of whistlers received at each station were on average centered at that station's
latitude.

1.3.3 Further development of whistlers as magnetospheric density probes

Storey's work stimulated VLF radio workers to plan certain tests of his ideas,
one being the observation of alternating appearances of multihop or echoing
whistlers in the northern and southern hemispheres. Such tests were planned
during an URSI conference at the Hague in 1954. Among the participants in the
planning were R. A. Helliwell of Stanford, M. Morgan of Dartmouth, G. McK.
Allcock of D.S.I.R in New Zealand, and R. Rivault of the University of Poitiers.
The conjugate hemisphere tests were successfully performed between Alaska
and Wellington, New Zealand (Morgan and Allcock, 1956), as were recordings
at Punta Arenas, Chile of magnetospheric signal propagation from the NSS

Figure 1.7 (a) Map showing the locations of the Whistlers West IGY stations
at Stanford (ST), Seattle (SE) and Unalaska (UN). (b) Map showing the locations of Antarctic stations Byrd (BY) and Eights (EI). Circles of 500 km radius indicate the distance from the ionospheric exit point of a whistler or VLF
emission within which a signal of average amplitude could be detected on the
ground (after Carpenter, 1986).

VLF transmitter at Annapolis, Maryland in the US (Helliwell and Gehrels, 1958). The observed delays at 15 kHz were of order 0.5 s, significantly longer than the mere tens of milliseconds required for direct propagation to the receiver in the Earth-ionosphere waveguide.

An important step in the development of whistlers as a diagnostic tool was the discovery in 1956 by Helliwell, Crary, Pope and Smith (1956) of the 'nose whistler'. This whistler was recognized as a higher latitude version of the signals that had previously been recorded by Storey and others. Its dispersion properties were readily explained by an extension of the basic propagation theory to higher normalized wave frequencies, that is, to frequencies that were closer to the minimum (equatorial) path electron cyclotron frequency f_{ceq} than those considered previously under the low-normalized-frequency or 'Eckersley' approximation (i.e. $f \ll f_{ceq}$). Most importantly, the whistler's nose-like form on spectrograms, with a minimum in travel time, was found to occur because of asymptotically large travel time as both zero frequency and f_{ceq} were approached. Since the nose frequency f_n was found to be approximately proportional to f_{ceq}, the nose effect opened the way to estimates of the values of f_{ceq} for the multiple components of a whistler through fitting of theoretical curves. And, since f_{ceq} provided a measure of B_{eq}, the magnitude of the Earth's magnetic field at the equator, one could therefore estimate the radii of the corresponding field-line paths of propagation. In fact, path radii extending to more than $4R_E$ geocentric distance were inferred from these initial events, which were recorded in 1955 at College, Alaska. Thus it was possible to extend to greater altitudes the region apparently penetrated by whistler paths and, by inference, characterized by dense plasmas.

The form of the nose whistler is evident from the integral expression for whistler travel time t versus frequency f, obtained from the equations for propagation in a cold, essentially collisionless plasma (e.g. Helliwell, 1965). The expression is

$$t = 1/c \int \mu_g \mathrm{d}s = 1/2c \int f_p f_c / (f^{1/2}(f_c - f)^{3/2}) \mathrm{d}s \tag{1.1}$$

where μ_g is the group refractive index in the direction of energy propagation, f_p is the plasma frequency, f_c is the electron gyrofrequency, $\mathrm{d}s$ is an element of path length, and the propagation vector k is assumed to be aligned with the geomagnetic field. It is also assumed that the effects of ions are small and that $f_p/f_c \gg 1$. It is clear from (1.1) that the travel time becomes asymptotically large as either zero frequency or f_{ceq}, the minimum value of electron gyrofrequency along the path, is approached. It is also clear that if f_{ceq} is known and the form of the variation of electron density (or f_p) along the field-line path is assumed, the observed travel time t at frequency f can be used to estimate the scale factor for the electron density.

In Storey's original work (1951, 1953), which concerned data at relatively low latitudes, f/f_c was relatively small over the entire field-line path and for diagnostic purposes it was reasonable to use the relation

$$t = 1/2c \int f_p/(f^{1/2}(f_c)^{1/2}) ds \tag{1.2}$$

Hence the expression for whistler dispersion, defined as $tf^{1/2} = D\{f\}$, became

$$D\{f\} = 1/2c \int f_p ds/(f_c)^{1/2} = D_0 = \text{constant} \tag{1.3}$$

In this situation the path radius or magnetic shell parameter could not be inferred, except from considerations of the location of the receiver.

1.3.4 The hypothesis of whistler propagation on discrete field-line paths

Another important advance in the late 1950s was the development of the concept of whistler-mode propagation along discrete field-aligned paths, or ducts. In studying the anisotropic nature of whistler propagation, Storey (1951, 1953) had recognized that the energy flow at the lower whistler frequencies was constrained to be within 19 degrees of the direction of the magnetic field, thus apparently undergoing the guiding necessary for hemisphere-to-hemisphere propagation. However, Smith (1961a) argued persuasively that propagation of whistlers received at ground points must occur along discrete geomagnetic-field-aligned paths. Preliminary evidence for such paths had appeared in the form of the multicomponent nose whistler reported by Helliwell et al. (1956) and in the occasional detection in 1957 of two closely spaced but separate time delays of whistler-mode signals propagating to South America from the NSS transmitter in Annapolis, Maryland (Helliwell and Gehrels, 1958). Smith noted that a single whistler could contain multiple components, apparently the result of propagation on separate paths; the f–t properties of successive multicomponent events were similar, irrespective of the location of the lightning source; and the time intervals between successive echoes of multihop whistlers were identical. He noted further that, as Yabroff (1961) had demonstrated, the wave-normal angles of whistlers with respect to the geomagnetic field direction should increase rapidly during initial upward propagation, thus ruling out the possibility of later ionospheric penetration from above. It was postulated that the discrete paths were in fact tubes of enhanced ionization or 'ducts', in which the wave energy was trapped and guided. Waves with frequencies below half the local gyrofrequency (or cyclotron frequency) would remain within a limiting cone of angles, the size of which depended upon the density enhancement factor of the duct. As the trapped whistler-mode energy moved between duct endpoints in the conjugate ionospheres, it would follow a snake-like path back and forth across intra-duct peaks in the refractive index. The total travel time would be approximately the same as that of a wave with propagation vector oriented strictly along the magnetic field.

The duct hypothesis greatly increased the diagnostic potential of the whistler method, since it essentially freed the observed dispersive properties of the whistler from dependence upon a highly variable and poorly known parameter, the lightning source location.

1.3.5   Carpenter's initial work with whistler data

In 1956, D. L. Carpenter enrolled as a graduate student in electrical engineering at Stanford University, his intention being to obtain a masters degree and then seek employment in industry. Previously a student of political science and international affairs, he had decided to change fields after experiencing two years of difficulty in getting the security clearance required for positions in the Federal Government (this was the infamous McCarthy era). While taking introductory courses in calculus and physics, he was hired by R. A. Helliwell to act as a data aide in the developing International Geophysical Year (IGY) Whistlers West Program (Helliwell and Carpenter, 1961; 1962). The project was in full swing; the senior graduate students R. L. Smith, J. H. Crary and W. Kreiss were occupied with various theoretical studies and with the problems of building, testing, and installing VLF receivers at a number of sites. Figure 1.7a shows the locations of three of the principal stations of the Whistlers West Network, Stanford (ST), Seattle (SE), and Unalaska (UN), with circles indicating radii of 500 km around the stations. As the first regular Stanford recordings began in mid-1957, Carpenter was assigned to study the dispersion properties of the recorded whistlers, which were typically displayed in a format of frequency 0–8 kHz versus 2.4 s in time on grey-scale paper records produced by a Kay Electric Sonagraph.

An important part of the Stanford data analysis program was the comparison of whistlers received at neighboring stations. One objective was to estimate the ground distance within which a whistler or VLF noise event emerging from the ionosphere could be detected. In most cases, whistler components recorded at Stanford or Seattle were found to be independent of one another, the two stations being separated by about 1500 km, more than twice the typical ~ 500 km detection radius that was eventually found in the spaced-station comparisons (Helliwell and Carpenter, 1961).

A problem immediately encountered in the data work was that of identifying the time of the causative atmospheric of a whistler. If that time could be known, the dispersion $D\{f\} = tf^{1/2}$ of a whistler component could be readily obtained from a single measurement of travel time t at frequency f. A rough estimate of the electron density near the top of the component's propagation path could then be made, based on an estimate of the endpoint latitude or magnetic shell parameter of the path. When the time of the causative spheric was not known, Storey (1951, 1953) had shown that it could be estimated, albeit with some difficulty, from slope measurements at the lower normalized whistler frequencies ($f \ll f_{ceq}$), where $dt/d(f^{-1/2})$ approaches a constant value D_0.

It was found that, in the spectrograms from Stanford (ST) and Seattle (SE), the causative atmospherics of one-hop whistlers were often easily recognized, either through simple visual inspection or through comparison of spectra from several successive events (Carpenter, 1960). This was fortuitous, two contributing factors being a low-loss propagation path over sea water from the southern-hemisphere lightning source region and the lack of thunderstorm activity near the SE and ST receivers. Elsewhere, for example in Europe, spheric identification of one-hop whistlers was inhibited by the strong spheric backgrounds that were often present on the records. And although the spherics of two-hop (long) whistlers could often be recognized, coming as they did from thunderstorms in the hemisphere of the receiver, the two-hop whistlers were often so diffuse as to be difficult to analyze (Laaspere et al., 1963). Thanks to the ability to identify causative spherics, the puzzling observation noted above was made in early 1958.

1.3.6 The unusual whistler events of 13 January 1958

Figures 1.8a and 1.8b show in coordinates of frequency 0–16 kHz versus time a weak whistler recorded at both ST and SE on 13 January 1958. The time scale is referenced to the time of detection of the causative atmospheric. One of the SE components (arrow above the record) exhibited travel times comparable to those of the only component at the lower latitude station ST. A better-defined example of this effect, recorded in 1959, is shown in the 0–8 kHz spectrograms of Figs 1.8c and 1.8d (note the difference in the time scales from those of Figs 1.8a, 1.8b). In this case the whistlers exhibited multiple components, several of which appeared on both the SE and ST records. An arrow above the SE record indicates energy that arrived at SE prior to the appearance of the earliest components at Stanford. Carpenter considered the 13 January 1958 event so unusual that it was shown in a Stanford Technical Report written in 1959 and issued in March 1960 on the topic of methods of identifying causative atmospherics (Carpenter, 1960). This was the first 'knee whistler', one of a type used to support the earliest hypotheses about the knee phenomenon, or abrupt density decrease in the equatorial electron density profile.

Throughout 1958 and into 1959, statistics were accumulated on mean monthly whistler dispersion D at 5 kHz at a number of IGY stations, this being essentially a measure of mean monthly travel time at 5 kHz ($D\{5\,\text{kHz}\} = t(5000^{1/2})$). There was a good deal of scatter in the daily values, but attention was focused on effects such as an unexplained annual variation in the monthly average values, with a peak in December and minimum in June or July (Smith, 1961b), and the course of day-to-day fluctuations was not followed in detail. Some apparent anomalies were seen; J. P. Katsufrakis, a fellow student and later manager of Stanford's highly successful Antarctic field programs, found that whistlers with dispersion a factor of 3 to 4 lower than usual had been received at SE following the great magnetic storm of 11 February 1958.

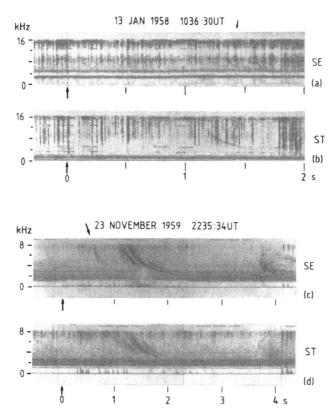

Figure 1.8 Simultaneous spectrograms from Seattle (SE) and Stanford (ST) showing evidence of the 'knee' effect, i.e. anomalously short whistler travel times on the longer, higher latitude paths. (a)–(b) SE and ST records from 13 January 1958 showing a weak one-hop whistler that originated in the southern hemisphere. Frequency (0–16 kHz) is displayed versus time from the reception of the causative impulse (vertical arrows). The first of two weak components received at SE (arrow above (a)) was delayed only slightly with respect to the main component detected at ST (from Carpenter, 1960). (c)–(d) SE and ST records from 23 November 1959 showing a multicomponent whistler, several components of which were detected at both stations. Frequency 0–8 kHz is displayed on a time scale differing by a factor of 2 from that in (a) and (b). An arrow above the SE record indicates energy that arrived at SE prior to the appearance of the earliest components at Stanford (after Carpenter, 1963b).

1.3.7 Investigation of density decreases following magnetic storms

In August 1959, Alex Dessler, then of the Lockheed Company in Palo Alto, visited Stanford and suggested that the VLF group look at whistler dispersion as a function of time during a magnetic storm that had begun on 16 August. The ST and SE records were examined, and a pronounced decrease in whistler dispersion, by a factor of order 3, was noted in the aftermath of the storm. Other storm periods were investigated, and decreases were again observed. Fig. 1.9, top, shows two elegant, closely spaced multicomponent nose whistlers recorded at Seattle during a period of low to moderate geomagnetic activity in June 1959, while the bottom panel shows a whistler recorded at SE on 18 August, only a day after the geomagnetic K_p-index had reached 8 + for the second time within 24 hours. The travel times of the August whistler components, referenced to the time of the first causative atmospheric (termed the spheric) in the June case (arrow below the panel) were notably shorter than those of the June event.

Around this time, interest at Stanford developed in extracting whistler path radius (or endpoint latitude) information from each specific whistler component that was to be used in electron density studies. This was a departure from the common practice (Allcock, 1959) of assuming that at lower and middle-latitude receiving locations, the ionospheric exit points of detected whistler components were centered around the position of the station. At middle latitudes the nose

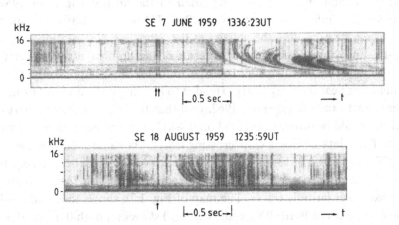

Figure 1.9 Spectrograms of whistlers illustrating the large travel time differences between whistlers recorded during geomagnetically calm conditions and those recorded in the immediate aftermath of magnetic storms. (a) Spectrograms of two closely spaced, exceptionally well-defined multicomponent whistlers received at SE after propagating through the plasmasphere. The record is from a magnetically calm period in June, 1959. Vertical arrows mark the two causative radio atmospherics. (b) Spectrogram of a multicomponent whistler received at SE after propagating through the plasma trough region. The record is from a magnetic storm period, and only a day after the K_p index had reached 8 + for the second time within 24 hours (after Carpenter, 1962a).

frequencies of many whistlers were above 10 kHz and were not directly observable on the records, partly due to a high frequency falloff in the spectrum of the whistler lightning sources (Helliwell, Jean and Taylor, 1958). It became clear that, in order to estimate the path radii of such whistlers, a means of extrapolating upward from the observed lower-frequency portions of the traces was needed. R. L. Smith, then involved in his PhD thesis work (1960), developed an approximation for the whistler travel time integral in terms of elliptic functions, so that travel time measurements at two reasonably well separated whistler frequencies could be used to estimate the whistler path radius and associated electron density level (Smith and Carpenter, 1961). Thanks to this method, one could clearly separate temporal variations in density from the effects of temporal shifts in the path radii of whistlers. The latter could strongly modify the observed dispersion, even in the absence of density variations.

1.3.8   Whistler data from Byrd, Antarctica

From work in 1960 with the extrapolation method, it soon became clear that there were two classes of magnetic-storm-related density reductions, decreases by factors of up to 3 and decreases by roughly an order of magnitude. Statistically, both types could be observed at any given latitude, but now there were two important developments that rapidly clarified the picture. One involved data from Byrd Station, Antarctica, recorded in the austral winter of 1959 and received at Stanford in the spring of 1960. The whistler activity at Byrd was surprising, far exceeding the activity at the Whistlers West stations in the US, Alaska, and New Zealand both in terms of the number of hours per day of activity as well as the whistler rates per minute. Most importantly, the paths terminating near Byrd regularly extended to geocentric distances of from ~ 4 to over 6 Earth radii (Carpenter, 1963a), so that the upper, nose-like part of the whistler could be directly observed on 0–8 kHz records. Soon there were found in the Byrd data whistlers with 'crossing traces,' the same effect observed in the SE/ST comparison of 13 January 1958, but now with the earlier arriving, higher latitude, components appearing on the same record with the conventionally delayed lower latitude traces. The map of Fig. 1.7b shows the southern-hemisphere location of Byrd (BY), while Fig. 1.10 shows on both 0–8 and 0–16 kHz records an example of a 1959 Byrd whistler with traces crossing one another in the part of the record above ~ 7 kHz. The high rates of activity at Byrd made it possible to compare the details of several events detected within a standard two-minute recording interval, and thus to confirm that, in a given event, the crossing traces were all from a single lightning source.

The analysis of Byrd data was challenging, in that Byrd was now providing a high-latitude 'view' of ground observed whistler activity, while stations such as ST and SE had provided a low-latitude perspective. At ST and SE VLF emission activity was relatively uncommon and the spectral records were usually dominated by spheric activity and relatively simple whistler forms. At Byrd, the spheric backround was low, and the multiple components of individual whistlers were often irregularly distributed with respect to one another, or shared the spectrum with a variety of VLF noise bands and wave bursts.

1.3.9 The knee effect as a regular feature of the equatorial density profile

The second development noted above came as the result of applying the extrapolation method to whistlers received at middle latitude stations such as SE and UN (see map of Fig. 1.7a). In many cases, all of the observed components exhibited evidence of the deeper type of density depression (as in Fig. 1.9b), so that it was not possible to identify the location or equatorial radius of a transition from one density regime to the other, as one now could with the Byrd data. However, the transition was well defined in a few remarkably clear cases (the SE spectra of Fig. 1.8c are an example), and it became increasingly evident that a sharp density decrease could exist at one time or another at any of a wide range of geocentric radii, and that whenever both low and high levels were observed simultaneously, the low-density regime was located exterior to the high-density one, in terms of geomagnetic field lines or magnetic shell parameters.

By mid-1961 Carpenter had completed an initial statistical study of density variations during magnetic storms as well as measurements of monthly average $D\{5\,\mathrm{kHz}\}$ at ST from 1957 to 1961. The results were reported in papers on the effect of magnetic storms on magnetospheric electron density and on secular variations in electron density (Carpenter, 1962a, 1962b). In these papers attention was limited to measurements of the size of the density variations and to the connection of certain variations with periods of magnetic activity. Meanwhile,

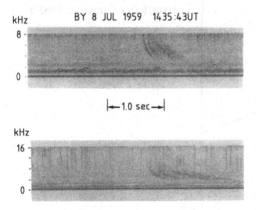

Figure 1.10 Spectrograms of a multi-component whistler whose earlier-arriving traces cross over some of the later arriving ones, i.e. the 'knee' effect. The spectrograms, representing 0–8 kHz and 0–16 kHz (note the different time scales), represent recording at Byrd, Antarctica on 8 July 1959.

the same material was included in the data section of his PhD dissertation (1962c), along with discussion of two observed types of post-magnetic-storm density effects, 'homogeneous' and 'inhomogeneous' depressions. It was suggested that the inhomogenous depression, exhibiting an inner region of moderately depressed electron density followed by a steep density drop and an outer region of still lower concentration, was in fact the more general state of the profile following disturbances. The homogeneous depressions were simply ones in which the observations had been limited to portions of the profile on one side of the steep fall off or the other. Fig. 1.11 shows the profiles in the schematic form presented in the thesis. The full curve was provided as a reference for quiet conditions. Profile segments A_1 and A_2 were homogeneous depressions, showing uniform logarithmic reductions by different amounts, while B showed the inhomogeneous depression, with the knee effect.

1.3.10   The search for interpretations of the knee effect

Having thus completed the data work, Carpenter began in mid-1961 to investigate possible explanations of the results. It was around this time that he applied the word 'knee' to the region of steep density gradients and the words 'knee whistler' to the particular type of whistler whose multiple components defined the location of the knee in geocentric Earth radii or L value. His interpretive efforts were strongly influenced by the seminal work of Axford and Hines (1961), who had postulated the existence of two principal magnetospheric cold-plasma

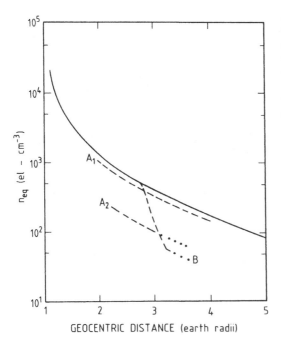

Figure 1.11   Carpenter's initial diagram of a reference or quiet day profile of equatorial electron density (solid curve) and the various forms of the profile observed by means of whistlers during magnetic storm periods (dashed curves) (after Carpenter, 1962c).

flow regimes, an outer, convection-driven one in which the circulation did not enclose the dipole and an inner, dipole-enclosing regime in which the plasma motions were dominated by the Earth's rotation. Carpenter considered it likely that the two essentially different density regimes, separated by the region of steep density gradients, were related to the corresponding flow regimes discussed by Axford and Hines. Since the high-density inner region appeared to be in a kind of quasi-equilibrium with the underlying ionosphere, it was the tenuous nature of the plasma beyond the knee that appeared to require explanation. However, Carpenter was unable to suggest an appropriate physical mechanism. The model of Axford and Hines involved a closed magnetosphere; a viscous solar-wind/magnetosphere interaction drove the plasma near the magnetopause antisunward to form the outer part of the convection loop, but the idea of a direct magnetic connection with the tenuous solar wind and the (as yet undiscovered) magnetotail were not included. Although Carpenter was content to associate the low density region with a corresponding high-latitude regime of enhanced convection, without further explanation, his supervisor R. A. Helliwell urged that a density-reducing mechanism be proposed. Carpenter (1962c) then suggested that the mechanism was latitude-dependent heating of the ionosphere, which might be expected to cause the light ions in the higher latitude, preferentially heated region to be redistributed throughout a very large high-latitude reservoir.

In 1961 Carpenter had received further encouragement from work at Poitiers, France by Y. Corcuff (1961), who reported a dramatic reduction in the dispersion of whistlers recorded during the great magnetic storms of the IGY. One such storm was that of 11 February 1958, during which whistlers of unusually low travel time had been observed by the Stanford group. Not having the advantage of the extrapolation method at the time, Corcuff had interpreted the change as predominantly due to an equatorward displacement of paths, an effect which may indeed have occurred in some of the cases studied. Although Corcuff did not stress the density changes that Carpenter believed to occur, her work nevertheless provided independent evidence of large reductions in whistler dispersion during magnetic storms, and thus contributed to the recognition of the plasmapause phenomenon (Carpenter, 1962c).

1.3.11 Initial publication on the knee

During 1962, Carpenter continued to accumulate examples of knee whistlers. The problem of demonstrating the essentially permanent nature of the knee was similar to ones faced by other IGY researchers whose data had been acquired during the highest levels of solar activity in ~ 200 years. Partly as a consequence of this high solar activity, the knee had frequently been at $L < 3$ during that period, such that its recognition required the use of the extrapolation method.

Furthermore, the data for such cases had to be obtained from middle-latitude northern-hemisphere stations where the whistler activity was sporadic. Already persuaded that the knee was a permanent feature of the magnetosphere, Carpenter finally decided that there was enough material on hand to persuade others. In the fall of 1962 an article on the knee was prepared and submitted to the *Journal of Geophysical Research*; it was published soon thereafter (Carpenter, 1963b).

Figure 1.12, a diagram from that paper, shows how in effect the plasmapause was discovered in whistler data. At the top are two idealized spectrograms of multicomponent whistlers. The second panels show the distribution in f–t space of the measured parameters, the nose frequencies and travel times (f_n, t_n) of the several whistler components. The bottom panels show equatorial electron density profiles that were estimated from the loci of the f_n, t_n values just above. The whistler at the left represented propagation entirely within the region corresponding to the high-density part of the profile, while in the case of the knee whistler (as it was now called) at right, propagation had occurred on both sides of a knee in the profile. The travel times of the components propagating outside the knee were shorter than those of their high-density counterparts (at left) by approximately the square root of the difference in density level. Since the normalized shape of the whistler dispersion curve, i.e. the variation of f/f_n versus t/t_n, was insensitive to the scale factor of the density distribution along the path and was only weakly sensitive to the functional form of that distribution (Smith, 1961b), the nose frequencies of the components propagating outside the knee were changed only slightly when the travel times of the components were reduced. The locus of f_n, t_n points folded back on itself, and on the spectra the traces cross one another, giving rise to the effect that was first noted in the SE/ST case of 13 January 1958 and had been described by Carpenter as crossing traces beginning in 1960.

1.3.12 News about the LUNIK results

Around the time of submitting the article on the knee, Carpenter heard for the first time of results from the USSR Lunik probes published by Gringauz, Kurt, Moroz and Shklovskii (1960c) and Gringauz (1963). He obtained one of the articles originally published in the USSR in 1960 and, being unfamiliar with particle instruments, discussed the *in situ* measurements with R. Mlodnosky, a young Stanford researcher who was managing a Stanford experimental VLF program on the upcoming OGO satellite series. Mlodnosky found the LUNIK results credible from the point of view of the instrument involved and Carpenter found them reasonable from the point of view of the density profile.

Invited by R. A. Helliwell, then Chairman of Commission IV of URSI, to present his findings at the 1963 URSI Assembly in Tokyo, Carpenter prepared a figure comparing the LUNIK results with an idealized equatorial density profile exhibiting a knee, of the kind obtained from whistlers (see Fig. 1.6). The comparison was complicated by the fact that the LUNIK trajectory had begun at ~60 degrees magnetic latitude, moving with altitude toward lower latitudes. At low altitudes the LUNIK numbers were much smaller than the whistler values, possibly because the rocket had initially been poleward of the plasmapause, and had then entered the plasmasphere as it moved equatorward. Carpenter was particularly pleased to meet Gringauz at the URSI conference and to find that they were in good mutual agreement about the knee effect. There were rumors about disagreements among Soviet scientists concerning the LUNIK profiles,

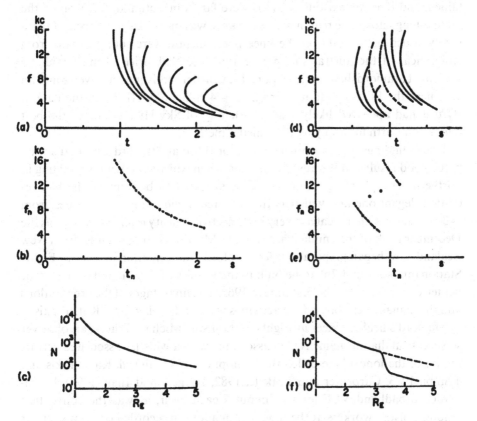

Figure 1.12 Diagram showing how the knee or plasmapause effect was identified in multicomponent whistler data. (a) Whistler spectra representing propagation on multiple paths in the plasmasphere, but not in the plasmatrough or plasmapause region. (b) The distribution in frequency-versus-time space of the whistler nose frequencies and travel times associated with the spectra of (a). (c) The equatorial electron density profile inferred from the data of (b). (d) Whistler spectra representing propagation on both sides of the plasmapause. The dashed curves represents propagation at intermediate density levels in the region of steep gradients. (e) The distribution in frequency-versus-time space of the nose frequencies and travel times associated with the spectra of (d). (f) The equatorial electron density profile inferred from the data of (e) (adapted from Carpenter, 1963b).

but Carpenter did not learn much about this at the time, having friendly but limited personal contacts with his Soviet colleague during the meeting.

1.3.13   The need for better data: Eights, Antarctica, as its source

While Carpenter regarded his 1963 paper as convincing on the subject of the knee as a permanent feature of the magnetosphere and on the negative correlation of the radius of the knee with magnetic activity, he believed that only a limited view of the behavior of the knee phenomenon had thus far been obtained, and that the existing stations were far from optimum for probing the middle-magnetosphere region where the knee was most likely to appear. Fortunately, much better data on the knee phenomenon were soon to come from Antarctica. In the austral summer of 1961–62, Neil Brice, then a graduate student at Stanford, had carried a portable whistler receiver on a traverse into an area near the Ellsworth mountains at $L \sim 4$ called Sky-Hi. From the data of Byrd it had appeared likely that the vicinity of Sky-Hi was one of the best regions on Earth for applying the whistler method.

Brice's findings were surprising, even for those at Stanford who had anticipated good results. In the austral summer, when whistler activity originating in northern-hemisphere lightning would be expected to be minimal, he had recorded elegant one-hop whistlers that covered a wide range of L values, from ~ 3 to 6, and which indicated a very high electron density level (recognized as the December peak of the annual variation). L. Martin, visiting scientist from New Zealand, now began preparations for Stanford's participation in work at Eights Station (map of Fig. 1.7b), to be built in the vicinity of Sky-Hi and occupied for winter-over science for the first time in 1963. The final stages of the preparations and the management of the field activities were undertaken by J. Katsufrakis.

Stanford's field engineer at Eights Station was Michael Trimpi, an observer so successful that his name is now associated with a whistler-associated perturbation of the ionosphere called the 'Trimpi event' (Helliwell, Katsufrakis and Trimpi, 1973; Carpenter and LaBelle, 1982; Burgess and Inan, 1993). He recorded broadband VLF for two minutes each hour, as was the convention among whistler workers at the time, but in addition recorded continuously for three hours each day in parallel with similar recordings at Byrd, advancing the start time of the recordings by three hours every 7 days.

In this period, the analysis of whistlers was increasingly done with a real-time spectrum analyzer, the Raytheon Rayspan, which utilized 420 magnetostrictive filters and a capacitive commutation system (Helliwell, Crary, Katsufrakis and Trimpi, 1961). The output was displayed on the Y axis of a CRT and intensity modulated, so that a 35 mm film could be exposed to create a continuous display of frequency versus time, with a time scale that depended upon the camera drive speed. The film was then projected onto a viewing screen, so that cross-hairs could be positioned and the coordinates of a location recorded digitally.

1.3.14 Initial whistler studies of the worldwide structure and dynamics of the plasmasphere

The recordings of knee whistlers from the austral winter of 1963 at Eights were astonishing in their extent and detail. The year was one of low solar activity, and the magnetic activity consisted principally of weak magnetic storms, characterized by maximum values of the K_p-index near 6 and then relatively steady but moderate substorm activity, with K_p near 3, during multiday recovery periods. The circumstances for whistler propagation were ideal. The magnetic storms were weak; storm-related interruptions in the whistler activity, well known from the great events of the IGY period, were minimal, and the ensuing recovery periods, known to be favored intervals for whistler activity (Yoshida and Hatanaka, 1962), were long in duration. For much of the June–July–August 1963 period, it was possible to obtain density profile information and to track the position and occurrence of the plasmapause for many hours each day.

The Eights data provided clear evidence of an inward displacement of the plasmapause during magnetic storms. Once this initial phase was complete, a distinctive diurnal variation in plasmapause radius appeared and was observed for several days in succession as relatively steady substorm activity continued into the recovery period. Fig. 1.13 shows plasmapause measurements near the beginning of a storm period; the upper panel shows hourly values of R_E, the equatorial geocentric range in Earth radii within which, or very near which, the knee was inferred to be located. The lower panel shows K_p, with 3-hour values increasing downward to illustrate the inverse relation between the knee position and magnetic activity. The geocentric range scale was made linear in R_E^{-2} so that, on the assumption that $\mathbf{E} = -(\mathbf{v} \times \mathbf{B})$, the first derivative of the plotted data time series $d(R_E^{-2})/dt$ would be approximately proportional to $\mathbf{E_e}$, the eastward component of the electric field.

Figure 1.14 shows the remarkably repeatable diurnal variation of the knee position that was observed for several days in the aftermath of a weak magnetic storm in early July 1963. The nightside was characterized by inward displacement from a relatively large radius; on the dayside the radius increased slightly, on average, to a minor peak at noon, and then in the late afternoon shifted abruptly outward by of order one Earth radius to form a bulge-like extension.

1.3.15 The use of whistlers to track cross-L plasma motions

While studying the Eights data on the knee position, Carpenter noted a phenomenon that was to play an important role in further studies of plasmasphere shape and dynamics. On one occasion, whistlers recorded for several successive nighttime hours exhibited unusually long trains of multihop echoes. The path radius could not easily be estimated in this case, but there was a steady, hour-by-hour decrease in the echo period. A density decrease did not appear to be the predominant effect involved, since magnetic activity was not changing sharply at the time. However, an inward displacement of the path would be consistent with the manner in which the plasmapause radius had been observed to change during such nights. Having read the work of Axford and Hines (1961) on the subject of magnetospheric convection, Carpenter concluded that the whistler path, in the form of a tube of enhanced ionization, was participating in the bulk $\mathbf{E} \times \mathbf{B}/B^2$ drift motions of the surrounding plasma. The duct hypothesis seemed to protect solidly against any possible interpretation in terms of a simple movement in time of the whistler lightning sources, without significant plasma drifts.

From this point on, the idea of tracking the cross-L motions of whistler paths developed rapidly as an aid to interpreting data on plasmasphere structure.

One of the first clear findings on plasma drifts covered a 3-hour continuous recording near midnight on 15 May 1963. A well-defined nose-whistler component, propagating near $L = 4.5$, was observed repeatedly; its nose frequency was found to increase relatively steadily as the nose travel time decreased (Fig. 1.15). The inferred inward displacement was ~ 0.3 Earth radii, and the overall change

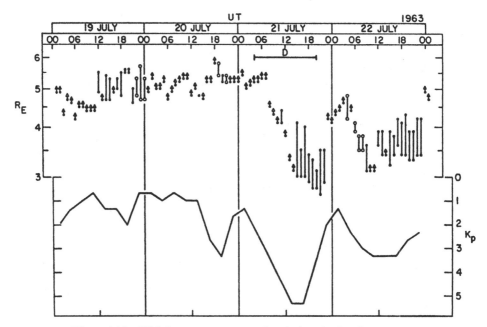

Figure 1.13   Whistler measurements of variations in the plasmapause equatorial radius with time during the onset of a weak magnetic storm. Lines connecting data points show the range within which the plasmapause was found to be located, while data points with arrows indicate points in the plasmasphere beyond which the plasmapause was inferred to lie. The K_p index is plotted with values increasing downward. The whistler data were recorded at Eights, Antarctica during a four-day period in July 1963 (after Carpenter, 1966).

was consistent with data on the change in the average position of the plasmapause during the same local time period.

Another finding on cross-L motions concerned whistler path radii near dusk. As local time advanced in the late afternoon and the region of larger plasmasphere radius began to be observed, there was usually no indication of a cross-L flow outward into the bulge region. Instead, whistler components observed inside the afternoon plasmapause continued to exhibit the same nose frequencies, as if at unchanged radii. This is indicated by arrow 'c' in Fig. 1.16, which is a

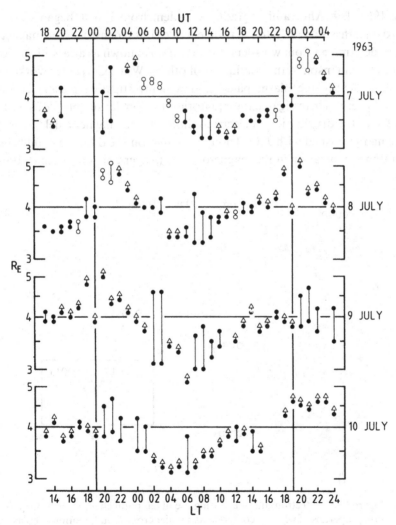

Figure 1.14 Whistler measurements of diurnal variations in the plasmapause position during a four-day period of relatively steady, moderate geomagnetic agitation that followed a weak magnetic storm. The whistler data were recorded at Eights, Antarctica in July, 1963 (after Carpenter, 1966).

summary figure from the paper on the plasmapause published in 1966. Meanwhile, the bulge-like region of high-density plasma that extended beyond the afternoon plasmapause radius was represented in the whistler data by a newly observed set of components. As indicated in Fig. 1.16, Carpenter interpreted this as indicating that, in terms of the origin of its plasma, the region of the bulge differed from the region interior to it, or main plasmasphere.

1.3.16 Angerami and studies of the magnetospheric electron density distribution

During 1965, J. J. Angerami, a graduate student from Brazil, began work at Stanford on a thesis devoted to the problem of electron density in the magnetosphere as determined from whistlers. Angerami was known as a careful, meticulous worker who inspired the confidence of others. While Carpenter worked to locate and identify the plasmapause and was investigating plasma cross-L motions, Angerami investigated the equatorial electron density profile as well as the profile of the distribution of electrons along the geomagnetic field lines. In 1964 he had published with J. O. Thomas a paper on the diffusive equilibrium distribution of ionization in the magnetosphere (Angerami and Thomas, 1964).

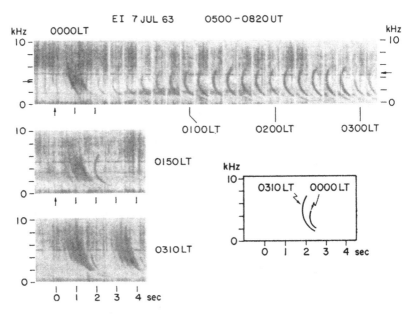

Figure 1.15 Spectrograms illustrating use of temporal changes in the dispersion properties of whistler components to infer cross-L bulk motions of plasma in the outer plasmasphere. The series of spectrogram segments in the top panel shows the appearance of a particular whistler component at intervals of approximately ten minutes over a three-hour period. At the left are spectrograms of the entire multicomponent whistler at three times during the period of interest (after Carpenter, 1966).

Previous workers such as Storey (1953, 1956) and Smith (1961b) had recognized that the whistler travel time was weighted in favor of conditions near the equatorial region. Hence relatively good estimates of the equatorial electron density could be made, within a factor of 2 or 3, in spite of the need to make assumptions about the functional form of the field-line model of plasma density.

Angerami sought to improve these already good density estimates through a combination of theory, whistler data on equatorial electron density, and topside ionosphere electron density data from the highly successful Alouette 1 topside sounding experiment (Schmerling and Langille, 1969). He showed that, while a diffusive equilibrium-type field-line model was consistent with the data from the plasmasphere, a model varying much more rapidly with radial distance was required in the region of the plasmatrough, in particular one such as the exospheric model calculated by Eviatar, Lenchek and Singer (1964), which varied approximately as R^{-4}. Angerami presented equatorial profiles with bars indicating the estimated range of uncertainty in the density levels, that uncertainty being due largely to lack of knowledge of the functional form of the field aligned density distribution. He also showed plots of the corresponding electron content in a tube of ionization above $1\,cm^2$ at $1000\,km$ altitude. As might be expected from the integral nature of whistler measurements, he found that the model-related uncertainty in the tube content was smaller than the corresponding uncertainty in the equatorial density. Angerami made clear what had been

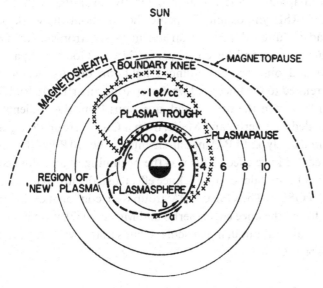

Figure 1.16   Sketch showing Carpenter's (1966) original model of the equatorial distribution of thermal plasma in the magnetosphere. The crosses bounding the region marked 'plasma trough' were not intended to represent a physical boundary, but rather indicated the region where significant amounts of whistler activity beyond the plasmapause had been observed (after Carpenter, 1966).

left uncertain in Carpenter's 1963 paper, namely that the density depletion beyond the plasmasphere could be assumed to extend essentially all along the field lines between the conjugate hemispheres.

1.3.17 Presentation of the new data on plasmasphere structure and dynamics

In the Spring of 1965, because of the wide scope of the new findings, Carpenter and an associate T. R. Jewell prepared a stop action movie film for presentation at a special American Geophysical Union (AGU)-URSI symposium on Solar–Terrestrial Relations to be held in Washington DC during the Spring, 1965 AGU meeting. One section of the film contained a series of 24 whistlers from 5 August 1963, one from each hourly synoptic recording. The whistler spectra, each displayed with respect to the time of its causative spheric, showed a diurnal sequence that had been found characteristic of periods of moderate, steady geomagnetic agitation, such as the ones illustrated in Fig. 1.14. Sketches of some of the spectral forms shown in the film sequence are presented in Fig. 1.17. The film also included the cross-L inward drift event illustrated in Fig. 1.15, as well as spectra showing the apparent tracking of plasmasphere whistler components over a ~ 20 hour period that terminated in the late afternoon. The hour-by-hour changes in travel time of the components suggested that their paths had undergone radial motions associated with the diurnal inward and outward displacements of the plasmapause, while also approximately corotating with the earth.

At the AGU–URSI symposium, Carpenter heard about the work of H. A. Taylor, who had acquired data from an ion mass spectrometer on OGO 1, launched in 1964. Taylor and his colleagues at NASA Goddard Space Flight Center (GSFC) had observed pronounced light ion density decreases that appeared to be related to the knee, and were preparing an article for publication (Taylor *et al.*, 1965a). On the other hand, questions from the audience at the symposium revealed that Serbu and Meier, from a GSFC group doing electron retarding potential analyzer (RPA) measurements on the IMP 1 and 3 satellites, had not observed the knee effect. This was the beginning of a prolonged, often lively, debate about the evidence for the knee effect, usually involving Carpenter and S. J. Bauer of GSFC. The GSFC RPA group were confident of their results, and not having found the knee effect (Serbu and Meier, 1967), were understandably skeptical of claims about it, in particular claims that it was a regular feature of the magnetosphere.

1.3.18 Introduction of the terms plasmapause and plasmasphere

Until mid-1965, Carpenter and Angerami worked steadily to complete two major papers, one on the new findings from Eights about the knee position and

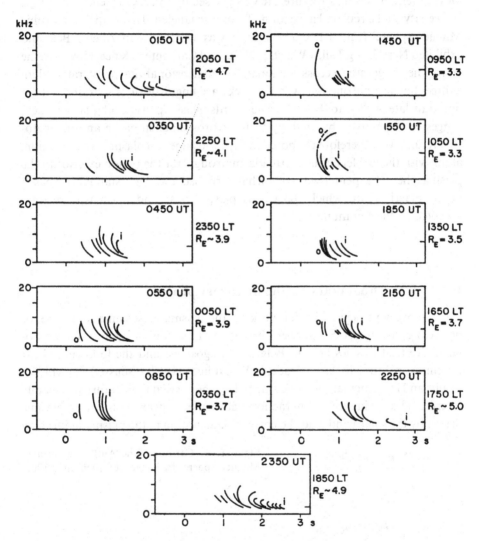

Figure 1.17   Tracings of whistler spectra showing an example of the diurnal variation in whistler form and in the inferred plasmapause radius during a period of moderate and steady geomagnetic agitation. Time is shown with reference to the time of the causative atmospheric. The approximate plasmapause radius deduced from each record is indicated at the right in Earth radii. The whistler component found to be closest to the plasmapause on the inside is marked 'i', the component closest on the outside with an 'o'. The recordings were made at Eights, Antarctica on 5 August 1963 (after Carpenter, 1966).

another on the equatorial electron density profile. These were then submitted to the *Journal of Geophysical Research* and were published together (Carpenter, 1966; Angerami and Carpenter, 1966) in early 1966. As the papers were being prepared, Carpenter had decided that some new descriptive terms were needed for the thermal plasma structure. He was pleased that B. J. O'Brien, then at Rice University, had circulated a cartoon of a mouse labeled 'The Anthropomorphic Magnetosphere' (Fig. 1.18). The mouse was wearing Van Allen's Belt and exhibited Ness' Long Tail, a Whistler Nose, and Carpenter's Knee. However, the word 'knee' suggested a cross-sectional, two-dimensional viewpoint, rather than a three-dimensional one, and was also awkward linguistically; for example, it did not translate well into French. The words plasmapause and plasmasphere suggested themselves, by analogy to the words magnetopause and magneto-sphere, and were therefore proposed in the 1966 paper on plasmapause location, along with the less felicitous word 'plasmatrough' for the region exterior to the plasmasphere. Carpenter was somewhat surprised and understandably pleased at the alacrity with which the words plasmapause and plasmasphere were accepted by the community.

1.3.19 Some final notes on the discovery phase

The discovery phase of whistler work on the plasmapause came to a close, or near close, in August–September 1966 at the Inter-Union Symposium on So-lar–Terrestrial Physics held in Belgrade, Yugoslavia and the following URSI Assembly in Munich. At Belgrade, J. W. Dungey (1967) discussed the whistler results and in particular the dusk-side bulge, finding that that feature was readily interpreted as evidence of interaction between a sunward convecting plasma, driven by the solar wind, and the corotational motion in the plasma induced by

ANTHROPOMORPHIC MAGNETOSPHERE

CHARACTERIZED BY

NOSE WHISTLERS

(VAN ALLEN'S) BELT

(CARPENTER'S) KNEE

AND

(NESS') LONG TAIL

Figure 1.18 Cartoon of the 'Anthropomorphic Magnetosphere', by Brian J. O'Brien, *circa* 1965.

the Earth. Later, at the URSI Assembly, during a Commission IV session on new developments, there was a debate between Carpenter and Bauer about the knee effect. The session, for which D. Gurnett and F. L. Scarf acted as reporters (Gurnett and Scarf, 1967), was well attended by those who had recently participated in the Belgrade symposium. Carpenter summarized the evidence that had been obtained from Eights, which now included material from an additional year of recording in 1965. He briefly reviewed the variety of evidence that appeared to support the existence of the knee, including data on ion density from satellites, slant columnar electron content to the ground at middle latitudes, and abrupt spatial changes in satellite observations of whistlers and of VLF noises such as triggered chorus. Then Bauer defended the GSFC electron trap data from IMP 1 and 3, pointing out that there was some evidence for a fairly rapid decrease in density, but not for one as pronounced as that reported from whistlers or from Taylor's OGO 1 instrument. It was noted, however, that uncertainty in the effective collecting area of the GSFC instruments could modify the inferred densities downward by a factor of 2–3. In the ensuing discussion R. L. Smith and F. L. Scarf questioned the estimated effective area of the GSFC electron traps, since the whistler results would indicate a discontinuous increase in the sheath size by almost an order of magnitude as the spacecraft crossed the region of steep density gradients. Questions about the possible effects on the whistler results of geomagnetic field distortions and persistent magnetospheric inflation were raised by Bauer and N. F. Ness. Finally, in support of the knee measurements, M. J. Rycroft indicated that the ionospheric trough (determined from Alouette 1) and Carpenter's knee appeared to fall on the same L-shells, for a wide range of K_p values. At the end of the debate there was an unofficial show of hands, for or against the knee. The outcome was not recorded, but Carpenter recalls that Alex Dessler, who had started him on the road to magnetic storm discoveries seven years before, stood up and said 'I vote for Carpenter.' It was one of those moments that a young researcher does not forget.

At the time of the debates on the knee, Carpenter considered Bauer to be a friendly but serious critic who was constantly probing for weaknesses in the knee picture. In retrospect, particularly in the light of an excellent talk given by Bauer (1970) at the Ottawa URSI Assembly in 1969 on the occasion of the final debate, it appears that Bauer's persistent efforts to get at the physics at issue were extremely helpful. They clarified the extent to which questions in the debate had been resolved, and held up for scrutiny those issues that needed more attention, one of which was the abiding question of the extent and nature of the coupling of the plasmapause region to the underlying ionosphere.

Chapter 2

Electromagnetic sounding of the plasmasphere

2.1　Introduction

The previous chapter contains contributions to the history of science in the field of space plasma physics. It explains how the plasmapause, this peculiar and unexpected magnetospheric frontier, was discovered independently in the late 1950s and early 1960s by two scientists from the two leading countries involved in space exploration. The discoveries were made by using two totally different technical methods of measurement: *in situ* spacecraft observations and electromagnetic sounding of the magnetosphere. These techniques were both in their infancy at the time.

The main results of electromagnetic sounding of the plasmasphere, from the ground and from satellites, will now be described. *In situ* satellite particle observations will be outlined in Chapter 3.

In both this chapter and the next the most relevant results will be presented without emphasis upon technical aspects of the experiments. Such aspects are well described in the specialized literature, examples in the case of the whistler method being works by Smith (1961a); Carpenter and Smith (1964); Helliwell (1965); Carpenter and Park (1973); Rycroft (1974a); Y. Corcuff (1975); Tarcsai (1975); P. Corcuff (1977); P. Corcuff, Y. Corcuff and Tarcsai (1977); Park and Carpenter (1978); Bernhardt (1979); Daniell (1986) and Rycroft (1987). An extensive review of the use of whistlers for magnetospheric diagnostics is given by Sazhin, Hayakawa and Bullough (1992).

2.2   **Initial results**

As noted above, Storey (1953) used whistlers for the initial identification of the
dense plasmasphere, and Carpenter (1962b) used evidence of unusually low
whistler travel times to infer the occurrence of deep, factor-of-∼ 10 depressions
in electron density during the severe magnetic storms of the IGY. The relations
between the ostensibly incompatible concepts of a dense plasmasphere and of
pronounced density depressions were clarified when, in 1963, Carpenter's research led him to conclude that 'the distribution of ionization in the magnetosphere may show a pronounced departure from smoothness. At low heights, the
equatorial density profile may show normal density levels and a relatively
smooth decrease with increasing altitude, but at a geocentric distance of several
Earth radii the density values may drop sharply, returning to a gradual rate of
decrease only after a substantially depressed level is reached' (Carpenter, 1963b).
This sharp drop in the equatorial plasma density became known as the 'knee in
the magnetospheric density profile'. It was only later, in 1966, that this peculiar
boundary was called the 'plasmapause' by Carpenter (1966). The word 'plasmapause' was originaly used when the three-dimensionality of the knee phe-

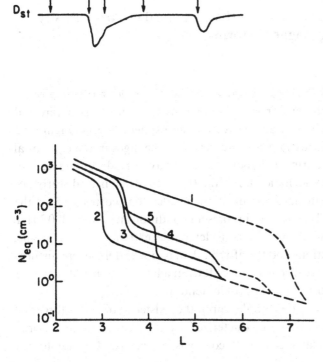

Figure 2.1   Schematic illustration of changes in
the equatorial profile of
the electron concentra-
tion in the magneto-
sphere expected during a
10-day period that in-
cludes two magnetic
storms. Above is an in-
dication of the D_{st}
geomagnetic activity in-
dex (ring current). Below
are electron concentra-
tion profiles that corre-
spond to the times
marked along the D_{st}
curve. These idealized
profiles represent condi-
tions near the 0400 MLT
meridian. They show es-
tablishment of well de-
fined plasmapause pro-
files during disturbance
periods and suggest the
complexities associated
with periods of recovery
(after Carpenter and
Park, 1973).

nomenon was emphasized, but that restriction has not been maintained in practice. It should also be mentioned that in this same 1966 paper Carpenter introduced the term 'plasmasphere' for the dense region inside the plasmapause and 'plasmatrough' for the tenous region outside.

In the following years much additional information about the plasmasphere was deduced from ducted whistler observations and from other forms of radio probing. Highlights of these results will be reviewed in the following sections.

In order to emphasize the essentially dynamic nature of the plasma-sphere–plasmapause system, we shall use the changes in the density structure of the plasmasphere during magnetic storms as an initial framework for discussion, many of the early insights about the plasmasphere having been obtained by focusing attention on changes during disturbed periods. We begin with those changes as they are registered in the radial profile of total density. This is followed by a review of data on the worldwide effects of magnetic storms on the plasmasphere. Coupling of the plasmasphere and ionosphere is then discussed as well as other aspects of plasmasphere structure and dynamics, including the equatorial electron density profile, irregularities, plasma temperature, and quiet-day electric fields. Finally, there is a discussion of certain aspects of wave activity related to the plasmasphere.

2.3   Plasmasphere dynamics

2.3.1   An overview of changes in the equatorial electron density profile during magnetic storms

Introduction

In 1973 Carpenter and Park published a review paper designed to provide ionospheric workers with recent results from whistlers about the structure and dynamics of the region overlying the regular ionospheric layers. Figure 2.1 (taken from that paper) illustrates schematically the changes in the equatorial profile to be expected during a hypothetical 10-day period containing two magnetic storms. Above is an indication of the D_{st} index of magnetic disturbance activity. Below are equatorial concentration profiles that correspond to the times marked along the D_{st} curve and represent conditions near the 0400 MLT meridian. Profiles 1 and 2 may be considered the extremes of the profile, representing quiet conditions and the aftermath of a moderately severe disturbance, respectively. They are discussed more fully in a later section and represent limits between which most observed profiles tend to lie.

The depicted changes in the profile imply the existence of a fast (order of hours) plasma removal process, characterized by a 'knee' in the equatorial profile and a slow, multiday recovery. Recovery is characterized by day-to-day increases in density in the depleted region and the continuing presence of steep density gradients at L values near or slightly beyond those of the stormtime knee development. (A plasmapause radius at time t_3 larger than at time t_2 is attributed to rotation into the meridian of observation of a portion of the plasmasphere with larger radius, rather than cross-L diffusion.) Since the process of plasmapause formation has been found to be an essentially permanent one, retreating to higher L as disturbance levels subside but continuing to be active, a new plasmapause or knee begins to form (at $L = 5.5$ in the case illustrated) as recovery proceeds. Its effects in terms of a knee in the density profile may not be identified for some time, however, until densities in the L range just inside the point of incipient knee formation approach plasmasphere levels.

During the second storm indicated in Fig. 2.1, a new knee is formed at $L \sim 4.3$, but because of the 'vestigial' plasmapause at lower L and the as yet only partially recovered density near $L = 4$, the resulting profile exhibits a step-like structure.

Examples from experiments

Figure 2.2 shows a series of nighttime profiles determined by Corcuff *et al.* (1972) from a combination of ground whistler and satellite data acquired in 1968. The profiles reflect the larger-scale features that were estimated to be present at the times noted along the plot of K_p values. The density levels in the trough region indicated in the figure are lower by factors of up to 10 than are now considered to be typical, but in general the curves illustrate well the idealized situation depicted in Fig. 2.1.

ISEE 1 satellite data on near equatorial electron density, acquired by the University of Iowa Plasma Wave Instrument (Mosier, Kaiser and Brown, 1973; Gurnett *et al.*, 1979) along a series of orbits at post-noon local time, are shown in Fig. 2.3 (from Carpenter and Anderson, 1992). The profiles, obtained from measurements by a sweep frequency receiver (SFR) of resonance phenomena in the local plasma (Gurnett and Shaw, 1973), exhibit changes similar to those illustrated in Figs 2.1 and 2.2. Following the appearance of a well-defined plasmapause on day no. 215, the profile is shown in two successive states of recovery, one only partial, with evidence of reduction beyond $L = 5$ (no. 217), and the other (no. 219) complete to at least $L = 8$. On day 224, following renewed disturbance activity, the profile (dashed line) again showed a well-defined knee effect, remarkably similar to that of day 215.

JIKIKEN (EXOS-B) electron density profiles obtained at $L > 4$ on the nightside by the Stimulated Plasma Wave (SPW) experiment are shown in Fig. 2.4 (Oya and Ono, 1987). The profiles show the result of a buildup of dense plasma in previously depleted regions during quieting, and also indicate a disappearance of dense plasma beyond a newly developed plasmapause as a consequence of increased disturbance.

The general pattern indicated in Fig. 2.1 has been supported by experiments with the satellite radio beacon technique (Almeida, 1973; Davies, Hartmann and Leitinger, 1977; Davies, 1980). Group delay and Faraday rotation measurements on signals propagating from synchronous orbit to a ground station allow estimates of the plasmasphere contribution to total electron content along the essentially linear path. This content has been found to be enhanced on the first day following a sudden commencement (see later discussion), then decreases for about 2 days, and then recovers slowly over periods as long as 14–20 days, depending upon the subsequent magnetic activity (Soicher, 1976; Degenhardt, Hartmann and Leitinger, 1977; Poletti-Liuzzi, Yeh and Liu, 1977 ; Kersley, Hajeb-Hosseiniem and Edward, 1978). Figure 2.5 shows an example of the variation in the inferred plasmasphere content N_p versus time measured at Graz and Lindau during a storm that began on 10 January, 1976. At the end of the main phase, N_p was $\sim 50\%$ of its pre-storm value.

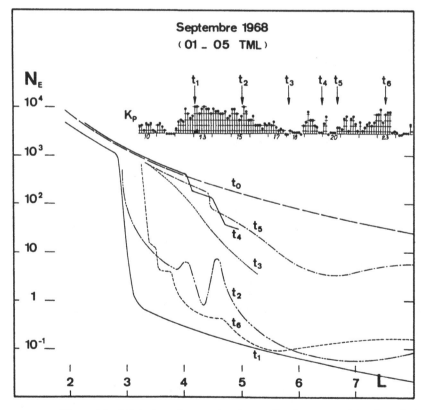

Figure 2.2 Nighttime electron density profiles from a combination of satellite and ground whistler data illustrating recovery effects similar to those indicated in Fig. 2.1. The indicated density levels in the trough region are lower by factors of up to 10 than are now considered to be correct (after Corcuff *et al.*, 1972).

Observations of plasmapause radius variations with magnetic activity

One of the properties of the knee most readily detected from ground whistler and satellite data is the tendency for its radius to vary inversely with the level of geomagnetic activity, much in the manner of correponding variations in magnetopause radius, auroral oval location, and ring current position (Carpenter, 1963b; Corcuff and Delaroche, 1964; Binsack, 1967; Chappell, Harris and Sharp, 1970a; Rycroft and Thomas, 1970). During some of the great storms of the IGY, the knee was found to be near $L = 2$, and it was probably at $L < 2$ at times (Corcuff and Delaroche, 1964). On the dayside on quiet days it may be observed beyond $L = 8$ (Gringauz and Bezrukikh, 1976; Carpenter, 1981; Nagai, Horwitz, Anderson and Chappell, 1985), if it can be identified at all (see, for example, the profile for day 219 in Figure 2.3).

Figure 2.6 shows some results from a whistler study, near sunspot minimum, of the observed radius of the knee at ~ 0600 MLT versus the maximum K_p value in the preceding 24 hours (Carpenter, 1967). A negative correlation coefficient of -0.67 was found, and an even more significant coefficient, -0.76, was cal-

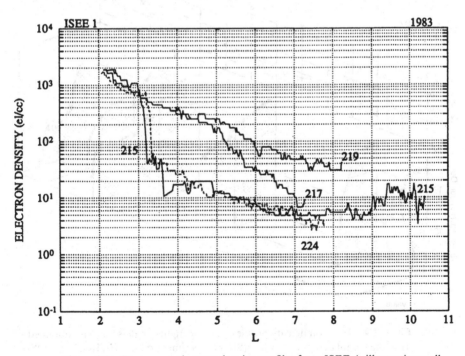

Figure 2.3 Post-noon electron density profiles from ISEE-1, illustrating well-defined plasmapause effects as well as states of recovery. The profiles were obtained from sweep frequency receiver (SFR) information about resonance phenomena in the local plasma (after Carpenter and Anderson, 1992).

culated when the K index for Byrd Station was used instead of K_p, Byrd being at the approximate longitude of the whistler measurements.

As an approximation to the data of Fig. 2.6, Carpenter and Park (1973) proposed the relation:

$$L_{pp} = 5.7 - 0.47K_p \tag{2.1}$$

as a predictor of the plasmapause position in the midnight-dawn sector, where K_p represents the maximum value of the 3-hour index in the preceding 12 hours. The formula was considered by the authors to be most useful during periods of increasing or persistent disturbance, when the effects of recent increases in activity should still be well defined in the profile.

A relation remarkably similar to (2.1) was obtained by Carpenter and Anderson (1992) from a study of 208 near-equatorial plasmapause crossings by the ISEE-1 satellite in 1977, 1982, and 1983. The crossings were identified on electron density profiles obtained from measurements of the upper hybrid

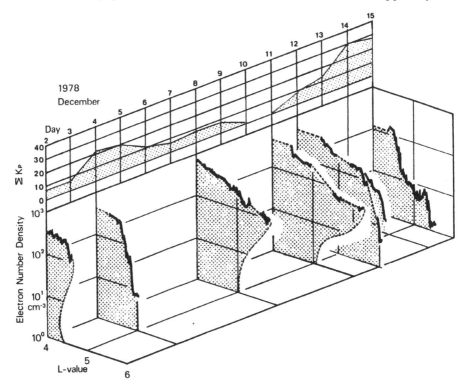

Figure 2.4   Electron density profiles near the nightside plasmapause obtained from the stimulated plasma wave (SPW) experiment on the JIKIKEN (EXOS-B) satellite. The dashed lines between the tick lines represent interpolations during times when the experiment was not operating. The sequence of data illustrates how the density profile responds to changing geomagnetic activity conditions indicated in the top panel by ΣK_p (after Oya and Ono, 1987).

resonance and plasma frequencies on sweep frequency receiver (SFR) records. A least squares linear fit to the data produced the relation:

$$L_{ppi} = 5.6 - 0.46K_{p,max} \qquad (2.2)$$

where L_{ppi} is the L value of the last point measured interior to a steep plasmapause falloff. $K_{p,max}$ is the maximum value of K_p in the preceding 24 hours, but in the case of periods centered at 09, 12, and 15 MLT, respectively, the values for one, two, or three immediately preceding 3-hour periods were ignored to account for observed delays in the response of the dayside radius to enhanced convection activity (Chappell, Harris and Sharp, 1971a; Décréau, Béghin and Parrot, 1982, 1984). The fact that the relation deduced by Carpenter and Anderson for data widely distributed over the 0000–1500 MLT agreed with the result of Carpenter and Park (1973) for times near local dawn appears to be related to the nearly circular shape of the statistically averaged boundary within the 0000–1500 MLT period (see the solid part of the curve of Fig. 1.16 and the discussion below of the distinction between the main plasmasphere and the bulge region).

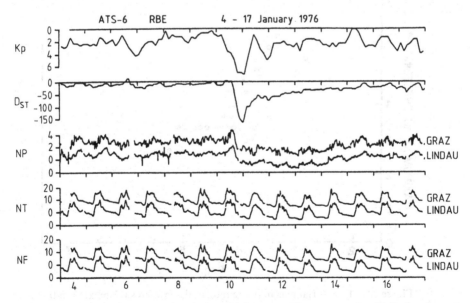

Figure 2.5 Temporal variations of electron columnar content N_p in the plasmasphere and N_F in the ionosphere along linear paths from ground stations at Graz and Lindau to a synchronous satellite. $N_T = N_P + N_F$ is the total columnar content. The K_p and D_{st} indices are shown above and ionospheric data for Graz and Lindau below (after Degenhardt *et al.*, 1977).

Comparisons of whistler and satellite data

The description of the plasmapause based on ground whistler data led to *in situ* searches for related effects. As one example, ALOUETTE 1 broadband radio data recorded at 1000 km altitude were compared with simultaneous ground whistler data as a means of cross-checking diagnostic methods and extending the early whistler results on magnetic disturbance variations in plasmapause position (Carpenter, Walter, Barrington and McEwen, 1968). At what was interpreted as the plasmapause at 1000 km altitude, the broadband VLF records from the polar orbiting ALOUETTE were often found to exhibit a sharp change or 'breakup' in a band of quasi-electrostatic noise associated with the lower hybrid resonance (LHR) frequency (Barrington, Belrose and Deely, 1963; Brice and Smith, 1965). At that location there was often a second effect, a cutoff with increasing latitude in whistlers propagating from sources in the conjugate ionosphere. The invariant latitude of these changes was found to agree well with

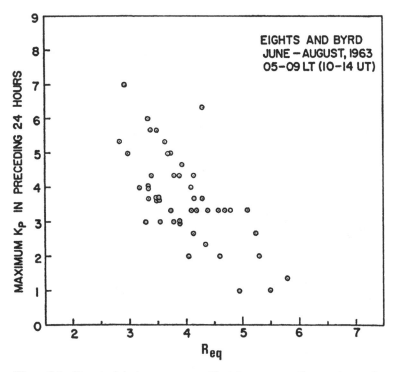

Figure 2.6   Equatorial plasmapause radii at dawn versus the maximum three-hour K_p value in the 24 hours preceding the measurement. The whistler data were recorded in the Antarctic at Byrd and Eights, and represent a roughly 30° range of longitudes near the prime geomagnetic meridian. Centered dipole coordinates were employed in the calculations of geocentric distance (after Carpenter, 1967).

simultaneous results from whistlers recorded at a ground station, and to vary inversely with magnetic activity. Figure 2.7 illustrates a sudden VLF change detected by ALOUETTE 1 on a pass near Byrd Station, Antarctica on 26 August, 1965 at ~0400 MLT (Carpenter et al., 1968). VLF spectra in the range 0–10 kHz are shown on a compressed time scale (above) and on an expanded scale below. At left, well-defined whistlers appeared following propagation through the outer plasmasphere from lightning sources in the northern hemisphere. An intense noise band at and above the LHR frequency appeared and fell slowly and steadily in frequency until a point of sudden change or breakup, after which the noise activity became irregular and burst-like and the evidence of one-hop whistlers became fragmentary. The transition from a smooth LHR noise band to another wave environment occurred within less than ~1 second, or within a distance of < ~6 km along the satellite orbit!

Multiday tracking of the plasmapause position from polar orbit

The scatter of ~ ± 0.5R_E in the values of R_{eq} in Fig. 2.6 for a given value of K_p was partially explained by data from another polar orbiting satellite, OGO 4. Fig. 2.8, from Carpenter and Park (1973), shows plasmapause positions at ~0100 MLT measured at roughly 90-minute intervals during a several-day period in September 1967. Below is a plot of the AE index. In this case of operations with a loop antenna, the plasmapause was detected from both the latitudinal whistler cutoff and abrupt latitudinal changes in propagating VLF hiss and chorus emissions. As in Figs 1.13 and 1.14, the L scale is linear in L^{-2} to facilitate recognition of effects due to convection electric fields. Dashed lines connecting either pairs of symbols or single symbols and arrows indicate the R_E or L range within which the plasmapause was estimated to lie. There is evident a general trend toward lower L values with increasing substorm activity, but also

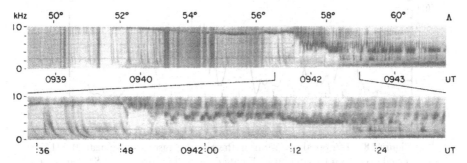

Figure 2.7 ALOUETTE 1 VLF frequency–time spectrograms illustrating abrupt changes in whistler and VLF noise activity at what appears to be the projection of the plasmapause at ~1000 km altitude. The changes occurred at ~57° invariant latitude, and are shown on an expanded scale on the second panel (after Carpenter et al., 1968).

rapid variations in the data, on a time scale of $\sim$90 minutes, such as near 0000 UT on 19 and 22 September. If such data were plotted versus a maximum 3-hour K_p value in the preceding 12 or 24 hours, there would clearly be scatter such as that in Fig. 2.6. This scatter and its possible causes are further discussed in the section below on global aspects of the plasmasphere response during magnetic storms.

Density recovery in the trough region

A more detailed view of the recovery stages of the profile than the one presented schematically in Fig. 2.1 was provided from the whistler measurements of Park (1974a). Fig. 2.9 shows plots of the day-to-day changes in the profile of flux tube electron content (2.9a) and equatorial electron density (2.9b) during an 8-day recovery period in June 1965. The tube content data represent the number of electrons in a tube with a $1\,cm^2$ cross-section at 1000 km altitude, extending from 1000 km to the magnetic equator. Plotted on a linear scale, they show approximately constant day-to-day electron content increases from 18 to 25

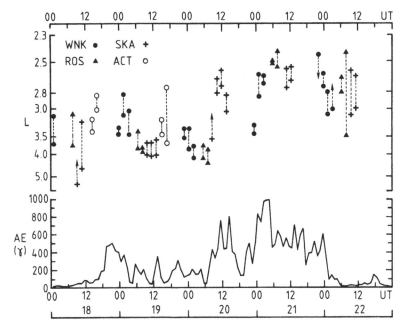

Figure 2.8 Temporal variations in the plasmapause position near 01 MLT detected from OGO 4 satellite VLF data during the period 18–22 September 1967. The auroral electrojet (AE) index is plotted at the bottom of the figure. The figure illustrates the rapid inward displacements of the plasmapause that are usually observed at times of increased substorm activity. It also illustrates the persistence during quieting of the plasmapause effects established during periods of increasing disturbance (after Carpenter and Park 1973).

June. The flattening of the tube content profiles in the regions of continued filling illustrates the dependence of the recovery process on upward fluxes from the ionosphere, which if roughly constant with latitude should cause the tube content in the L range of continued refilling to be roughly constant with L (see below for additional details).

Figures 2.1 and 2.2 indicate another feature of the disturbance/recovery cycle, a decrease and subsequent recovery of the electron density within the storm-time plasmasphere. The decrease was first identified by Carpenter (1962b); later Park and Carpenter (1970) showed that the outer plasmasphere could exhibit localized regions where the density is lower than quiet day levels by as much as a factor of 3. In a remarkable case study involving both whistler and ionosonde data, Park (1973) showed evidence of dumping of outer plasmasphere plasma into the dayside ionosphere during a sudden burst of substorm activity (see below).

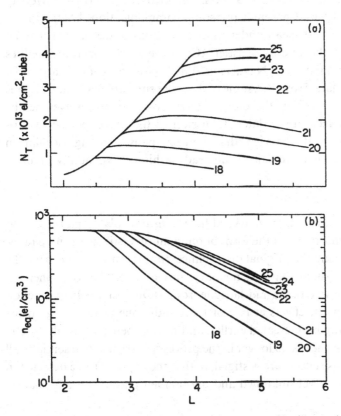

Figure 2.9 (a) Flux tube electron content profiles illustrating day-to-day refilling of the plasmasphere during an extended quiet period following a magnetic storm. The profiles were obtained by drawing smooth curves through the data points of whistlers. The numbers indicate Universal Time days in June 1965. (b) Equatorial electron concentration profiles corresponding to the tube content profiles (after Park, 1974b).

The permanence of the knee phenomenon

Early whistler observations such as those summarized in Figs 1.13 and 1.14 indicated that a knee effect was regularly present in the plasmasphere, at least in the ~ 00–17 MLT sector, provided that the K_p-index had previously increased to 2 or more and that the intervening period had not been one of exceptionally deep quieting. The question of the persistence of quiet-time knee effects at $L > 5$ remained open; studies were complicated by a falloff in lightning source activity with increasing latitude. However, the falloff was least restrictive in its effects at the longitude (~ 90 degrees W) of Byrd Station, Antarctica, $L \sim 7$, thanks to the ~ 11 degree offset of the spin and geomagnetic axes near that meridian. In Byrd Station records, Carpenter (1981) found evidence of ducted whistler propagation to $\sim 8R_E$ on the dayside, and was able to conclude that a plasmapause or plasmapause-like decrease is almost always present near or within $L = 5.5$ on the nightside. A similar impression has been obtained from study of University of Iowa ISEE profiles by Carpenter (personal communication, 1992). On the dayside the picture is less clear; under quiet conditions dense plasma may be observed to $L \sim 8$ or beyond by whistlers (Carpenter, 1981) or from satellites (Gringauz and Bezrukikh, 1976; Nagai *et al.*, 1985), with no clear indication of a plasmapause in the data. However, some of the more sunward extending regions of dense plasma observed in the dayside afternoon may themselves be the consequence of convection (see later discussion of the bulge), which in turn would be consistent with the view that a plasmapause-forming mechanism functions (albeit in its characteristically unsteady fashion) at essentially all times.

'Vestigial plasmapauses'

As noted in Figs 2.1 and 2.2, the marks of the storm-time plasmapause tend to remain on the plasmasphere in the form of remnant or vestigial plasmapauses, often for days following the original disturbance. In this sense, as in others, the plasmasphere is a system that has memory, just as does the ionosphere in notable cases such as that of the mid-latitude trough (Quegan, 1989). In practice, if there are plasmapause effects at two or more radii, one tends to regard the inner one(s) as having been created earlier, and thus as being vestigial. Strictly speaking, if 'vestigial' means no longer in the process of formation, essentially all observed plasmapause effects are vestigial, in that they appear to be detected in the aftermath of an as yet unknown and unobserved plasmapause formation process.

2.3.2 An overview of changes in the worldwide shape of the plasmasphere during magnetic storms

Introduction

The foregoing results, while in some cases illuminating about the plasmasphere as a global phenomenon, tend to emphasize the state of the radial density profile, much as ionosonde data emphasize the states of the height profile in the ionosphere. We now consider those measurements which emphasize the structure and dynamics of the plasmasphere on a global scale.

The global nature of the plasmapause phenomenon became evident through the multiplicity of early rocket, whistler and satellite measurements (Gringauz, 1963; Carpenter, 1963b; 1966; Taylor, Brinton and Smith, 1965a,b; Binsack, 1967). Strong additional evidence of a global extent came through simultaneous comparisons of ground whistler data and near-equatorial OGO 3 satellite data acquired in 1965 at widely separated longitudes (Carpenter, Park, Taylor and Brinton, 1969).

Observations of changes in the plasmasphere boundary during weak magnetic storms

Figure 1.13 illustrates changes in the observed plasmasphere radius during the onset of a weak magnetic storm, as determined from whistlers recorded in a coordinate system rotating with the Earth (Carpenter, 1966). The K_p histories of two such July 1963 events were similar and the plasmasphere responses were found to be similar as well. Because the whistler observations were made from a single ground station, magnetic local time (MLT) effects were folded into the general pattern of inward plasmapause displacement following the increase in the K_p-index. However, the comparative steadiness of magnetic activity for several days following the weak magnetic storm onsets, coupled with whistler activity that often persisted throughout the 24 hours, made it possible to interpret the MLT variations in terms of the spatial structure of the plasmasphere.

Figure 1.14 shows examples of the day-to-day repeatability of the observed MLT variation. The solid curve in Fig. 2.10 shows an average position of the plasmapause versus MLT for 1963 periods such as that of Fig. 1.14, when the K_p-index was relatively steady and in the 2–4 range during the long recovery phases of weak magnetic storms. The curve exhibits an inward trend across the nightside, a broad minimum near dawn, and then a slight outward trend in the morning, leading to a secondary peak at noon. Following the noon peak the radius is relatively steady until near dusk, where the boundary exhibits a relatively abrupt increase in radius, by of order $1R_E$, this being the westward or sunward 'edge' of the bulge region.

Quite different trends in the data, observed during periods of changing, as opposed to relatively steady, activity are illustrated in Fig. 2.10 by the dotted and dashed sequences. The dots represent the period marked 'D' in Fig. 1.13, when a weak magnetic storm was developing. The observed plasmapause radius underwent inward displacement well into the dayside hours, behavior not evident in the average curve. The dashed sequence in Fig. 2.10 represents a quieting period (28–29 July, 1963). In this case the detected plasmapause radius exhibited a nearly constant value across the nightside instead of an inward trend.

Dawn–dusk and noon–midnight asymmetries in plasmasphere radius

The dawn–dusk asymmetry indicated in Fig. 2.10 was found to become more pronounced with increasing magnetic activity (Carpenter, 1966). OGO 2 and 4 polar satellite data acquired during magnetically disturbed periods indicated ratios of $\sim 3:2$ in plasmapause radii detected nearly simultaneously near the ~ 2100 and 0900 MLT meridians (Carpenter, 1971).

A noon–midnight asymmetry much larger than the one indicated in Fig. 2.10 was reported by Gringauz and Bezrukikh (1976) (see Chapter 3) from PROGNOZ satellite observations. However, their data represented magnetic conditions quieter than those represented in the solid curve of Fig. 2.10, and indeed they found that the asymmetry was reduced under more disturbed conditions. This question will be discussed below in a section on distinctions between the main plasmasphere and the bulge region.

Penetrating east–west electric fields in the outer plasmasphere

While data on plasmapause position were interpreted as a kind of integral measure of preceding convection activity (Carpenter, 1970; Chappell et al.,

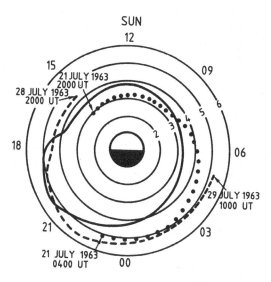

Figure 2.10 Equatorial radius of the plasmapause. The solid line represents the average equatorial position of the knee versus local time, during periods of moderate, steady geomagnetic agitation ($K_p = 2$–4). The observations were made in July and August, 1963 at Eights, Antarctica. The dots show a particular example involving increasing magnetic activity; the dashes illustrate an example of decreasing agitation (after Carpenter, 1966).

1971a), the development of a second whistler technique, involving tracking the radii of whistler paths, made it possible to estimate the instantaneous value of the east–west electric field in the outer plasmasphere (Carpenter, 1966; Carpenter, Stone, Siren and Crystal, 1972). Whistler time series data from an isolated substorm on 15 July 1965, illustrated in Fig. 2.11, provided the first experimental tracking of the bulk flow of the low-energy magnetospheric plasma, showing that entire field-line paths could retain their identities for hours and move in a manner consistent with the presence of a penetrating, large-scale electric field (Carpenter and Stone, 1967; Carpenter et al., 1972).

The whistler method was able to resolve east–west field components with periods of 15–20 minutes or longer. Block and Carpenter (1974) showed evidence that the bulk of the detected changes in whistlers during a substorm could be attributed to potential as opposed to induced electric fields. A comparison was made of Millstone Hill radar data on ion drifts poleward of $L = 4$ and Antarctic whistlers propagating in the outer plasmasphere near $L = 4.4$ (Gonzales et al., 1980). The whistlers were received near the meridian of the radar during a multi-hour period on 9–10 July 1978 that included a small, temporally isolated substorm. It was found that the north–south ion drifts at ionospheric heights were consistent with the equatorial radial drifts from whistlers, provided

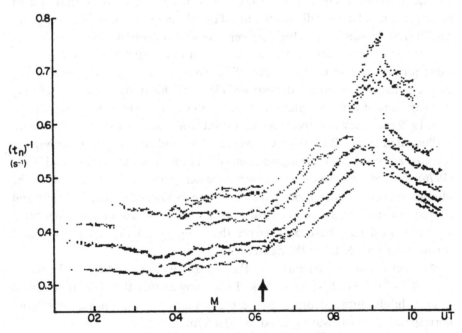

Figure 2.11   Whistler data that provided the first evidence of cross-L inward bulk motions in the outer plasmasphere during a substorm. An arrow marks the onset of an isolated substorm at $\sim$0610 UT. Fast inward drifts occurred between $\sim$0610 and $\sim$0900 UT. The quantity plotted is inversely related to the length of the whistler's field-aligned path (after Carpenter et al., 1972).

that magnetosphere-to-ionosphere mapping occurred along essentially equipotential field lines. For the same substorm period, the east–west field deduced from whistlers agreed with a suitably averaged version of the fluctuating east–west electric field detected near the equator by the double probe antenna on ISEE 1 (Maynard, Aggson and Heppner, 1983).

Among the early findings reported by Carpenter *et al.* (1972) were the following:

The observed penetrating east–west electric fields were largely unsteady, with durations comparable to those of substorms and peak amplitudes in the range 0.3–0.6 mV/m, which at $L \sim 4$ corresponds to cross-L displacements of $\sim 0.4 R_E$ /hour.

Substorm fields could penetrate one or more Earth radii inside the plasmapause, giving rise to significant departures from simple corotation of the plasmasphere with the Earth (the corotation electric field at $4R_E$ is ~ 1.8 mV/m).

With a notable exception near dusk (see below), the sense and magnitude of cross-L motions in the outer plasmasphere were consistent with statistical data from whistlers on the changes with time in the radius of the plasmapause. In one whistler study, the plasmapause position, although not tracked continuously, was displaced inward to an extent consistent with the drift pattern of nearby whistler ducts (Carpenter *et al.*, 1972). It was therefore concluded that at least some of the rapid inward displacements of the plasmapause observed from polar satellites near midnight (see Fig. 2.8) were the result of actual cross-L motions of the magnetospheric plasma, and not the result of an earthward surge of some other process (Carpenter and Park, 1973). Thus interpreted, these decreases in plasmapause radius indicated westward electric fields in the range 0.3–0.6 mV/ m, in agreement with the values reported from the tracking of whistler paths.

A 1979 summary of whistler results on substorm-associated cross-L motions is illustrated in Fig. 2.12 (from Carpenter, Park and Miller, 1979). Among the features are outward morningside and afternoon drifts. Carpenter and Seely (1976) showed case studies in which whistler paths drifting inward in the afternoon sector under quiet conditions (see below) reversed direction and moved outward during isolated substorms. Hence the associated penetrating westward field may have been larger than implied by the radially outward arrows near 1500 MLT in the figure.

Another feature, noted earlier, is the lack of fast inward drifts in the sector prior to ~ 2300 MLT. In fact, Park (1978) showed that this local time region often exhibits outward drifts in the outer plasmasphere during substorms, corresponding to eastward fields of ~ 0.2 mV/m.

A remarkable substorm-associated effect, not shown in Fig. 2.12, is a flow reversal, such that whistler paths observed to move inward during a temporally isolated substorm reversed direction afterward and moved outward at a roughly comparable speed and for a comparable period (Carpenter *et al.*, 1972; Carpenter and Akasofu, 1972; Carpenter and Seely, 1976). Figure 2.11 shows an example of this effect, the reversal having occured at ~09 UT. The quantity plotted is the inverse of travel time at the whistler nose frequency, which can be visualized as a measure of the inverse length of the whistler's field-aligned path.

The outflow effect has not been modeled or otherwise interpreted. Its full spatial and temporal signature has not yet been explored because of limitations on the perspective of the single observing ground station as it advances in MLT during the substorm and its aftermath. A possible clue to its physical basis is the tendency for the effect to be more pronounced after those substorms that follow periods of extended quiet, rather than those occurring during periods of separated but relatively regularly repeated events (Carpenter et al., 1979).

Penetration of substorm fields deep within the plasmasphere

Blanc (1978, 1983a) used the incoherent scatter radar at St Santin to study substorm-associated perturbations of ion drifts near 45 degrees magnetic latitude. The perturbations were found to consist of generally westward flows at all local times, the largest velocities, reaching ~200 m/s, being observed in the pre-midnight sector. The pattern was interpreted as the result of a high latitude convection source field, acting in the presence of globally non-uniform iono-

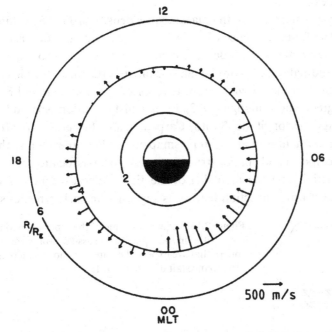

Figure 2.12   Average westward substorm electric field in the outer plasma-sphere at $L = 4$, represented in terms of cross-L flow velocities at the equator. The values were taken from a curve fitted to data representing 34 days of multi-hour cross-L whistler-path drift measurements (after Carpenter et al., 1979).

spheric conductivity and supplemented by a disturbance dynamo field. The latter field develops in response to the deposition of heat in the high latitude nightside thermosphere. A model of the fields (see Chapter 4) was developed and was found to be consistent at $L = 4$ with the average east–west component of substorm fields in the plasmasphere reported by Carpenter *et al.* (1979).

Irregularities in the plasmasphere radius

The solid curve of Fig. 2.10 was intended to show average positions as determined within a rotating coordinate system by a particular observing method, not a 'snapshot' of the instantaneous plasmasphere radius. This fact notwithstanding, curves such as that of Fig. 2.10 or even smoother teardrop-shaped theoretical curves have been widely used in interpretive works in which a 'snapshot' viewpoint was implied (Brice, 1967; Kavanagh, Freeman and Chen, 1968; Chappell 1972; Higel and Wu, 1984). Such usages have been convenient and were often appropriately qualified, but as Carpenter *et al.* (1993) point out, they have probably tended to obscure the actual complexity of the boundary and its dynamics. As an earlier reminder of this, Carpenter (1983) published the sketch of Fig. 2.13 showing some of the types of spatial irregularities that might be expected to be present in the plasmasphere radius at a given time (exclusive of outlying features).

In studying whistler data on the plasmapause density profile, Angerami and Carpenter (1966) found evidence of variations in plasmapause radius of order $0.5R_E$ within the estimated ~ 30-degree longitudinal field of view of the whistler station. Longitudinal structure in the plasmasphere surface obtained from VLF goniometer observations of whistlers was reported by Bullough and Sagredo (1970) and Sagredo and Bullough (1973). From whistler stations spaced by ~ 30 degrees in magnetic longitude, Smith, Carpenter and Lester (1981) observed differences of a few tenths of an R_E in plasmapause radius, interpreting them as the result of temporal and spatial structure in substorm-associated convection electric field activity on the nightside of the Earth. The existence of spatial structure in substorm convection fields was suggested in early whistler studies

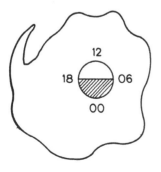

Figure 2.13 Sketch of some of the types of spatial irregularities in the equatorial cross-section of the plasmasphere that are expected during a period of increasing disturbance (after Carpenter, 1983).

which, as noted above, showed an absence of fast inward substorm-associated drifts in the sector westward of the ~ 2300 MLT meridian, as well as a reversal of fast, substorm-associated flow from inward to outward following temporally isolated substorms (Carpenter et al., 1972).

As remarked previously, variations up to as much as $1R_E$ in observed plasmapause radius have been detected between successive polar satellite observations, or within about 20 degrees in longitude (Carpenter and Park, 1973, Carpenter and Chappell, 1973). These variations were interpreted both in terms of unsteady local cross-L drifts and in terms of rotation past the orbital plane of the satellite of structure imposed in earlier local time sectors by non-spatially uniform electric fields.

Variation with local time in the delay between substorm activity onset and inward plasmapause displacement

The concept of a nightside process that tends rapidly and locally to reduce the plasmasphere radius during substorms was supported by studies of satellite data on plasmapause radii by Chappell et al. (1971a) and by Décréau, Beghin and Parrot (1982). They found that while the plasmapause radius in the midnight sector correlated well with current levels of magnetic activity, the dayside radii correlated best with earlier disturbance levels, the time delay being approximately equal to the time for the observed plasma to rotate with the Earth from the midnight sector to the region of observation. The OGO 4 plasmapause crossing data of Fig. 2.8, acquired near 0100 MLT, illustrate the comparative immediacy of the radius reduction response on the nightside. The inward displacements at the beginning of AE surges on 20 and 21 September are examples.

Specific examples of the delay effect have been reported. A whistler case study showed that a station that was located in the morning sector when magnetic storm activity began did not detect substantial reductions in plasmasphere radius until the station had rotated past the dusk meridian (Carpenter, Park, Arens and Williams, 1971). R.R. Anderson (personal communication, 1992) reports evidence of dayside delays in plasmapause reduction, based on comparisons of ISEE 1 electron density profiles acquired along inbound orbits in the late morning sector with the immediately following (~ 2–3 hours later) outbound orbits in the predawn sector. If the outbound pass occurred closely following a sharp increase in the K_p index, the predawn plasmapause radius tended to be as much as $2R_E$ less than that observed ~ 2–3 hours earlier in the late morning sector. In contrast, if both orbital segments occurred well into a period of relatively steady agitation, the plasmapause radii at the two locations tended to be roughly equal, as expected from the data of Fig. 1.14 and the solid curve of Fig. 2.10.

Observations pertaining to erosion of the plasmasphere

Observations of plasmasphere erosion in process tend to be fragmentary because of the great spatial extent of the region being affected. Carpenter *et al.* (1972) found that when the overall plasmasphere size was significantly reduced, as implied in the data of Fig. 1.13, this tended to occur during occasional periods of sustained substorm activity that lasted for periods of order 10 hours. In such cases there was no indication of the type of compensating outward drifts observed immediately after some temporally isolated substorms. Thus it appeared that in the many situations in which substorm activity was not continuous enough to produce a large integrated boundary displacement effect, distortion of the plasmasphere could apparently occur without significant changes in its overall plasma content. Carpenter (personal communication, 1992) has suggested visualizing the distorted but not seriously eroded plasmasphere as a kind of rotating waterbag, whose boundary is temporarily driven inward in a region near midnight, but which then reacts by expanding outward in a region eastward of the region of transient inward displacement.

Several observations have been made of what appears to be sunward entrainment of the plasma in the outer duskside plasmasphere during periods of enhanced magnetic activity. In addition to the outward afternoon flows reported by Carpenter and Seely (1976) and noted above, Carpenter *et al.* (1993), who studied the data of the Retarding Ion Mass Spectrometer (RIMS) on DE 1, found evidence of sunward and outward flows of thermal light ions in the outer duskside plasmasphere during substorms. Yeh, Foster, Rich and Swider (1991), using Millstone Hill radar data, found that during periods of enhanced activity, leading to a substantially increased ring current, a region of intense westward ion drifts appeared well within the duskside plasmasphere (see also Okada *et al.* (1993) for corresponding AKEBONO satellite observations). Fig. 2.14 displays the observed eastward component of ion drift velocity versus magnetic latitude under conditions of disturbance ranging from $K_p = 9$ (top panel) to $K_p = 2$ (bottom panel). In the top and middle panels there were two local minima (maxima in westward velocity), the higher latitude one being identified with the subauroral ion drift (SAID) phenomenon (Galperin, Ponomarov and Zosinova, 1973a,b; Smiddy *et al.*, 1977; Spiro, Heelis and Hanson, 1978; Anderson, Heelis and Hanson, 1991; Anderson *et al.*, 1993) that has been associated with the vicinity of the plasmapause. The lower-latitude peaks in Fig. 2.14 show westward velocities of $\sim 1500\,\text{m/s}$. Sharp inner limits of the flow were located below 50 degrees invariant latitude in one case and 60 degrees in the other. The effect was observed to last for at least 6 hours, and to be most pronounced at ~ 18 MLT.

The duskside bulge sector during and following periods of plasmasphere erosion

Much of the currently available evidence on plasmasphere dynamics during substorms comes from the duskside bulge sector. As noted above, the bulge was at an early date interpreted by theorists as evidence of solar-wind-induced sunward plasma flow activity in the middle magnetosphere (Nishida, 1966; Brice, 1967; Dungey, 1967). East–west motions of the outer plasmasphere in the dusk sector were inferred from whistler observations of the apparent position in local time of the distinctive westward or sunward shoulder of the bulge (Carpenter, 1970). The edge of the bulge tended to be observed:(1) in the afternoon sector during and immediately following periods of enhanced substorm activity as measured by the AE index; (2) near 1800 MLT during more or less steady activity, and; (3) near 1900–2000 MLT, or sometime later, during periods preceded by 6–8 hours of quieting. From this type of data it was inferred that the outer plasmasphere near dusk was displaced sunward during substorms and that it tended to move in the direction of the Earth's rotation during quieting, approaching corotation if quieting were deep and prolonged.

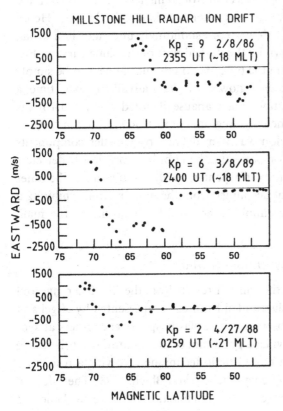

Figure 2.14 Latitudinal profiles of the east–west ion convection velocities observed with the Millstone Hill incoherent scatter radar in the evening MLT sector for three K_p activity levels. The convection pattern shifts equatorward and expands as K_p increases. Dual maxima form as K_p reaches moderately disturbed conditions ($K_p > 4$) (after Yeh et al., 1991).

Detached plasmas

A major step in the investigation of the bulge region was the identification by Chappell *et al.* (1971a) of 'detached plasmas', regions of dense plasma interpreted as being effectively disconnected from the main plasmasphere. The regions were located predominantly in the afternoon–dusk sector, and were interpreted as having been displaced sunward during disturbed periods. The mechanism of detachment was not identified, although Chappell (1974) suggested that it could be due to the effects of localized structure in electric fields. Barfield, Burch and Williams (1975) discussed it in terms of rapid reconfiguration of the magnetosphere during disturbed periods, and Lemaire (1974, 1975, 1985, 1987) proposed a detachment process involving the gravitational interchange instability (see Chapter 5). Furthermore, it was not clear how apparently detached outlying regions were related to the bulge effects detected from whistlers. Figure 2.15b shows an example of outlying dense plasmas detected along an afternoon inbound pass of ISEE 1 on 30 October, 1983 (adapted from Carpenter *et al.*, 1993). An inset shows an equatorial projection of the orbit.

Decoupling of the bulge from the inner region

Although the existence of the bulge was clear from whistlers, the limits or radial extent of the main body of the bulge were often not defined in the data. Hence the shape of the curve in Fig. 2.10 was correspondingly uncertain in a broad range near 20 MLT (this is shown by the dashed part of the same curve in Fig. 1.16). Furthermore, the plasma in the bulge region did not appear to be simply connected to the main plasmasphere flow lines. With advancing local time, a whistler path just inside the afternoon plasmapause (denoted 'c' in Fig. 1.16) was not observed to drift out into the bulge region, but instead stayed at a roughly constant radius. The bulge region was then defined by whistler components propagating at higher L values on paths that had not been previously detected. And, as noted in the whistler study (Carpenter, 1970), the plasma in the outer region often appeared somewhat decoupled from the inner region, exhibiting a density profile that was not a simple extension of the profile in the main plasmasphere.

Bulge observations at synchronous orbit

Information complementary to the whistler results from the $L \sim 3$–5 range and to the OGO 5 data on widely distributed features was presented by Higel and Wu (1984) through GEOS 2 observations of plasmasphere encounters at synchronous orbit. Figure 2.16 provides a polar plot of the equatorial density on 30 January 1979, obtained using a pulsed radio technique involving a resonance response close to the plasma frequency (Etcheto and Bloch, 1978). The values of the densities (dots) for a full day are plotted as a function of Universal Time and

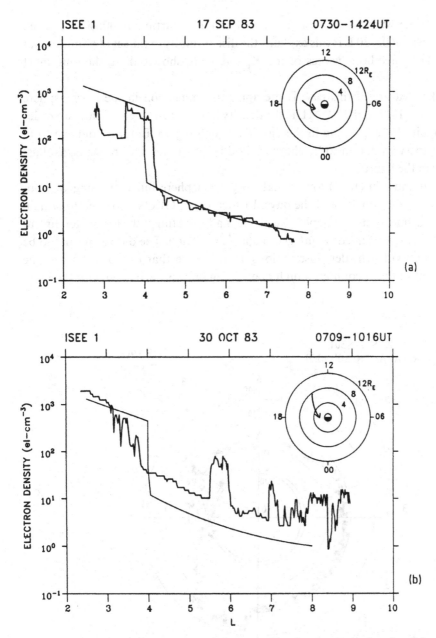

Figure 2.15 ISEE-1 electron density profiles showing examples of outlying high-density regions in the afternoon-dusk sector. Insets show the satellite orbit in terms of radial distance versus MLT. The dashed curves represent Carpenter and Anderson's (1992) empirical model of the nightside equatorial profile with a plasmapause arbitrarily set at L ~ 4 (adapted from Carpenter *et al.*, 1993).

Local Time with a time resolution of 1 minute, starting at 00:00 UT when GEOS was at 02:30 LT; this explains the discontinuity in the set of points at that time. The three-hour values of the K_p index are shown along the outermost circle.

A key feature of this case is the apparent immersion in the plasmasphere between $\sim$ 1815 and 2040 MLT, the density in that sector being nearly an order of magnitude higher than in the dayside trough region. Entry to and exit from the region were found to be characterized by sharp density gradients, as illustrated in the figure.

Higel and Wu (1984) found that no plasmasphere bulge crossing was observed at $6.6R_E$ on 20% of the days. In such cases, GEOS appeared to remain either outside or inside the plasmasphere at all local times, the former tending to occur during sustained strong geomagnetic agitation. The data appeared to be consistent with whistler observations, in the sense that the local time at the midpoint of the encountered high density region was correlated with the mag-

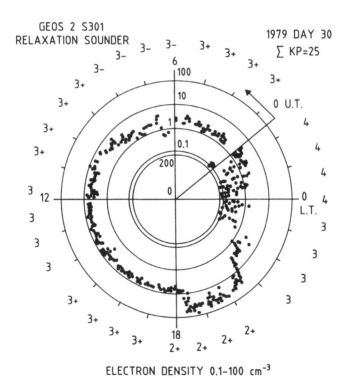

Figure 2.16 Distribution of electron densities measured along the geostationary orbit of GEOS 2 on 30 January 1979 with the relaxation sounder experiment (after Higel and Wu, 1984), showing a $\sim$ 2-hour encounter with the plasmasphere bulge near dusk.

netic disturbance level, tending to be earlier for higher K_p values. These results are also confirmed by the Magnetospheric Plasma Analysers on three LANL geosynchronous satellites (McComas *et al.*, 1993).

The bulge as observed by GEOS 2 was used by Fontaine *et al.* (1986) to test the use of the EISCAT incoherent scatter radar as a tool for estimating the plasmapause position. In four case studies, thermal plasma flow patterns at ionospheric heights in the $L \sim 4$ to 10 range were deduced from 24-hour periods of measurements, and were compared with the MLT range of the bulge encounter reported from GEOS 2 and projected down the geomagnetic field lines to ionospheric heights. Under conditions of steady K_p values, the mapped bulge (from GEOS) appeared to be at the latitude and MLT in the ionosphere where a regime of essentially corotating plasma penetrated from lower (unobserved) latitudes into the radar field of view. This was interpreted as a validation of the belief that, under steady conditions, the plasmapause represents a spatial transition from corotation-dominated to convection-dominated flow, and that the radar can under such conditions be used to estimate certain large-scale features of magnetospheric convection. It was evident, however, that the mapped relations were not always simple, in particular during periods of changing magnetic activity.

In situ three-dimensional plasma data from Los Alamos magnetospheric plasma analyzers (MPAs) (McComas *et al.*, 1993) on board two geosynchronous satellites, 1989–046 and 1990–095, have confirmed previously reported general features of the bulge region but have also indicated a much higher level of complexity in plasmasphere morphology than had previously been reported (Moldwin *et al.*, 1994). In particular, the authors found many brief, order of an hour, plasmasphere encounters during quiet periods (K_p less than 2), and also found that bulge-like plasma encounters, while few in the midnight–dawn sector, were not strongly dependent upon local time. Most remarkable was their finding that two synchronous satellites, six to eight hours apart in local time, can encounter plasmasphere plasma at different local times, even during periods of relatively steady geomagnetic activity. The authors concluded that 'the outer plasmasphere is often highly structured even during steady geomagnetic conditions, and that the simple teardrop model of the bulge rarely, if ever, adequately describes the duskside plasmasphere'.

A new perspective on the bulge as detected from whistlers

New information about that part of the bulge near the main plasmasphere was obtained by Carpenter *et al.* (1992) using ground whistler stations spaced in longitude and the DE and GEOS 2 satellites. Evidence was found of a narrow dense plasma feature extending sunward from the bulge region, of the general type indicated schematically in Fig. 2.13. It was concluded that the abrupt westward edge of the bulge detected from whistlers is often the topological consequence of an inward (toward the main plasmasphere) spiralling effect of dense plasma streamers during recovery periods, the streamers having developed during brief periods of enhanced convection.

Results of a multiplatform study of the bulge

Carpenter *et al.* (1993) have conducted several multiday case studies of plasma structure and dynamics in the bulge sector. They worked with a combination of ground whistler data, DE 1 thermal ion data, and GEOS 2 and ISEE 1 electron density data. Several of their conclusions are as follows:

During active periods the plasmasphere tends to become divided into two topologically distinct entities, a main plasmasphere and a duskside bulge region. Carpenter and Anderson (1992) had reported that when all the innermost detectable plasmapause crossings on 208 ISEE orbits were plotted versus MLT, the plasmapause radius appeared on average to be quasi-circular. These results are shown in Fig. 2.17. Carpenter *et al.* (1993) concluded that 'the bulge is essentially the plasma that has originally been entrained by penetrating convection electric fields and displaced sunward and outward from the duskside plasmasphere, while the main plasmasphere is the bulk of the remaining dense plasma. The latter, through approximate rotation with the earth during quieting, assumes a quasi-circular shape, with a duskside radius only $\sim 0.5 R_{\mathrm{E}}$ greater than the radius at dawn, and thus a mean radius close to the one established on the nightside during the main erosion period'.

The authors found this distinction between the main plasmasphere and the bulge region useful in explaining differences between earlier OGO 5 satellite (Chappell *et al.*, 1971a) and ground whistler (Carpenter, 1966) reports on the average plasmapause position versus local time. The smaller plasmapause radii found from whistlers in the late afternoon and in the dusk sector were considered to be due to the more effective probing by whistlers of the 'innermost plasmapause' present along a given meridian, while in the case of OGO 5, an 'outer plasmapause' when present, was likely to be identified as the plasmasphere boundary along a particular orbit.

Dense plasma patches may exist near the afternoon magnetopause for several days after a period of enhanced convection. Carpenter *et al.* (1993) stated that 'the patches may at times cover a significant fraction of the outer afternoon–dusk magnetosphere, and are estimated to represent from ~ 10 to 30% of the outer plasmaspheric plasma entrained by the convection electric field during a weak magnetic storm'. Figure 2.18 shows three examples of afternoon profiles, one (2.18a) during a period of deep quieting, a second (with data gaps) within 24 hours following a recurrence of moderate substorm activity (2.18b), and a third two days later, when the substorm activity had been fairly steady for about three days.

The regularity with which outlying dense plasma structures were found led to the suggestion that there is a quasi-permanent circulation of irregular plasma

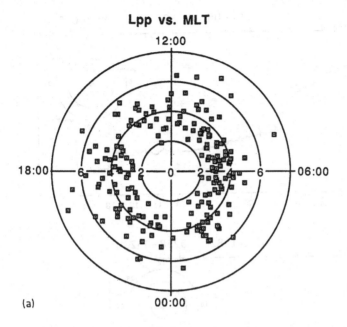

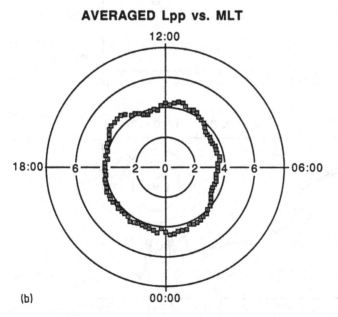

Figure 2.17 Plots of plasmapause *L* value versus MLT, illustrating the tendency of the main plasmasphere to become roughly circular in the aftermath of plasmasphere erosion events in which a 'new' plasmapause is formed. (a) Scatter plot of 208 plasmapause crossings identified in ISEE-1 SFR data. (b) Two-hour running average of the data of (a) (after Carpenter and Anderson, 1992).

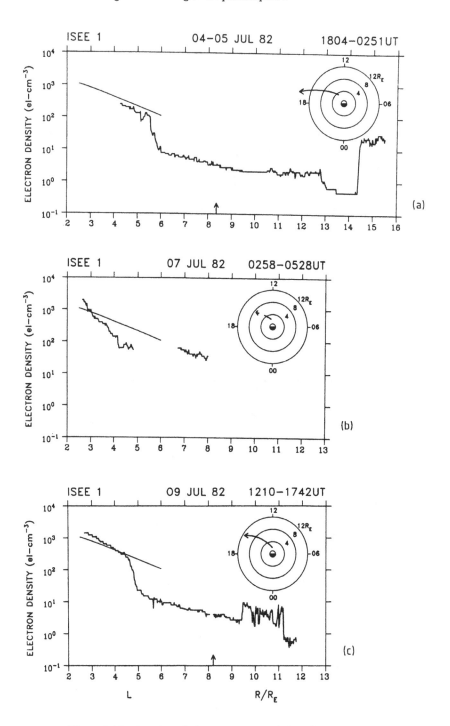

Figure 2.18 A series of afternoon equatorial profiles observed from ISEE-1, illustrating first a lack of outlying high-density features during deep quieting (a), and then, following renewed substorm activity, the detection of outliers (b), (c) (from Carpenter *et al.*, 1993).

structures in the dusk sector, these features disappearing only under conditions of very deep quieting.

On the question of detachment of outlying plasmas, Carpenter *et al.* (1993) concluded that 'the properties of outliers observed near the plasmapause and out to synchronous orbit suggest that many of these are rooted in or attached to the main body of the plasmasphere. On the other hand, the distribution and occurrence of dense plasmas observed at $L \sim 6$ and beyond, and in particular their observation several days or more after an erosion event, suggest that many of those regions are effectively isolated from or detached from the main plasmasphere'. While detachment may develop in the aftermath of entrainment and outflow, velocity shear effects observed in the duskside ionosphere as well as various properties of pre-dusk substorm-associated convection surges detected by auroral radar (Freeman *et al.*, 1992) suggest that detachment may occur to some extent when the plasma first becomes entrained. An example of an outlier considered likely to be rooted in the main plasmasphere is shown in Fig. 2.15a, while Fig. 2.15b shows the other type, particularly at $L > 7$.

To the extent that duskside bulge sector plasmas are detached from the main plasmasphere, their occurrence appears to be inconsistent with MHD models that predict continuity of all outlying dense plasma structures with the main plasmasphere.

2.3.3   Post-storm refilling of the plasmasphere

There has been strong interest among theorists in modeling the earliest stages of plasma flow from the ionosphere into depleted flux tubes (Singh and Horwitz, 1992 and references cited therein). However, there is essentially no experimental data indicating the density conditions that actually exist at the time and location of plasmapause formation. Carpenter and Anderson (1992) have been able to define nightside trough density levels from ISEE 1 data, but they interpreted these as representing the profile after some initial filling process, perhaps of less than an hour's duration, had already been at work. That there exist some type of lower density 'initial conditions' for filling is suggested by the GEOS 2 plots of Higel and Wu (1984), which often showed irregular densities between 0.1 and 1.0 el-cm^{-3} in the pre-midnight sector. Figure 2.16 shows an example of this effect.

The plasmasphere refilling process has been documented in radio probing data to a limited extent. The most extensive work on the recovering trough region profile is by Park (1974a). Figure 2.9, noted above, shows the day-to-day variation of the equatorial electron density and tube electron content profiles during an exceptionally quiet eight-day recovery period in 1965. Smooth curves were drawn through data points from whistlers. As recovery proceeded, the point at which the tube content curve changed from a rising characteristic to one constant with latitude moved from $L = 2.5$ on 19 June to $L = 4$ on 25 June. The plasmasphere inside this point appeared to saturate in terms of daily interchange of plasma with the ionosphere, while the outer plasmasphere continued to fill, the movement of the point to larger L with time being attributed to the increase with L in tube volume. (This and similar effects led Park (1974a) and other workers (Titheridge, 1976) to refer to the 'inner plasmasphere' and the 'outer plasmasphere', the inner one being at or near saturation.) Since refilling started on 17 June, the recovery time of the plasmasphere may be said to have varied from one day at $L = 2.5$ to more than eight days at $L = 4$. In his recovery period study, Park (1970) found that, at $L \sim 5$, the flux tube electron content increased at an average rate of 5×10^{13} el/cm^2/day throughout the period of observation. This net increase was the result of a daytime filling flux of $\sim 3 \times 10^8$ el/cm^2/s and a nighttime draining flux of $\sim 1.5 \times 10^8$ el/cm^2/s.

The question of whether net daily filling stops at low L values after one or two days has been raised by Tarcsai (1985), who has shown whistler data indicating that, at $L = 2$, net filling can continue for at least five days. Further evidence of prolonged net filling deep within the plasmasphere came from Doppler shift and group delay measurements on fixed-frequency VLF whistler-mode signals by Saxton and Smith (1989). They found a constant upward directed flux of $(0.26 \pm 0.19) \times 10^8$ el/cm^2/s at $L = 2.5$ in the inner plasmasphere throughout an extended period of quiet conditions, up to nine days. This represents the net filling rate of the plasmasphere averaged over all local times and over the nine days studied. The rate is small compared with the daytime filling rates noted above, but is nevertheless roughly consistent with the results on filling near $L = 4$ by Park (1970) and between $L = 2$ and $L = 2.8$ by Tarcsai (1985).

Flux estimates from other radio probing methods have been dependent, as were those of Park, on case studies, and have been generally supportive of the results from whistlers. Reported values from radio beacon measurements have been quite varied, ranging from 5×10^8 el/cm^2/s (Soicher, 1976) to 1.4×10^5 el/cm^2/s (Poletti-Liuzzi et al., 1977). However, from beacon measurements Kersley et al. (1978) found upward fluxes only slightly smaller than those of Park and fluxes 1.5×10^8 el/cm^2/s downward. Meanwhile, with incoherent scatter, both Evans and Holt (1978) at Millstone Hill and Vickrey et al. (1979) at Arecibo found interchange fluxes in rough agreement with the whistler results.

2.4 Coupling of the plasmapause and plasmasphere regions to the ionosphere

2.4.1 The concept of a plasmasphere boundary layer

Carpenter (1992, personal communication) has suggested that the concept of a 'plasmasphere boundary layer' could provide a useful basis for discussing certain particle and wave phenomena that apparently occur because of the presence of the plasmapause 'boundary'. Among these phenomena are effects of the intrusion of the ring current into the plasmasphere (La Belle *et al.*, 1988), especially in the pre-midnight region of larger plasmasphere radius (also discussed in Chapter 3). One occasional effect of this intrusion may be the above-noted penetration of fast westward ion drift activity to latitudes well below those where subauroral ion drifts (SAID) normally occur (Yeh *et al.*,1991).

2.4.2 SAR arcs

The abrupt density change at the plasmapause appears to be a key factor in a process that leads to the appearance during magnetic storms of stable auroral red (SAR) arcs in the mid-latitude ionosphere (see Barbier, 1958). The arcs involve radiation at 630 nm from excited atomic oxygen near 400 km altitude; they last for hours and may be detected on two or three consecutive nights after a major magnetic storm onset. Their distribution in L versus LT is similar to that of the plasmapause, being tilted across the nightside from higher to lower latitudes, with a minimum latitude near dawn (Glass *et al.*, 1970; Hoch and Smith, 1971; Hoch and Lemaire, 1975). A close spatial link between reported SAR arcs and the plasmapause was indicated in studies of OGO 5 high altitude ion density data by Chappell, Harris and Sharp (1971b) and OGO 2 and 4 polar satellite data by Carpenter (1971).

2.4.3 Ion and electron troughs

The existence of light ion troughs in the topside ionosphere, characterized by a sharp drop in H^+ and He^+ concentration with increasing latitude, has been correlated with the position of the overlying plasmapause (Taylor *et al.*, 1969). Electron temperature in the topside ionosphere has been found to show a latitudinally narrow peak that coincides with the outer edge of the plasmasphere (Brace and Theis, 1974; Brace *et al.*, 1988). In the case of the light ions, the density change is explained as a consequence of the upflow of the light ions into the low density trough regions, an upflow sometimes referred to as the polar wind. In the case of the subauroral electron temperature peak or SETE, it is considered to be the result of downward heat conduction from a region just inside the plasmapause, where heat is deposited.

The connection between the plasmapause and the main electron density trough in the ionosphere has been the subject of much investigation. At the time of the early reports on the knee effect and the plasmapause phenomenon, attention was drawn to satellite data showing a mid-latitude electron density trough in the ionosphere (Muldrew, 1965; Sharp, 1966). Initial statistical studies of the locations of these trough positions seemed to reveal that they occurred on the same geomagnetic field lines as the knee in the equatorial density distribution (Rycroft and Burnell, 1970; Rycroft and Thomas, 1970). However, as topside ionosphere measurements were made by the OGO series satellites, evidence was found that a light ion trough in H^+ and He^+ was probably more directly related to the plasmapause than the trough in total electron density (Taylor, Brinton, Pharo and Rahman, 1968b; Taylor et al., 1969; Morgan, Brown, Johnson and Taylor, 1977).

The situation was found to be complex; under some circumstances with chemical, convective and neutral-wind-induced effects, there may exist a significant trough in electron density and in O^+ as the dominant ion, and in H^+ as the second prominent ion. But in other cases there may exist pronounced troughs in H^+ and He^+ with very little or no evidence of a trough in O^+ and n_e. This complicates the study of manifestations of the plasmapause in the topside ionosphere, as indicated by Taylor and Walsh (1972) and Taylor and Cordier (1974).

Smith, Rodger and Thomas (1987) used a combination of ground whistler and ionosonde observations to confirm that the plasmapause and the mid-latitude trough often occur on the same field line. The two features are fairly closely co-located in the morning hours, but there is often a significant difference (up to 2 L-shells) in the evening hours. These results supported those obtained earlier by Rodger and Pinnock (1982). In the pre-midnight region of substantial separation of the trough and plasmapause, Yeh et al. (1991) found that the low-latitude limit of the plasma sheet electrons is a better signature of the equatorial plasmapause than is the ionospheric trough. Earlier observations by Galperin et al. (1977), Jorjio et al. (1978) and others, already showed that the soft electron boundary (SEB) where plasmasheet electrons are precipitated, often coincides with the high altitude plasmapause in the nightside.

In a recent review of F-region troughs, Rodger, Moffett and Quegan (1992) discussed the physical processes that could lead to trough formation, ion drifts being of paramount importance in such a case. A key point here is that the ionosphere density is subject to decay in a manner that the overlying plasmasphere is not; hence the wide divergence in the effects before midnight, when ionosphere decay might be expected to be relatively important, and the greater agreement near dawn, when convection tends to cause the movements of separate mid-latitude thermal plasma features to converge and the action of the protonosphere as a plasma reservoir may have a bigger effect on the underlying region.

From an analysis of a very large number of electron density–height profiles derived from the ALOUETTE 1 topside sounder, Titheridge (1976b) concluded that an ionospheric projection of the plasmapause can be defined by (1) the latitude at which a peak in electron temperature occurs, (2) the latitude down to which the ion transition height remains high, and (3) the latitude below which

H^+ densities start to increase rapidly. He found statistical agreement among these three criteria to within about 2° of latitude (i.e. approximately 0.5 of an L unit at $L = 4$). He also found agreement with the statistical position of the equatorial plasmapause determined by whistlers except, notably, in the dusk sector, where, like Brace and Theis (1974), he found no bulge. The disagreement in the bulge sector appears to have been explained in the report by Carpenter and Anderson (1992) noted earlier. From ISEE density profile data, they concluded that, on a statistical basis, the plasmapause radius does not exhibit a pronounced duskside bulge. Later, Carpenter *et al.* (1993) found that, topologically, the bulge and the main plasmasphere are really two different entities, again as noted above.

2.4.4   Ionospheric effects of cross-L plasmasphere drifts

The whistler results showing a reversal of the substorm electric field from eastward to westward near midnight agree with results of ionosonde observations of the nightside F-layer during substorms (Park and Meng, 1971; 1973). The observations showed a large-scale distortion, with the F layer lifted upward in the pre-midnight sector and pushed downward in the post-midnight sector. Park and Meng suggested that this distortion was the result of $\mathbf{E} \times \mathbf{B}$ drifts, involving lifting by the eastward electric field before midnight and lowering by the westward field after midnight. The observed ionospheric effects were more pronounced in the winter hemisphere than in the summer hemisphere because the winter ionosphere decays continuously throughout the night and is controlled primarily by recombination and diffusion. ATS 6 radio beacon satellite measurements of nighttime winter increases in total electron content were interpreted by Davies *et al.* (1979a,b) as evidence of increases in ionosphere content associated with westward electric fields, essentially the mechanism proposed by Park and Meng (1973).

2.4.5   Ionization interchange effects

There is some evidence that ionization interchange processes between the ionosphere and the plasmasphere are enhanced during magnetic disturbances. One such effect may be the apparent increase in total plasmasphere content (along a linear path to synchronous orbit) during the first day following a magnetic storm sudden commencement (Soicher, 1976; Degenhardt *et al.*, 1977). During the main phase of a storm, the plasmasphere is depleted both inside and outside the plasmapause (see Fig. 2.1), but there are essentially no experimental data documenting detailed aspects of the depletions.

It is not well known what fraction of the 'lost' plasma is carried to interplanetary space by convection and what fraction is dumped into the ionosphere.

From the satellite beacon technique, Degenhardt *et al.* (1977) found that the noted increase in (line of sight) plasmasphere electron content could not be explained entirely as the result of outward $\mathbf{E} \times \mathbf{B}$ drifts prior to dusk, which effectively move topside ionosphere plasma upward into the range embraced by the plasmasphere content measurement. Interestingly, model calculations showed that the later-observed ~50–75% loss of plasmasphere content could not be explained as the result of a decrease in radius of the plasmasphere and the removal of the plasma beyond the new plasmapause, but by implication involved interchange with the ionosphere.

Park (1971) suggested that the normal nighttime flow of plasma from the plasmasphere into the ionosphere is enhanced during substorms and causes the enhancements in ionospheric concentrations frequently observed during mid-latitude winter nights. Furthermore, in a study of a several-hour burst of substorm activity that followed many days of quiet, he found that significant amounts of plasma were dumped from the dayside plasmasphere into the ionosphere and that this may be an important process in understanding the storm time behavior of both regions. The solid curve B in Fig 2.19 (Park, 1973) shows the total electron content versus L in the forenoon sector prior to a substorm that started at 1430 UT on 25 June 1965. The dashed curve A shows the content profile after the substorm activity; various symbols denote specific measurements made more than 11 hours later. Park identified two ways in which the profile could change from the solid to the dashed curve. (a) Tube content could decrease by the dumping of plasma into the underlying ionosphere without involving any plasma motions across L shells; in this case the amount of plasma dumped into the ionosphere would have to be latitude dependent, as illustrated in Fig. 2.20a. (b) Inward convection of plasma across L shells and a uniform draining into the ionosphere of about 10^9 electrons $\mathrm{cm}^{-2}\mathrm{s}^{-1}$ could be combined as illustrated in Fig. 2.20b to produce the same

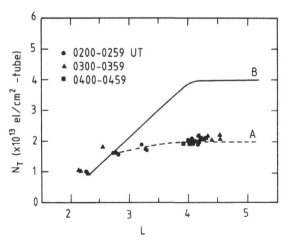

Figure 2.19 Plots of flux tube electron content versus L deduced from whistlers received at Eights before (B, solid profile) and after (A, dashed profile) the substorm of 25 June, 1965. The data are grouped into three-hourly intervals (after Park, 1973).

final profile as in (a). Whistler observations made at various times following the onset of substorm activity could be used to support this second scenario.

These whistler observations also indicated that the substorm caused a depletion of the plasmasphere in a limited local time sector, leaving tube content profiles at other local times unperturbed. Longitudinal structures were thus produced, which persisted after the substorm was over. Ionosonde records from a number of stations showed that the peak F2 ionospheric electron density was enhanced over the monthly median level during that same substorm of 25 June 1965, but only in a limited local time sector (Park, 1974b).

A remarkable aspect of the dayside events studied by Park is the inferred occurrence of cross-L inward plasma drifts, a flow direction opposite to the one usually observed on the dayside during substorms (see Fig. 2.12). However,

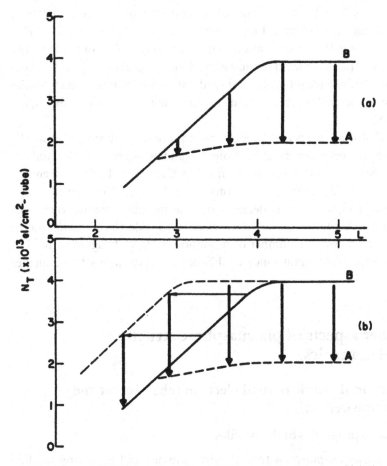

Figure 2.20 Illustration of two different mechanisms by which the total flux tube electron content changed from the solid line (B) immediately before a substorm to the dashed profile (A), which corresponds to observations afterward (from Park, 1973).

Carpenter (1966) had reported substantial inward displacement of the plas-mapause on the dayside during the onset of a weak magnetic storm (see dots in Fig. 2.10 and also Fig. 1.13). It is possible that these effects are peculiar to the erosion phase of the plasmasphere, when it is first disturbed following an extended period of quiet.

Several mechanisms have been suggested for the stormtime draining of the plasmasphere through downward flow of ionization. Hanson and Ortenberger (1961) and Hanson (1964) suggested mechanisms based on heating of the iono-sphere and of the protonosphere respectively. A third mechanism more consist-ent with whistler observations was suggested by Park (1971, 1973). In the topside ionosphere where the O^+ concentration decreases with increasing alti-tude a downward drift results in a reduced O^+ concentration at a given altitude. Below the critical level where H^+ is in chemical equilibrium a reduction in O^+ must be accompanied by a corresponding reduction in H^+ concentration. Indeed, in the region of chemical equilibrium one has : $O^+ + H \rightleftarrows O + H^+$. This reduction of the H^+ concentration induces downward flow of H^+ from the protonosphere. The downward draining mechanism is expected to be most prominent in the nightside ionosphere, since daytime flow of ionization from the ionosphere into the protonosphere tends to overcome the draining effect (Que-gan *et al.*, 1982).

During substorms, a special type of precipitation or drainage from the plasmasphere has been observed by Potemra and Rosenberg (1973) and re-ported in the context of other substorm effects by Carpenter, Foster, Rosenberg and Lanzerotti (1975). During substorms on 2 January 1971, when wave-induced burst precipitation was detected outside the plasmapause, there was also sufficient precipitation of >40 keV electrons inside the plasmapause to cause phase anomalies on a number of sub-ionospheric VLF signal paths. The nature and geophysical significance of this precipitation needs to be further explored.

2.5   Other aspects of plasmasphere structure and dynamics

2.5.1   Equatorial profile of total electron tube content and electron density

General properties of the profiles

Figure 2.21 shows examples of tube electron content and the corresponding equatorial electron density in the plasmasphere obtained in a whistler case study by Angerami and Carpenter (1966). The total content N_T profile shows a characteristic feature, noted earlier (Fig. 2.9), involving an increase with radial distance followed by a region of nearly constant content, the break point being considered a transition from an inner region that is in a quasi-equilibrium with the underlying ionosphere and an outer region that is continuing to fill on a day-to-day basis (Park, 1974a).

Notes on the plasmasphere profile

Figure 2.22 shows an empirical model of the equatorial density profile developed on a piecewise basis by Carpenter and Anderson (1992) from ISEE 1 data of the kind illustrated in Fig. 2.3 and from published whistler results. The plasmasphere segment is linear, corresponding to an exponential falloff with radial distance. A similar type of behavior was found from statistics of whistler data by Park, Carpenter and Wiggin (1978) and from DE 1 thermal ion data by Gallagher and Craven (1988). Inside $L \sim 2.5$ the profile has been found to fall off more steeply, approximately as L^{-4}. This was shown in the whistler data of Fig. 2.23, obtained by Tarcsai, Szemerédy and Hegymegı (1988).

The ISEE/whistler model was intended to represent the state of the profile approached under quiet conditions, a state of 'saturation' in the sense that the electron density n_e has increased to the point such that any further daily increases Δn_e give rise to $\Delta n_e/n_e$ that are only marginally detectable.

If an empirical model such as that of Park et al. (1978) is prepared based upon plasmasphere data acquired during various stages of recovery, the slope of the

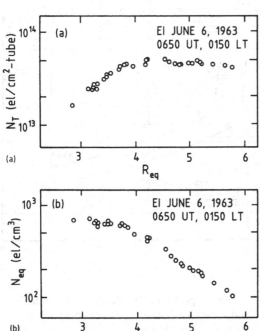

Figure 2.21 Electron total flux-tube content N_T (a) and corresponding equatorial density N_{eq} (b) versus equatorial geocentric radius R_{eq}, as deduced from multipath nose whistler recorded at Eights, Antarctica on a magnetically quiet night (after Angerami and Carpenter, 1966).

plasmasphere profile segment tends to be steeper than that in Fig. 2.22 because of the tendency for the outer plasmasphere densities to be reduced by as much as a factor of 3 and then to increase from day to day during cycles of disturbance and recovery (see Fig. 2.1). (Based upon the fact that density depressions by up to a factor of 3 can be observed in the outer plasmasphere, it may be suggested that any equatorial region where the density is within a factor of 3 of the saturation profile level be regarded as part of the plasmasphere, densities at lower levels being considered representative of the plasmatrough. This criterion is roughly similar to the one used by Chappell (1974) to identify outlying or 'detached' regions whose origin could be inferred to be in the plasmasphere.)

The profile in the plasmapause region

Relatively little work has been done on the details of the plasmapause profile. Its apparently greater steepness in the immediate aftermath of its formation and on the nightside has been noted by Chappell (1972), Nagai *et al.* (1985) and by Oya and Ono (1987), among others, while Higel and Wu (1984) have commented on

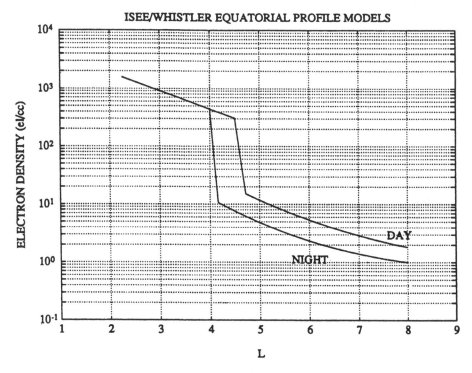

Figure 2.22   Equatorial electron density profile models developed from ISEE-1 sweep frequency receiver data and whistler data by Carpenter and Anderson (1992). The models are presented in piecewise fashion. The night trough segment represents 02 MLT and the day trough segment 12 MLT. Plasmapause *L* values were chosen for convenience in illustration.

the occasional abruptness of the density drop when GEOS 2 exits a region of bulge plasma near or after dusk (see the example of Fig. 2.16). From their study of ISEE plasmapause crossings, Carpenter and Anderson (1992) found that the scale width of the plasmapause, defined as the fraction of an earth radius within which the equatorial electron density dropped by an order of magnitude, varied widely, but could be approximated as $0.1R_E$ on the nightside and by values near $0.2R_E$ on the dayside. Figure 2.24 shows a plot of their data on 'clean' profiles, ones without the irregularities mentioned above. The smallest values, limited by the ~ 150 km spatial sampling interval of ISEE 1, are near 150 km; these were mostly on the nightside. Figure 2.25 shows six examples of steep plasmapause profiles recorded over a range of L values, including one case (D) containing an irregular feature in the region of steep gradients as well as structure in the outer plasmasphere. These data suggest that, for scale widths greater than 150 km, the steepness of the observed profile is not a strong function of L value (at $L < 5$), and hence of the severity of the causative disturbance.

The trough profile

In the bulge region the trough profile was found by Chappell, Harris and Sharp (1970b) to fall off as L^{-4} and, in their ISEE 1 study, Carpenter and Anderson (1992) found that $L^{-4.5}$ provided a good fit to the region just beyond the plasmapause. At greater distance, especially beyond L $\sim$ 7, the profile was found to flatten out and approach a near-constant level near 1 el-cm^{-3}. This effect was

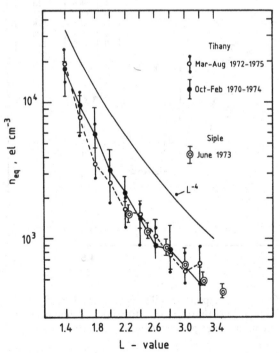

Figure 2.23 Median equatorial electron density determined from 985 whistler observations at Tihany (Hungary) as a function of L. The observations were made during two different seasons (March–August 1972–75 and October–February 1970–74). The vertical bars indicate the upper and lower quartile values. For comparison, June 1973 data from Siple station reported by Park et al. (1978) are also shown. An L^{-4} profile has also been added in this figure (after Tarcsai et al., 1988).

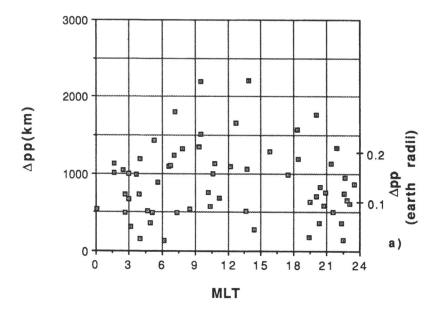

Figure 2.24   Distribution versus MLT of plasmapause scale width Δpp in kilometers and in Earth radii, where Δpp is defined as the distance within which the density changed by a factor of 10. These data represent plasmapause profiles that were identified by Carpenter and Anderson (1992) as 'smooth'.

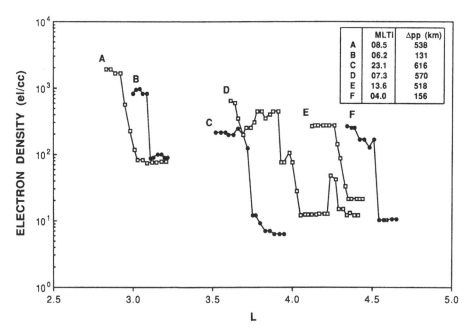

Figure 2.25   Examples of steep plasmapause profiles recorded on ISEE 1 over a range of L values. In case D there was an irregular feature in the region of steep gradients as well as structure in the outer plasmasphere (after Carpenter and Anderson, 1992).

clearest on the dayside but not well documented at night. Figure 2.18a (from Carpenter *et al.*, 1993) shows a trough profile of this kind, while Fig. 2.15a shows a trough distribution beyond the outlying peak that follows an $\sim L^{-4.5}$ curve (dashed curve) very closely.

The distribution of ionization along magnetic field lines

Whistlers are particularly useful for estimating the near equatorial electron density, both because the group velocity does not vary strongly with position along the field-line path, and because the weighting that there is favors the near equatorial region. This can be seen qualitatively by reference to equation (2) of Chapter 1 ($\sim 80\%$ of the group delay at regularly observed whistler frequencies occurs within ~ 30 degrees of the magnetic equator (Park and Carpenter, 1978)). Thus whistlers become very useful for measuring total electron content and estimating equatorial electron density, but are not readily used to infer the field-line plasma distribution (some interesting attempts have been made to infer the spatial variation of the near equatorial distribution from dispersion information at the higher normalized whistler frequencies, which are the most sensitive to near equatorial conditions (Bernhardt, 1979; Sazhin *et al.*, 1990). For this reason, the study of field line models has been based upon a combination of physical arguments, certain propagation tests of opportunity, and comparisons of results from different methods or from spaced locations. Comparisons of equatorial electron density estimates from whistlers with Alouette satellite measurements at 1000 km (Angerami and Carpenter, 1966) and measurements of whistler dispersion along partial magnetospheric paths to the OGO 3 satellite (Angerami, 1970; Smith and Angerami, 1968) suggested use of a diffusive equilibrium (DE) model inside the plasmapause and a 'collisionless' model, behaving approximately as R^{-4} outside. Later, comparisons of plasmasphere data from whistlers with ISEE 1 near-equatorial SFR measurements in rendezvous situations showed agreement within $\sim 20\%$ if a diffusive equilibrium model were used in the whistler analysis (Carpenter, Anderson, Bell and Miller, 1981).

It should not be supposed that a DE model is considered a proper physical basis for understanding the plasmasphere field-line distribution. The distribution is known to exhibit many dynamic effects, including prolonged net upflow during recovery periods. Whatever the plasma dynamics may be, the plasma density distribution seems to be well approximated by DE- type models, where the electron density falls off quite slowly along the field lines. Indeed, as already indicated above, most of the whistler-mode group delay occurs within about 30 degrees of the magnetic equator, where any of the hydrostatic (DE) or hydrodynamic models show field-aligned density distributions which are rather flat, i.e. not drastically dependent on distance along the magnetic field line.

A more difficult problem has been the choice of a model to apply beyond the plasmasphere. The R^{-4} model suggested by Angerami and Carpenter (1966) had been recognized as an extreme case, likely of meaningful application only in the earliest phases of refilling (see also Roth, 1975a). Again, satellite/ground whistler comparisons proved of value; in two cases of dayside observations outside the plasmapause, whistler data were found to agree well with resonance and mutual impedance probe data from GEOS 1 (Corcuff and Corcuff, 1982). The comparison provided support for use of a 'hybrid' field-line model of the type discussed by Park (1972). This model leads to estimates of density and path radius intermediate between those from the DE and the R^{-4} models (see also Carpenter, 1983). Lemaire (1976) pointed out that the R^{-4} density distribution, illustrated in Fig. 2 of his article, is a relatively good approximation for the density distribution of escaping particles (see Chapter 5).

2.5.2   Irregularities

Large-scale irregularities in the plasmasphere

Information about the distribution, amplitude and spatial wavelength of magnetospheric density irregularities is fragmentary at best. Large-scale structures in the electron density in the plasmasphere have been identified from whistlers (Park and Carpenter, 1970). These structures have dimensions of the order of an Earth radius at the equator. Figure 2.26 shows an outlying peak in equatorial electron concentration at a radial distance of $4R_E$, with a peak-to-valley ratio of 2. Park and Carpenter also showed evidence of large variations in outer plasmasphere density level with longitude, up to a factor of 3, within the ~ 30 degree viewing range of a whistler station. Once formed, these structures could be observed for several hours, indicating that their lifetimes must be even longer. The irregularities were found to appear in the aftermath of brief enhancements of the magnetospheric convection activity. The authors suggested that they may be formed by the combined influence of non-uniform coupling fluxes into or out of the underlying ionosphere and convection electric fields that can create or modify already existing irregularities by mixing tubes of ionization with different content. A counterpart of these irregularities appears to exist in the ionosphere. Topside sounders have revealed finger-like regions of locally enhanced or reduced density that extend in the north-south direction and may persist for extended periods (Piggott et al., 1970).

Carpenter et al. (1993) have remarked that regions of irregularity established in the outer plasmasphere during a disturbed period often exhibit sharp low-L limits, and are approximately coincident with the outer plasmaspheric regions of depressed density.

Oya and Ono (1987) used the Stimulated Plasma Wave (SPW) experiment on the Jikiken (EXOS B) satellite to investigate departures from smoothness in electron density profiles sampled at spatial intervals of ~ 100 km. They reported examples of irregularities in the outer plasmasphere with scale sizes of $0.1-0.5 R_E$ (600–3000 km).

Field-aligned wave ducts

Plasmaspheric irregularities are manifested not only through density measurements, but also through their influence on propagating waves. A large and observationally elusive class of such irregularities is that associated with the guiding or ducting of whistler-mode and medium frequency waves along the geomagnetic field lines. The whistler duct is believed to be a field-aligned density enhancement, of order 5–10% over background (at middle latitudes) and of order 10 km in scale size at ionospheric heights, that guides the wave energy between the conjugate ionospheres (Smith, 1961; Helliwell, 1965). Much work has been done on the problem of ionospheric penetration by VLF signals and their trapping and guiding within ducts (Helliwell, 1963, 1965; James, 1972; Strangeways and Rycroft, 1980; Strangeways, 1991; Clilverd et al., 1992) but a large part of the early experimental evidence of the existence and properties of ducts was indirect, emphasizing those aspects of the observed waves that could be expected as a consequence of guided propagation (Carpenter, 1968; Angerami, 1970; Thomson and Dowden, 1977).

A still indirect but promising source of information on duct occurrence and

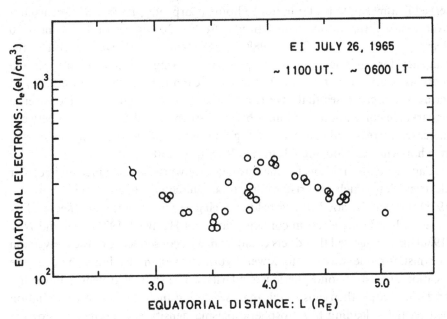

Figure 2.26 Equatorial electron density profile in the plasmasphere showing a large-scale enhancement near $4R_E$ geocentric distance in the dawn local time sector. The short bar near $2.8R_E$ represents uncertainty due to measurement error associated with that whistler component (after Park and Carpenter, 1970).

size is satellite or ground recording of the occurrence and spatial extent of charged particle bursts that are apparently driven by ducted whistlers (Rycroft, 1973; Imhof et al., 1989; Burgess and Inan, 1993; Walt, personal communication, 1993). In the case of quasi-continuous VLF emission activity it has been possible to make simultaneous measurements of a local density irregularity profile and wave intensity. Scarf and Chappell (1973) reported duct thicknesses ranging between 68 and 850 km and density enhancements ranging from 10 to 40%. Koons (1989) reported on whistler-mode hiss trapped in irregularities at the plasmapause, Chan and Holzer (1976) on ELF noise trapped in outlying or detached plasmas, and Beghin, Pandey and Roux (1985) on ELF hiss concentrated in density enhancements poleward of the plasmapause.

A new technique for study of density structure within the plasmasphere has been reported by Jacobson and Erickson (1993). Using a Very Large Array radio interferometer, they identified a type of irregularity extending 30 to 40 km perpendicular to the magnetic field with amplitude of order 100/cc, or approximately 5–10% of the background level. Possibly evidence of the whistler duct phenomenon, the irregularities were detected most frequently in the $L = 2$–3 range and were found to be approximately rotating with the Earth.

The physical basis of whistler ducts remains to be determined. It seems clear that from their regular occurrence at middle latitudes during relatively quiet periods (Laaspere, Morgan and Johnson, 1963), their occasionally long observed lifetimes of tens of minutes to hours (Carpenter, 1966), their occurrence over a wide range of invariant latitudes from ~20 degrees or less to near 70 degrees (Carpenter and Sulic, 1988; Tanaka and Hayakawa, 1985) and the tendency for their identity to be preserved during periods of bulk plasma motions (Carpenter et al., 1972), that they cannot be understood entirely in terms of the processes that give rise to ionospheric irregularities. However, a remarkable correlation was found by G. Carpenter and Colin (1963) between meter scale irregularities at E and F region heights and the occurrence of ducted two-hop whistler-mode signals from a VLF transmitter.

Candidate mechanisms for the formation of whistler ducts include irregular electric fields, which give rise to flux tube interchange (Cole, 1971; Thomson, 1978) and thundercloud electric fields, which give a rise to a stirring effect on flux tubes with differing electron content (Park and Helliwell, 1971). Bell and Ngo (1988) have suggested that ducts could form as a consequence of the conversion of whistler-mode waves into lower hybrid waves in the ionosphere at the location of existing ionospheric irregularities. The short-wavelength lower hybrid waves could then cause pitch angle scattering and precipitation of radiation belt particles, leading to ionospheric plasma density enhancements. Upward diffusion of cold plasma from the enhancements could then produce magnetic-field-aligned density irregularities.

Irregularities other than whistler ducts appear to be present in the regions overlying the ionosphere. Topside sounding satellites have indicated the presence of field aligned ducts of order 1 km in width that guide HF wave energy back and forth along magnetospheric path segments of varying length (Muldrew, 1967; Muldrew and Hagg, 1969). Sharma and Muldrew (1975) have suggested that ducts connecting conjugate points can occur as the result of magnetospheric currents flowing from one hemisphere to the other along geomagnetic field lines. Bell and Ngo (1988) found that the 10–100 m irregularities involved in the production of lower hybrid waves from whistler-mode waves occurred over a wide range of latitudes and at various altitudes along field lines extending to at least $2R_E$.

Irregularities near the plasmapause

Important information on irregular density structure near the plasmapause has been reported by LeDocq, Gurnett and Anderson (1994), who performed power spectral analysis on sweep frequency receiver data from the CRRES satellite. On some orbits, plasma density irregularities in the outer plasmasphere were found to have power spectral slopes near $-5/3$, thus suggesting the presence of well-developed two-dimensional magnetohydrodynamic turbulence. It was suggested that such turbulence could originate in a velocity shear at the plasmapause and associated Kelvin–Helmholtz instability. Density structure of a similar nature was found by Carpenter et al. (1993) to develop in outer plasmasphere regions where the mean density had decreased during disturbed periods. Such reduced-density, irregular regions were often found to have well-defined low-L limits.

A statistical analysis of the CRRES density data by LeDocq et al. (1994) showed that density fluctuations were largest between $L = 3$ and $L = 6$, in the region of the plasmapause. Fluctuation amplitudes greater than 10% of a local density average were found in only 25% of the data studied.

There is substantial evidence of electron density irregularities in the vicinity of the plasmapause. For example, Koons (1989) used the sweep frequency receiver on the AMPTE spacecraft to study irregular structure near the nightside plasmapause, finding cases of density enhanced by more than 40% over nearby levels and diameters of order 500 km. In ISEE 1 SFR data, Carpenter and Anderson (1992) found that within the region of steep density falloff at the plasmapause, roughly 50% of the 208 profiles examined exhibited either a density plateau or one or more local irregularities. Such irregularities appeared to be less likely to occur following onsets of disturbance activity that involved abrupt, step-like changes to higher levels. From JIKIKEN plasma wave measurements in the outer plasmasphere, Oya and Ono (1987) found evidence of strong irregularities in the plasmapause region, with variations up to 70% of background, the scale sizes varying from several 100 km to less than 100 km. The authors inferred that 'the irregularities are associated with the plasma instability closely connected with the formation mechanism of the plas-mapause'.

Recently, Moldwin *et al.* (1995), using plasma data from the Los Alamos satellite 1989-046, found the region of dense, cold plasma at synchronous orbit to exhibit much fine structure, with scale sizes on the order of 1000 km or less. The amount of variability in the density was found to generally increase with increasing K_p, and the cases of greatest variability tended to be associated with substorm activity.

Inan, Bell and Anderson (1977) used variations in the delay of signals from the Siple, Antarctica VLF transmitter as measured along the orbit of IMP 6 to identify a local density enhancement in the region of steep plasmapause gradi-ents. Kowalkowski and Lemaire (1979) studied the distribution of irregularities in the thermal ion density data from OGO 5. They used a less restrictive criterion than that employed earlier by Chappell *et al.* (1971a) to identify outlying plasma features, and found a peak in occurrence in the post-midnight sector. On individual orbits, the irregularities in that sector appeared to be concentrated near the observed plasmapause.

In the LeDocq *et al.* (1993) study of CRRES data, a case was found of quasi-periodic density fluctuations outside the plasmasphere in the post-mid-night sector. The spatial scales along the orbit were approximately 750 km.

In a case study of the dusk bulge region, Carpenter *et al.* (1993) found that large-scale irregularities, with scale sizes of order $0.1 R_E$, were present near the plasmapause in the post-dusk sector during periods of continuing low-level substorm activity ($K_p = 1-2$) that lasted for as many as seven days after a plasmasphere erosion event. Higel and Wu (1984) reported from GEOS 2 on substantial density structure on the sunward or westward edge of bulge regions encountered along the geosynchronous orbit. An effect of this kind is evident in the data of Fig. 2.16.

Irregularities beyond the plasmapause

Carpenter and Anderson (personal communication, 1992) have found that the plasma trough region, when observed by ISEE 1 prior to significant refilling, does not tend to exhibit depressions from the apparent large scale backround level, as does the plasmasphere. Instead, irregularities tend to take the form of localized increases, the structures in Fig. 2.15b being examples (from Carpenter *et al.*, 1993). They also noted that dense plasma structures such as those illustrated in Figs. 2.15b and 2.18c may exhibit much more fine structure, with peak-to-valley ratios of 5 or more, than is usually present in the outer plasma-sphere and in some of the less-distant outlying features such as the one shown in Fig. 2.15a.

2.5.3 Plasmaspheric temperatures

A number of attempts have been made to deduce plasmasphere electron temperatures from the dispersion properties of whistlers near their upper frequency limits of observation (Guthart, 1965; Sazhin et al., 1990). Upper limits on T_e of a few eV have been deduced in the few cases analyzed thus far, but the method, while promising, requires further development, as discussed by Sazhin et al. (1992). Plasmaspheric temperatures are more commonly determined from in situ particle flux measurements. These in situ spacecraft observations will be presented in Chapter 3.

2.5.4 Quiet day electric fields in the plasmasphere

Quiet day ion drifts in the plasmasphere have been investigated by the incoherent scatter technique from a number of locations, including Millstone Hill near $L = 3$ (Evans, 1972) and St Santin near $L = 2$ (Blanc, Amayenc, Bauer and Taieb, 1977). The early observations did not relate in a simple way to the predictions of existing ionospheric dynamo models, and Blanc and Amayenc (1976) pointed out a need to obtain improved measurements of ionospheric neutral winds and electric currents and fields, and in the case of the theoretical models to include considerations of the penetration of the plasmasphere by electric fields of magnetospheric origin.

Studies of the cross-L drifts of plasmasphere whistler paths in quiet periods were reported by Carpenter and Seely (1976) and by Carpenter (1978a). The largest drifts were found on the dayside, corresponding to east-west fields in the range ~ 0.1–$0.2\,\mathrm{mV/m}$. Outward drifts occurred in the morning sector and inward drifts in the afternoon, as indicated by the time series of measurements of path L value in Fig. 2.27. This effect should contribute to a quiet day maximum in plasmasphere radius at noon, as Gringauz and Bezrukikh (1976) have reported (although it does not explain the size of the noon–midnight asymmetry indicated by these authors). Harmonic analysis of each of two quiet day series showed diurnal and semidiurnal terms consistent with the findings from radar. However, the flow speeds near $L = 4$ in the magnetosphere were less than those predicted by early dynamo theories (Maeda, 1964; Schieldge, 1974). The d.c. terms were small compared with the fluctuating terms; that was to be expected to the extent that the effects of a time stationary potential field were being sampled. Quiet day electric fields at synchronous orbit were measured on GEOS 2 on the dayside by Baumjohann, Haerendel and Melzner (1985), who used an elegant technique involving determination of electric-field effects on the orbits of electrons fired transverse to the geomagnetic field from an electron gun. The inferred electric fields for $K_p \sim 0$–1 were consistent with (and an extension of) the findings from whistlers, and were interpreted as evidence of an ionospheric dynamo process.

In an analysis of whistler path drift rates versus L, Carpenter (1978a) showed a case in which the inferred east–west field component decreased with increasing invariant latitude, a behavior that would be consistent with an ionospheric dynamo source of the fields and quite inconsistent with the flow patterns expected from a solar-wind dynamo of essentially high-latitude origin.

The global quiet time electrostatic potential in the plasmaphere has been derived from multipoint incoherent scatter and whistler observations by Richmond (1976). The potential is represented by a finite series of spherical harmonic functions symmetric about the equator. Figure 2.28 illustrates the equipotential lines in the equatorial plane. It can be seen that the cross-L electric drifts predicted by this empirical model fit qualitatively those obtained from whistler observations reported by Carpenter and Seely (1976).

The latter whistler observations show a small dc eastward electric field component of $0.016 \, \mathrm{mV/m}$ and $0.001 \, \mathrm{mV/m}$, respectively for the two quiet days: 13 July 1963 and 7 July 1973. These two isolated cases tend to support the existence of an outwardly directed plasmaspheric wind, as proposed by Lemaire

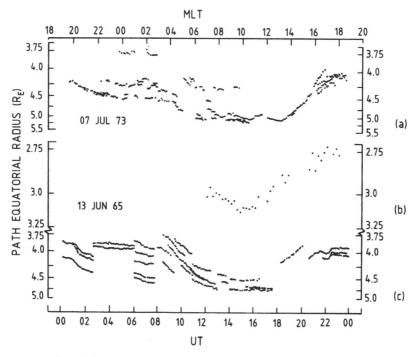

Figure 2.27 Two case studies showing variations with time in the equatorial radii of whistler paths in the outer plasmasphere during geomagnetically quiet periods. The dayside cross-L displacements are larger than these indicated at night. The whistlers were recorded at Siple, Antarctica (a) and at Eights, Antarctica (b).

and Schunk (1992, 1994); however, a more extensive survey would be needed to test this suggestion. In Richmond's (1976) and in other magnetospheric convection electric field models a net radial plasma flow was excluded a priori by setting the d.c. component of the east–west electric field arbitrarily equal to zero.

Doppler shift and group delay measurements on fixed-frequency VLF whistler-mode signals have been used by a number of workers to investigate cross-L drifts in the $L = 2$–3 range (Andrews, Knox and Thomson, 1978; McNeil, 1967). Saxton and Smith (1989) confirmed the presence at $L = 2.5$ of quiet day cross-L drifts similar to those reported from whistlers.

2.5.5 Temporal variations in electron density

The annual and semiannual variations

One of the first and most puzzling discoveries made from whistler dispersion measurements was an annual variation in equatorial electron density, with December values higher by a factor of roughly 2 than those in June. The effect was reported from IGY measurements in the western US by Smith and Helliwell (1960), Helliwell (1961) and Smith (1961a). An annual variation with a similar phase in other quantities such as neutral air density at ionospheric heights (Paetzold, 1962) was also being reported at that time. The large size of the plasmasphere electron density variation appeared to preclude the ∼3% eccentricity of the Earth's orbit around the sun as a cause, and tentative consideration was given by Helliwell (1961) to an explanation in terms of the offset of the Earth's spin and geomagnetic axes and the consequent annual variation in the geomagnetic latitude of the subsolar point at a particular geomagnetic longitude. If the geomagnetic asymmetry were the controlling factor, the annual variation in electron density should reverse its phase in the eastern hemisphere.

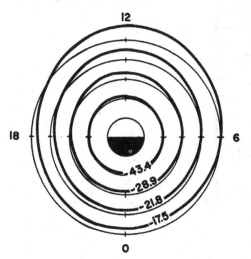

Figure 2.28 The electrostatic potential versus local time mapped into the magnetospheric equatorial plane from an ionospheric potential pattern that was inferred from observations. The circles are drawn at $L = 2, 3, 4$ and 5. The heavy solid lines are equipotential lines for -43.4, -28.9, -21.8 and -17.5 kV from inside to outside. Convection of cold plasma occurs counterclockwise along equipotential lines (after Richmond, 1976).

Such a phase reversal seemed unlikely, in view of early results showing the December maximum to exist over a ~ 180 degree range of longitudes from ~ 180 degrees West (New Zealand–Alaska) (Helliwell, 1961; Smith, 1961a) to 0 degrees (Poitiers, France) (Bouriot et al., 1967). However, more detailed assessments of the amplitude of the variation (Park et al., 1978) indicated the existence of a longitude effect, with a December–June ratio of ~ 1.5 near $L = 2.5$ in the data of both Stanford, near 120 degrees West (Carpenter, 1962b) and of Poitiers near 0 degrees (Bouriot et al., 1967), but a larger ratio, around 3 : 1, at intermediate longitudes, near the 75 degrees West meridian (Park, 1974a). Figure 2.29 compares the median monthly values of tube electron content N_T and equatorial electron density for November–December 1964 and June 1965 as determined by Park (1974a) from whistlers recorded at Eights, Antarctica. The difference between the November–December and June values of equatorial density (Fig. 2.29b) dropped substantially with increasing L.

Additional evidence of a falloff in amplitude of the annual variation with longitude away from ~ 75 degrees West was supplied by Tarcsai et al. (1988), whose data from Tihany at ~ 25 degrees East, illustrated in Fig. 2.23, showed northern winter densities near $L = 2$ to be $\sim 30\%$ above summer values; he found no evidence of a significant annual effect at $L > 2.4$.

Recently, VLF doppler experiments on fixed frequency transmitter signals

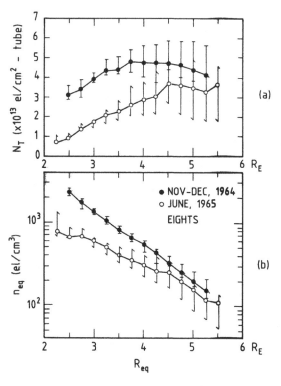

Figure 2.29 (a) Median monthly values of flux tube electron content N_T versus tube equatorial radius observed from whistlers near the 75°W meridian, showing a large annual variation. (b) Plot of equatorial electron concentrations corresponding to the tube content data (after Park, 1974a).

performed at Faraday, Antarctica, ~60 degrees West, and at Dunedin, New Zealand, ~170 degrees East, appear to have provided the basis for a clearer assessment of the annual effect (Clilverd, Smith and Thompson, 1991). The authors confirmed the factor of 3 difference in equatorial density near 60 degrees West and found a much smaller difference, by a factor of ~1.4, at Dunedin.

Clilverd *et al.* (1991) developed an explanation of the observations in terms of the observed values of ionospheric densities near the ends of the field lines of interest, as expressed in monthly foF2 medians. They found that the observed variations, including a reduction of the variation amplitude from 3:1 to ~2:1 between years of high and low solar activity, could be modeled by assuming diffusive equilibrium, on a time scale of order a month, from the F region peak to the equator. Shorter time scale variations (of the order of days or less) in the ionospheric density are unlikely to be reproduced faithfully in the plasmasphere because of the mismatch between time constants for ionospheric production and loss on the one hand (minutes) and plasmasphere refilling on the other (days). The larger size of the variation near the 60–75 degree West meridian was attributed to the large annual variation in the nighttime F region densities in the nearby southern hemisphere ionosphere, compared either with the corresponding variation in the northern conjugate region or at other longitudes.

A semi annual variation in equatorial density with equinoctial maxima was noted by Carpenter (1962b), Corcuff (1962, 1965) and Bouriot *et al.* (1967), the data of Bouriot *et al.* suggesting an amplitude of ~30%. The presence of such an effect is indicated in Fig. 2.30 by the comparative narrowness of the June minimum in monthly average values of electron density at $L = 2.25$ deduced from whistlers at Poitiers for the interval 1958 to 1965.

The solar cycle variation

It is well established that the F2 electron density peak, N_mF_2, has a significant and systematic solar cycle variation, varying nearly linearly with the mean

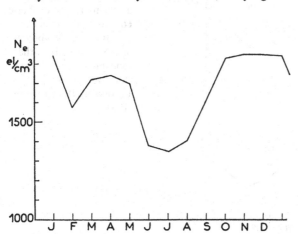

Figure 2.30 Averages of monthly values of electron density at $L = 2.25$ for the years 1958 to 1965, showing evidence of a semiannual variation. The associated whistler data were recorded at Poitiers and were reported by Bouriot *et al.* (1967).

sunspot number R (Ratcliffe, 1960). This variation is illustrated by the lower curve in Fig. 2.31, which shows annual mean values of N_mF_2 at 00 MLT. These values were determined from F_2 critical frequencies measured at Poitiers between 1957 and 1966. The upper curve shows the equatorial density N_e for $L = 2.25$, deduced from whistlers recorded at Poitiers between 00 and 02 MLT. It can be seen that both curves have similar trends between the 1958 solar maximum and the solar activity minimum in 1964. N_mF_2 decreased monotonically by 75% over this period, while N_e decreased by only 25%, but in a more irregular manner. There were important enhancements in the annual mean values of N_e in 1957, 1960, 1963 and 1965. These enhancements, superimposed on the overall solar cycle variation, were attributed by Bouriot *et al.* (1967) to the enhanced occurrence of solar proton events observed during those same years. Study of the total ionospheric electron content, N_T, deduced from the Faraday rotation and Doppler effects, led Bhonsle *et al.* (1965) and Yeh and Flaherty (1966) to similar conclusions.

Diurnal variation

Evidence of a diurnal variation in plasmasphere electron density at $L = 3$ with a post-midnight minimum, an afternoon maximum, and a range of ± 10–15% around the 24-hour average was reported from whistlers by Park *et al.* (1978). This effect is illustrated in Fig. 2.32, from a study of 985 whistlers by Tarcsai (1985), in terms of total electron content N_T at $L < 2.8$ versus LT. The figure shows that the total electron content had a maximum of 2.1×10^{13} electrons cm^{-2} around 17–19 LT and a minimum at 04–05 LT that was a factor of 2 lower. From study of the group delay of VLF transmitter signals propagating near $L = 2.5$ at 75 degrees West, Clilverd *et al.* (1991) reported ratios varying up

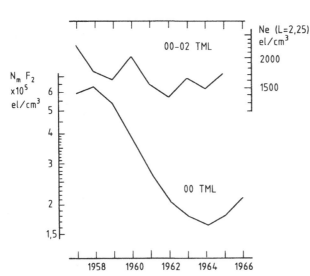

Figure 2.31 Comparison of solar cycle variations in magnetospheric equatorial electron density at $L = 2.25$ (above) and peak density in the ionosphere. The observations were made at Poitiers by Bouriot *et al.* (1967).

to $\sim 1.5:1$, depending upon season, between electron densities near dusk and near dawn.

These findings compare well with those of Park *et al.* (1978). As L decreases below $L \sim 2.5$, the tube content drops rapidly, and a given upward flux from the ionosphere may be expected to have a larger effect, as in the case of the low density trough region (Higel and Wu, 1984, Carpenter and Anderson, 1992). The amplitude of the diurnal variation compares well with that determined by Park *et al.* (1978). Park *et al.* had found that, beyond $L = 3$, diurnal effects in the dense plasmasphere could not be identified because of the large day-to-day variations associated with magnetic storms and substorms. In any case, such changes were expected to be small (percentagewise) because of the large values of flux tube electron content associated with the outer plasmasphere. Later, Tarcsai (1985) found that the magnitude of the diurnal variation of N_T is almost inversely proportional to L, at least from $L = 2$ to $L = 4$.

It is possible that the smallness of the observed diurnal variation in some parts of the plasmasphere, particularly near and beyond $L = 3$, is in part a consequence of the countering of H^+ downward nighttime flows from the protonosphere into the winter hemisphere near the solstices by continued upward nighttime flows from the summer hemisphere, as reported by Evans and Holt (1978).

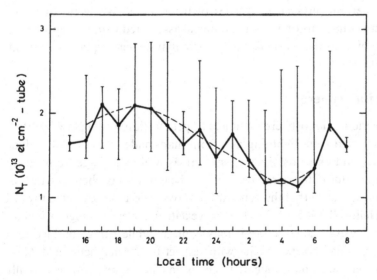

Figure 2.32   Hourly median values of the plasmaspheric electron content of a flux tube of $1\,cm^2$ cross-section at $1000\,km$ altitude, for L between 2.0 and 2.8. The data were deduced from Tihany (Hungary) whistler observations collected between 1970 and 1975. The error bars indicate the minimum and maximum value observed in each LT hourly period. The dashed line is a smoothed fit (after Tarcsai, 1985).

2.6   Plasma wave observations and the plasmasphere

2.6.1   Introduction

There is a vast literature on wave activity observed within the plasmasphere and in its near vicinity. Many of the relevant topics are either reviewed or treated in detail in papers such as those by Kurth and Gurnett (1991) on plasma waves in planetary magnetospheres, by Sonwalkar (1994) on magnetospheric LF, VLF, and ELF waves, by Anderson and Kurth (1989) on discrete emissions, by Hayakawa (1994) on whistlers, by Lanzerotti and Southwood (1979), Southwood and Hughes (1983), and Glassmeier (1994) on hydromagnetic waves, by Gurnett and Inan (1988) on DE-1 plasma wave observations, by Oya (1991) on EXOS-D plasma wave observations, by Kimura (1989) on ray paths in planetary magnetospheres, by Hayakawa and Sazhin (1992) on whistler-mode mid-latitude and plasmaspheric hiss, by Helliwell (1988) on active VLF wave experiments in the outer plasmasphere, by Kurth *et al.* (1981) and Morgan and Gurnett (1991) on terrestrial continuum radiation, by Sazhin and Hayakawa (1992) on whistler-mode 'chorus', by Sato and Fukunishi (1980) on quasi-periodic whistler-mode emissions, and by Sazhin *et al.* (1992) on auroral hiss.

Rather than attempting a wide-ranging discussion of wave observations, we propose to call attention to some of the wave phenomena that occur in particular regions and to the geophysical significance that these phenomena may acquire from certain physical properties of the regions. We begin our brief discussion with comments on wave phenomena associated with the ionosphere, and then continue to the plasmasphere, plasmapause, near-equatorial region, and plasmatrough.

2.6.2   The ionosphere

The highly refracting ionosphere is subject to penetration from below by ELF/VLF waves injected from lightning and from man-made sources that include fixed frequency and experimental transmitters as well as leakage from power grids. The large value of the refractive index in the lower ionosphere at whistler-mode frequencies is such that the penetrating waves tend to be refracted toward the vertical (Helliwell, 1965, 1969). They then excite the overlying region, including whatever density fine structure it may contain, while being confined to a limited range of wave normal angles with respect to the magnetic field. Conversely, the lower ionosphere appears to act as an efficient reflector of virtually all downcoming waves that have wave normal angles outside a 'transmission cone,' that is, a narrow (order of 3–10 degrees) cone of angles (with respect to the vertical) within which penetration of the lower ionospheric boundary should be possible according to Snell's law (Helliwell, 1963, 1965). Thus the ionosphere has the effect of limiting ground observations to waves guided in the 'ducts' discussed earlier, while ensuring that spacecraft are exposed to a wide variety of wave trajectories, most of which represent nonducted propagation. Ducts, while numerous in terms of the number that can be separately identified in the records of a single ground station (Lester and Smith, 1980; Hamar et al., 1990) are estimated to occupy only of order 0.001% of the plasmasphere/plasmatrough volumes that they penetrate (Burgess and Inan, 1993).

The regular ionosphere, while being both a source and sink for the overlying plasmasphere plasma (apparently being partially maintained at night by downward fluxes from above (Park and Banks, 1974)), is also perturbed and conditioned by wave action in the plasmasphere in which both ducted and nonducted whistler mode waves scatter energetic electrons from their orbits into the dense ionosphere (Helliwell et al. 1973; Rycroft, 1973; Voss et al., 1984; Hurren et al. 1986; Inan et al. 1990; Jasna et al., 1992; Burgess and Inan, 1993). Here the precipitating charged particles, electrons in the case of a wide range of whistler-mode signals, produce secondary ionization, often detected as density perturbations at the lower boundary of the nightside D region. The process of precipitation can be quasi-steady (Gail and Carpenter, 1984), but in its most readily observed form is driven by individual lightning whistlers and is burstlike. In subauroral ionospheric regions poleward of the plasmapause, wave induced electron precipitation effects tend to be even more intense (Imhof et al., 1989). Both whistlers and naturally occurring chorus and hiss emissions may be involved, and associated bursts of X-rays and optical emissions can be detected (Rosenberg et al., 1971, 1981; Helliwell et al., 1980; Doolittle and Carpenter, 1983; Hurren et al., 1986).

Density irregularities in the lower ionosphere have important effects upon signals penetrating the ionosphere from below. Irregularities with scale size of order 10 km to 100 km cause the wave normal angles of upgoing waves to be tilted at various small angles from the vertical, thus widening the distribution of nonducted ray paths that can be excited by waves entering the magnetosphere at a particular point (Sonwalkar et al., 1984). On the other hand, irregularities in the 10m to 100m range can cause profound changes in the wave modes observed. Whistler-mode waves incident upon such irregularities can excite short wavelength lower hybrid waves (Bell and Ngo, 1988, 1990; Titova et al., 1984; James and Bell, 1987) that are believed capable of strong interactions with ionospheric ions (Bell et al., 1991, 1993).

2.6.3 The plasmasphere

The collisionless aspect

Within the regular ionospheric layers, particularly in the dayside D and E regions, whistler-mode waves may suffer significant collisional damping (Helliwell, 1965). However, in the essentially collisionless region above $\sim 500\,\mathrm{km}$ altitude, collisional damping becomes unimportant in comparison with the effects of wave-particle resonance interactions. The spatial damping decrements are low; in the high-altitude regions one can observe waves that have propagated along ray paths extending over thousands of wavelengths. In such cases, a broadband signal can become highly dispersed, due to the frequency dependence of the refractive index. The amplitude of a given wave train may then depend upon the path-integrated effects of dispersion, of spreading losses, and of growth and damping due to Landau or cyclotron resonance. Landau

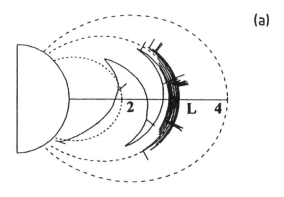

(a)

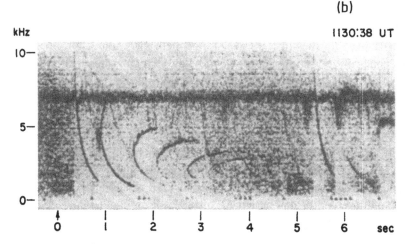

(b)

Figure 2.33   (a) Ray tracing of a nonducted ray path in the magnetosphere, illustrating some effects of magnetospheric reflection. Shown is the calculated trajectory of a 1 kHz signal injected into the topside ionosphere at 400 km altitude with vertical wave normal. Sample wave-normal directions are shown as line segments at various points along the ray path (from Draganov *et al.*, 1992). (b) Spectrogram of magnetospherically reflected (MR) whistlers received on the OGO-1 satellite (adapted from Smith and Angerami, 1968). The multiple signals between 0 and 4 s originated in a single lightning flash.

damping is presumed to become more important as the angle of the wave normal of the wave becomes progressively larger with respect to the direction of the geomagnetic field (Huang *et al.*, 1983; Church and Thorne, 1983). Increases in the wave normal angle from an initial zero value (at an ionospheric injection point) are illustrated in Fig. 2.33. Line segments show the wave vector orientation at various points along the ray path.

It is of interest that the dense 'cool' background electron plasma of the plasmasphere (~ 0.1–$5\,\text{eV}$; ~ 100–$1000\,\text{el/cc}$) dominates the real part of the refractive index and hence the velocity and direction of whistler-mode propagation, while the 'hot' but tenuous electron population ($\sim 5\ \text{eV}$–$500\,\text{keV}$; ~ 0.1–$1\,\text{el/cc}$) dominates the imaginary part and hence the growth or damping of the waves (Kennel and Petschek, 1966; Kennel and Engelmann, 1966).

Ultra-low frequency wave activity

In a cold, uniform magnetoplasma, the two MHD wave modes, the shear mode and the fast mode, are uncoupled (Lanzerotti and Southwood, 1979). However, in the non-uniform magnetosphere the two modes are coupled, typically such that the fast mode efficiently carries wave energy in directions transverse to the magnetic field, meanwhile exciting the shear mode so as to produce resonant oscillations of particular field-line regions. The waves are believed to originate in various ways, including generation within or upstream of the Earth's bow shock or through solar-wind-driven wave activity at the magnetopause (Greenstadt *et al.*, 1980; Odera, 1986), spreading from the cusp region at ionospheric altitudes (Engebretson *et al.*, 1991), anisotropy in the hot component of the magnetospheric particle distribution (Kennel and Petschek, 1966; Cornwall *et al.*, 1970), instabilities at the plasmapause (Hasegawa, 1971; Lakhina *et al.*, 1990), modulations of ionospheric conductivity by quasiperiodic energetic particle precipitation (Arnoldy *et al.*, 1982; Oguti *et al.*, 1984), and thunderstorm electric fields (Dejnakarintra and Park, 1974; Fraser-Smith, 1993).

ULF pulsations are classified as continuous (Pc) and irregular (Pi). Pc3–5 pulsations, with periods in the range 10–600s and wavelengths of the order of a few Earth radii, are predominantly dayside phenomena that fall into two classes: the first corresponds to harmonically structured and azimuthally polarized waves for which the fundamental frequency varies with local time and radial distance, and for which the spectrum is dominated by resonances of the local magnetic field lines (Takahashi and McPherron, 1982; Engebretson *et al.*, 1986; Potemra *et al.*, 1989). In the other class, the perturbation has a substantial compressional component and a large part of the magnetosphere may oscillate at a single frequency (Greenstadt *et al.*, 1986).

The action of the dense plasmasphere as a resonant cavity at ULF has been investigated by Yeoman and Orr (1989), who showed that observed magnetic pulsation periods are consistent with those expected for plasmaspheric cavity resonance. A cavity resonance mechanism for mid and low-latitude Pi2 signatures was suggested by Saito and Matsushita (1968): an impulse upon the magnetospheric cavity produces compressional cavity modes that stimulate field lines where the resonant frequencies are matched; this then leads to a low latitude enhanced signal.

Whistler-mode ray paths

If the plasmasphere were somehow homogeneous, the wave normal angle and the ray or energy flow direction of a whistler mode wave with respect to the geomagnetic field direction would not be expected to vary as the wave progresses. (At relatively low wave frequencies compared with the local electron cyclotron and plasma frequencies and for any allowed value of the wave normal angle, the ray direction tends to remain within ~ 20 degrees of the B field and thus tends to be weakly guided along it (Storey, 1953)). However, in the real magnetosphere, the effects of gradients in the B field and plasma density are such as to cause the wave normal angle to vary along the ray path, typically increasing until a limiting angle for propagation, or resonance cone angle, is reached (Edgar, 1976), as illustrated in Fig. 2.33a. Under such conditions the refractive index also becomes large, so that the wave can exchange energy with the denser, lower-energy particles within the distribution of radiation belt electrons. At large wave-normal angles, Landau damping eventually leads to the attenuation and effective disappearance of the waves.

The magnetospheric reflection phenomenon

When a whistler-mode ray path extends earthward into regions of increasing magnetic field, there may occur one or more magnetospheric reflections, as a consequence of which the ray direction is approximately reversed and propagation back into a region of weaker B field occurs (Kimura, 1966, 1989; Helliwell, 1969). This remarkable effect has been particularly well documented in spacecraft observations of the magnetospherically reflected, or MR, whistler, which has been found to propagate repeatedly back and forth between MR reflection points in opposite hemispheres (Smith and Angerami, 1968; Edgar, 1976). Figure 2.33b shows a spectrogram of an MR whistler received on the OGO-1 satellite. Also shown as Fig. 2.33a is a diagram of a ray path for a frequency of 1 kHz injected into the ionosphere at 46 degrees latitude with initial wave vector oriented along the magnetic field (from Draganov et al., 1992). The ray path shows a number of MR reflection points as well as a tendency to remain in a particular L region as the higher order hops occur. The phenomenon of magnetospheric reflection is essential to the action of the plasmasphere as a kind of cavity in which waves are trapped and throughout which they may spread widely from localized source regions.

Discrete emission band

An intense, usually structured band of whistler-mode noise with spatially vary-ing center frequency has been observed from satellites at various altitudes within the plasmasphere (Poulsen and Inan, 1988; Boskova et al., 1988, 1993a). From ~ 1 to 5 kHz in width, the band is found to fall in center frequency with increasing L, and may extend outside the plasmapause in slightly modified form. The individual elements comprising the band may fall or rise with time, often at rates substantially in excess of those of the structured 'chorus' observed beyond the plasmasphere (Poulsen and Inan, 1988). This remarkable noise phenomenon has been found to occur in the range $L = 2$ to 4.5 in the morning local time sector under magnetically disturbed conditions. Ray tracing and other analysis suggests that the noise is generated near the magnetospheric equator at frequen-cies of ~ 0.2–$0.3 f_{\text{heq}}$ (equatorial gyrofrequency) and that it propagates in a nonducted mode to low altitudes.

An apparently related noise band, only occasionally structured but similarly variable in frequency with L, was reported by Boskova et al. (1993a) from low-altitude satellite data. The band appeared over a wide range of local times under magnetically calm conditions. It was found to be located inside the plasmasphere and in a limited range of L at which ionospheric electron tempera-ture enhancements were observed.

The generation mechanisms of these noise bands remain to be determined. In view of the occurrence of the structured band in the dawn sector under disturbed conditions and hence in the aftermath of substorm particle injections, the possibility of geophysically significant wave–particle interactions, including particle scattering, is substantial. In one case of DE–1 data, the power spectral density of this type of noise was greater than that of any other wave activitiy in the 10 Hz–400 kHz range of the sweep frequency receiver!

Plasmaspheric hiss

Plasmaspheric hiss (Dunckel and Helliwell, 1969; Russell et al., 1969, 1972; Muzzio and Angerami, 1972; Thorne et al., 1973; Storey et al., 1991) is a well known and still controversial noise phenomenon that is believed to depend upon extended whistler-mode propagation back and forth within the plasma-sphere at frequencies generally in the ~ 100–1000 Hz range (Thorne et al., 1979). Figure 2.34, a sweep frequency record from the CRRES satellite, shows plasmas-pheric hiss as well as a variety of other wave phenomena that are observed within and beyond the plasmasphere in the frequency range 10 Hz–400 kHz. The record shows receptions during a ten-hour period centered on apogee, which in this case was at $L \sim 6.47$ and at ~ 0630 MLT. On this orbit the spacecraft emerged from the plasmasphere at ~ 0400 MLT and $L \sim 4.5$, and re-entered it at ~ 1000 MLT and $L \sim 3.8$. Plasmaspheric hiss between ~ 300 Hz and 3 kHz is faintly evident at the left between ~2200 and 2330 UT, or in the predawn plasmasphere. At the right, as CRRES crossed the plasmapause near 1000 MLT, an apparently sharp increase in plasmaspheric hiss intensity was observed, following detection of weaker activity in thë 300–800 Hz range since ~0430 UT.

Partly because of its widespread occurrence, plasmaspheric hiss has been proposed as a factor in the regulation of the radiation belts and in particular in the creation of the slot region near $L = 3$ (Thorne et al., 1979). This regulation is proposed to occur through cyclotron resonant scattering of energetic electrons by the hiss waves.

Although easily detected because of evidently efficient propagation in the plasmasphere, identification of the origin of plasmaspheric hiss has proven difficult, as discussed in a recent paper by Storey et al. (1991) and in an extensive review by Hayakawa and Sazhin (1992). As the consequence of a debate between Church and Thorne (1983) and Huang et al. (1983), it became clear that growth of the waves from ambient noise levels to observed levels was not possible along cyclic trajectories that repeatedly cross the equator. Solomon et al. (1988) found

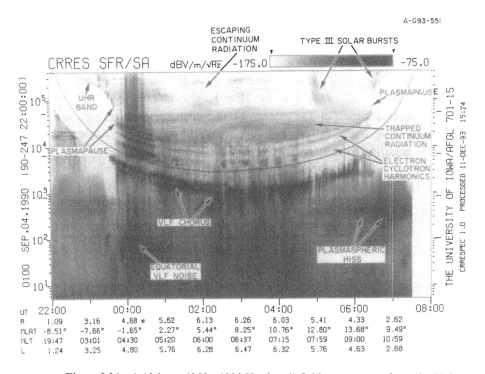

Figure 2.34   A 10-hour 40 Hz–400 kHz electric field spectrogram from the University of Iowa CRRES plasma wave experiment, showing examples of several forms of wave activity observed within the plasmasphere and in the nearby plasmatrough region. The thin continuous curve with minimum near apogee represents the electron gyrofrequency. The horizontal line structure in the 100–1000 Hz range is of instrumental origin (after Anderson, 1994).

in a particular case of GEOS data that the anisotropy of the hot plasma was sufficient to explain the growth of hiss to observed levels during a single crossing of the equator. However, this inference was based upon assuming a wave normal parallel to the magnetic field and, as Storey *et al.* (1991) point out, such wave-normal angles, previously thought to be required for wave growth, have not been found in wave-normal analyses that use data from multiple satellite antennas (Storey and Lefeuvre, 1979, 1980) and/or fading effects due to antenna spin and vehicle motion (Sonwalkar and Inan, 1989). Thus it appeared that an 'embryonic source' was needed, either in order for the limited amplification available along cyclic trajectories to have a significant effect, or to be consistent with the large wave normal angles observed. A proposal for such an embryonic source is the lightning whistler, evidence for which was offered by Sonwalkar and Inan (1989). A still more prominent role for whistlers was proposed by Draganov *et al.* (1992), who suggested that dispersed waves from an individual flash can form a diffuse continuous band of noise and that the collective effects of lightning worldwide could produce the levels of observed hiss activity, essentially without requiring the action of an amplification mechanism. However, this conclusion depended upon assumptions about the distribution function of low energy electrons, which is not well known experimentally.

Field-aligned plasma irregularities

The existence of whistler ducts of the kind discussed in Section 2.5.2 ensures that some of the wave energy of injected whistler mode waves propagates to the conjugate ionosphere rather than undergoing MR reflection before reaching the ionosphere. Part of the wave energy may then penetrate the ionosphere and be detected by ground receivers, but a significant fraction will be reflected and may then (1) re-excite the orginal duct, (2) excite neighboring ducts, and (3) propagate upward in the nonducted mode. This simple framework leads to an astonishing variety of effects. As an example, given strong wave growth in the aftermath of a magnetic storm, the energy may echo back and forth along one or more ducted paths for times up to minutes, exhibiting almost no damping decrement on individual hops (Helliwell, 1965). In addition, whistler energy that originally propagated in a duct at relatively low latitudes, say near $L = 2$, may spread poleward through interduct coupling at the time of reflection and excite paths with endpoints at substantially higher latitudes, near $L = 4$ (Carpenter and Orville, 1989). Furthermore, a whistler wave propagating just inside the plasmapause can cross the boundary during reflection and then propagate outside (Smith and Carpenter, 1982).

Many discrete wave events observed well above the ionosphere are found to have propagated in a hybrid mode, involving initial ducted propagation from a region of origin or of ionospheric penetration to a region of ionospheric reflection, followed by nonducted propagation upward to a point of observation

(Rastani *et al.*, 1985; Smith *et al.*, 1985). Thus both the MR process and the ionospheric reflection process play parts in the large-scale redistribution of whistler-mode wave energy within and beyond the plasmasphere. The ionospheric reflection, while it involves wave attenuation associated with passage through the regular ionospheric layers, can at times be highly efficient, as indicated by standing wave patterns in hiss-like noise observed in the lower ionosphere (Brittain *et al.*, 1983).

2.6.4   The plasmapause

Variations in resonance conditions at the plasmapause

Since the whistler-mode phase velocity is inversely proportional to the plasma frequency, the plasmapause marks a sharp drop (going inward) in the energy of particles that can resonate with a wave of a particular frequency. Partly as a consequence of this jump in resonance conditions, certain wave–particle interaction effects at whistler-mode frequencies can differ strongly from one side of the plasmapause to the other.

A predicted consequence of the intrusion of the disturbed-time ring current into the cool plasmasphere is the generation of ion cyclotron turbulence (Cornwall *et al.*, 1970). Although a number of ion cyclotron wave events inside but near the plasmapause have been reported (Kintner and Gurnett, 1978), the extent and geophysical importance of that process has not yet been well established.

The 'coherent wave instability'

A striking example of wave activity changes at the plasmapause is the so-called coherent wave instability, or CWI, which has been found to characterize the response of the magnetosphere to the injection of whistler-mode waves into ducted paths in the outer plasmasphere near $L = 4$ (Helliwell, 1988). This instability has been studied most extensively in data of the experimental transmitter at Siple, Antarctica (Helliwell and Katsufrakis, 1974; Paschal and Helliwell, 1983; Paschal, 1988), and has also been investigated by Dowden *et al.* (1978) using signals from a transportable VLF transmitter with balloon-borne antenna operating in Alaska near $L = 3$. The instability has been found to involve complex nonlinear processes that have yet to be fully explained (Helliwell, 1988), including the existence of an amplitude threshold for wave growth, growth by 30 dB to a saturation level, the development of sidebands, and the suppression of background broadband wave activity within $\sim 100\,\text{Hz}$ of an amplified carrier.

Figure 2.35 shows spectrograms of signals from the Siple transmitter that were received in the northern hemisphere conjugate region. A 'diagnostic' frequency–time format, part of which is shown at the bottom of the figure, was transmitted at five-minute intervals within a 1000 Hz band centered at 2.46 kHz. This format included frequency ramps and constant-frequency pulses transmitted at various power levels and with various polarizations. The spectrograms show three ∼ 20-s examples from a multihour transmission period and illustrate slow changes (time scale of tens of minutes) in the magnetospheric response to the same sequence of injected waves. At 0000 UT (top panel) a response to the initial 1-s ramps and pulses was barely detectable and there were well defined but temporally limited responses to the longer ramp and pulse. At 0040 UT (second panel) there was a relatively strong response to both the shorter and longer transmitted signals, including strong triggered emissions that initially rose and then fell in frequency with time. Then at 0150 UT the overall response was again weaker but now different in that there were a number of triggered rising tones.

The occurrence of the phenomena noted above was found to be dependent upon the phase coherence of the amplified signal incident upon a near-equatorial portion of the field line region called the interaction region. When simulated broadband noise was transmitted in a range around a constant frequency

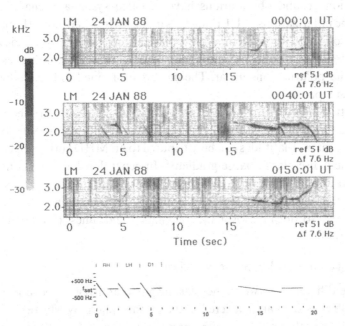

Figure 2.35 Spectrograms showing receptions in the northern hemisphere conjugate region of VLF signals transmitted from Siple Station, Antarctica. Below is shown a diagram of the transmitted frequency–time format. The format (not exactly to scale in frequency), is positioned in time to reflect an approximately 2 s transmission delay between conjugate hemispheres. The 1-s ramps and pulses at left were transmitted with right hand (RH), left hand (LH), and linear (D1) polarization, while RH was used with the longer ramp and pulse.

carrier, the growth of the carrier was found to be suppressed when the broad-band field strength of the noise exceeded a level believed to represent that of the ambient noise (probably nonducted hiss) present in the medium (Mielke and Helliwell, 1992).

The above-described phenomena have not been observed on a regular basis on paths outside the plasmasphere (Carpenter and Miller, 1983). In that region there is generally a higher level of natural wave activity, both of discrete (chorus) and diffuse (hiss) emissions. The lack of the kind of quiet noise background often evident in the outer plasmasphere during magnetically calm to quiet periods may be a factor in the difference in response.

Whistlers and ULF

In the data of ground stations located at $L > 3.5$, there is on average less evidence of whistler propagation outside the plasmasphere than inside, although at certain times and at certain distances beyond the plasmapause substantial whistler activity may be observed in the trough region (Carpenter and Sulic, 1988). The reasons for the reduced whistler activity beyond the plasmapause are not well understood.

Multistation ground observations have revealed systematic changes with latitude in the polarizations of ULF waves with periods ~ 200–600 s (Lanzerotti and Southwood, 1979), such that the polarizations on opposite sides of a 'demarcation line' are opposite, and at the demarcation line amplitudes are a maximum and polarizations linear. The demarcation line has been associated with the plasmapause.

Lanzerotti et al. (1975) combined observational data with the theoretical ideas of Chen and Hasegawa (1974) to suggest that damped sinusoidal oscilla-tions observed at mid-latitudes can be attributed to an MHD surface eigenmode excited at the dense plasmapause gradients. In general, it has been concluded from observations and theory that plasma density gradients, such as those observed in the plasmapause region, are important in coupling ULF driving energy sources to locally resonant magnetic field lines (Lanzerotti and Fukunishi, 1975).

Triggered whistler-mode emissions

One notable difference between the plasmasphere and region beyond is that when discrete emissions are triggered in the plasmasphere by whistlers, experi-mental signals, or other discrete emissions, the emission that develops is short in duration, of order 1 s, as illustrated in Fig. 2.35 (Carpenter and Sulic, 1988). However, in the trough region relatively long enduring bursts of hiss-like noise or discrete emissions can be seen, the discrete emissions sometimes lasting for many tens of seconds. Thus the amount of energy released from the energetic particles into wave activity during a characteristic event can differ strongly between the plasmasphere and trough regions.

The plasmapause as a 'source' region

In addition to serving as a kind of separatrix between wave regimes of the kind mentioned, the plasmapause itself can be a source region of waves, through instabilities associated with the steep density gradients, and through processes of mode coupling and mode conversion that can take place in a region of rapidly varying refractive index. Drift wave activity at the boundary, observed as electrostatic waves, Doppler shifted to a satellite-detected range of ~ 1.7–$178\,\mathrm{Hz}$, has been reported (Kintner and Gurnett, 1988).

Terrestrial continuum radiation, observed in the trough region at frequencies above the local plasma frequency, is believed to originate in the vicinity of the plasmapause (Jones, 1976; Kurth *et al.*, 1981). The component below the magnetosheath plasma frequency (often $\sim 50\,\mathrm{kHz}$) is observed to remain trapped in the cavity formed between the plasmapause and the magnetopause (see Fig. 2.34), while the component at higher frequencies up to several hundred kHz has been found to escape the magnetosphere (Morgan and Gurnett, 1991). To explain the continuum radiation, Gurnett and Frank (1974) suggested that low-energy electrons injected into the outer radiation belts excite electrostatic waves, which in turn produce continuum radiation via a mode conversion process. The mode conversion process involved has been the object of a number of studies (e.g. Jones, 1982; Morgan and Gurnett, 1991), but experimental work from DE–1 by Morgan and Gurnett (1991) indicates that it remains to be clearly identified.

2.6.5 Equatorial phenomena

The existence of an equatorial region, characterized by a local minimum in spatial inhomogeneity as seen by waves and particles moving along the field lines, has apparently fostered a number of special wave phenomena, several of which also depend upon the magnetospheric reflection phenomenon. These include ion cyclotron waves observed within the plasmasphere near the equator and near $L = 2$ (Kasahara *et al.*, 1992), enhanced upper hybrid resonance emissions in the plasmasphere at the equator (Oya *et al.*, 1991), and a concentration of quasi-static whistler-mode waves near the equator beyond the plasmapause (Russell *et al.*, 1970; Gurnett, 1976; Curtis, 1985; Olsen *et al.*; 1987, Boardsen *et al.*, 1992). The quasi-static whistler-mode waves, noted in Fig. 2.34 as equatorial VLF noise, are believed to be associated with heating of thermal light ions in the ~ 1–$100\,\mathrm{eV}$ range.

The equatorial region also appears to be a favored location for amplification of constant frequency or nearly constant frequency whistler-mode signals, as evidenced in VLF wave injection experiments near $L = 4$ (Helliwell, 1988).

2.6.6   Wave regimes just outside the plasmasphere

The remarkable differences in wave regimes inside and outside the plasmapause are well illustrated in sweep frequency receiver data from satellites. These have shown varied wave phenomena, generally above the local electron gyrofrequency, that are apparently confined to the region beyond the plasmapause (Gurnett and Shaw, 1973). These include auroral kilometric radiation, or AKR (Dunckel et al., 1970; Gurnett, 1974), electron cyclotron waves near half harmonics of the electron cyclotron frequency (Kennel et al., 1970; Shaw and Gurnett, 1975), type III solar radio bursts (Gary and Hurford, 1989), and a band of terrestrial continuum radiation (Gurnett and Shaw, 1973) that appears to be the result of wave trapping in the cavity or density trough between the higher density plasmapause and magnetopause boundaries (Kurth et al., 1981).

The spectrogram of Fig. 2.34 shows examples of several of these phenomena, as well as the corresponding wave activity above the gyrofrequency seen inside the plasmasphere, namely the upper hybrid resonance (UHR) noise band. This band, which occurs at approximately the frequency $f_{UHR} = (f_{pe}^2 + f_{ce}^2)^{1/2}$, is regularly observed along extended spacecraft orbits within the plasmasphere. Thanks to the fact that in such regions of relatively high plasma density $(f_p/f_c)^{1/2} > 1$ the Z mode emissions of the UHR band can only be generated locally within a narrow frequency range and do not propagate far without encountering regions of nonpropagation, they appear to a spacecraft as narrowband and hence are indicative of 'local' plasma conditions. Thus they represent a powerful means of determining the profile of electron density along a satellite path (Gurnett et al., 1979).

As satellites cross the plasmapause it is common to observe a transition from a plasmaspheric region dominated by hiss to a trough region in which whistler-mode chorus is observed (Dunckel and Helliwell, 1969; Anderson and Gurnett, 1973; Holzer et al., 1974; Burtis and Helliwell, 1976). Phenomenologically there are three types of whistler-mode hiss, plasmaspheric, mid-latitude, and auroral. Mid-latitude hiss differs from plasmaspheric hiss in that it (1) is often observed on the ground and is thus presumed to be ducted, (2) regularly occurs in the 1–4 kHz range and sometimes above, (3) often occurs in conjunction with chorus, and (4) may be observed on either side of the plasmapause (Sazhin and Hayakawa, 1992).

Of particular interest is the observation near but outside the plasmasphere of impulsive emissions, broadband signals at normally whistler-mode frequencies that do not show evidence of dispersive propagation (Ondoh et al., 1989; Sonwalkar et al., 1990). The signals have a small magnetic component, and thus appear to be electromagnetic in nature. While often observed, they are often ignored or unreported because of their apparently anomalous behavior. Figure 2.36b shows spectra from a DE 1 orbit (Fig. 2.36a) in which the satellite moved from outside to inside the plasmasphere. Chorus activity and hiss were first observed in the trough region (top panel). As the satellite approached the plasmapause, impulsive emissions appeared against a hiss backround (middle panel). Then, after penetration of the plasmasphere (bottom panel), only hiss was observed.

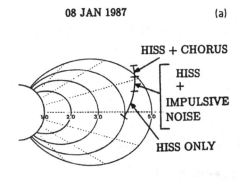

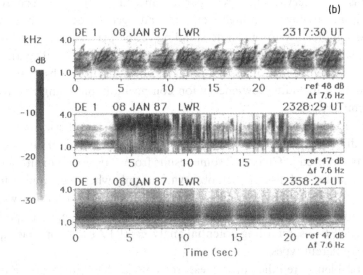

Figure 2.36 Spectrographic examples of apparent spatial variations in VLF wave activity observed as the DE-1 satellite moved from outside to inside the plasmasphere. (a) Meridional projection of portions of the DE-1 orbit showing the dominant wave types observed. (b) Spectrograms showing a mixture of chorus and hiss (top panel), followed by impulsive emissions against a hiss background (middle panel) and then hiss (bottom panel) as the satellite penetrated the plasmasphere.

Chapter 3

Plasmasphere measurements from spacecraft

3.1 Introduction

3.1.1 Preliminary remarks

Besides whistler observations, direct particle measurements from spacecraft (especially from satellites with highly elliptical and geostationary orbits) have contributed significantly to our understanding of the plasmasphere and of its outer boundary, the plasmapause. In particular, such measurements permit us to investigate a number of topics that are not subject to direct observation by radio techniques, including low-energy ion composition, pitch angle distribution, and temperature.

Satellite instruments which have contributed to plasmaspheric studies involve both direct particle flux measurements as well as wave observations. We have already reported in Chapter 2 some results from wave experiments, and in the present chapter will discuss such observations only when they appear to be complementary to direct plasma measurements. Most of our attention will be focused on direct particle measurements, obtained with Langmuir probes, charged particle traps, retarding potential analyzers (RPA), and ion mass spectrometers of different types.

Several problems are inherent in measurements made with the above-mentioned devices. The most serious problems arise when the instruments operate in a very low-density plasma and/or when the energy of the measured particles is very low. Indeed, it is difficult in practice to eliminate all the factors distorting the direct measurements, in spite of the care taken by their developers, including extensive preflight tests and calibration.

The most disturbing factor is electrostatic charging of the spacecraft. The magnitude of the spacecraft potential with respect to the ambient plasma depends on several factors: (i) the intensity and spectrum of solar UV radiation causing photoemission of electrons at the surface of the satellite, (ii) the ambient electron and ion fluxes impacting on the surface of the spacecraft, (iii) the spacecraft shape, the nature of materials and coatings forming the surface of the spacecraft, and even (iv) operational characteristics such as the power of the telemetry radio transmitter. The telemetry antenna can acquire a net negative charge due to electron flux from the plasma to the high-frequency antenna which exceeds, during the positive voltage phase, the ion flux to the antenna during the negative half-period. Even comparatively small spacecraft potentials (of the order one volt) distort the particle flux measurements in a substantial manner. The effects of electrostatic spacecraft charging have been examined by many scientists and were reviewed by Whipple (1981).

The retarding potential analyser (RPA) method overcomes these difficulties when operating in sufficiently dense plasmas; i.e. when the Debye length is comparatively small and the sheath of space charge around the spacecraft is thin. However, the efficiency of an RPA is drastically reduced in the case of thick Debye sheaths, and thus the thin sheath or zero spacecraft potential approximation is generally used in the analysis of RPA data.

Active and passive wave experiments used to determine the ambient electron density in the magnetosphere are not very sensitive to the spacecraft potential and thickness of the Debye sheath. Indeed, the Debye length in the magnetosphere is always much smaller than the wavelength of the radio waves used in this category of devices. This is the reason why plasma densities measured in wave experiments are much more independent of the spacecraft potential than are direct particle flux measurements. The most reliable results are therefore obtained when both type of instruments (direct particle flux measurements devices, and radio wave techniques) operate simultaneously on the same spacecraft. That ideal situation is not always met, because active wave experiments can perturb the operation of other instruments on board the spacecraft.

It seems clear that *in situ* particle and wave measurements can be made of most of the physical parameters required for study of the plasmasphere. Coordinated observations from different spacecraft operating in different parts of the plasmasphere would be ideal for study of the 3-D structure of the plasmasphere, its dynamical deformations, and its role in the dynamics of the ionosphere–magnetosphere system.

In this chapter the main results obtained since 1960 from direct particle measurements will be summarized and discussed. To keep these studies in their historical perspective as much as possible, and for the convenience of the reader, the material will be presented in chronological order, and in three parts corresponding, respectively, to the three decades : 1960–70, 1971–80 and 1981–90. A few scientific satellites equipped with instrumentation to study the cold plasma trapped in the plasmasphere have been launched during the last five years. APEX, ACTIVINY (INTERCOSMOS-24 and MAGNION-2), CRRES and INTERBALL are among them. The results of observations from these new missions are not yet fully analyzed at the time this manuscript is submitted. In the last part of Chapter 4 we have added some of the recent findings based on these missions, but without pretending to review this material comprehensively.

In Chapter 3 we emphasize mainly results from high altitude satellites. Some of the relevant findings from altitudes below ~3000 km are also discussed in Chapter 2 and in Chapter 4.

3.2   Experimental results from the decade 1960–70

3.2.1   Density distribution; plasmapause position

Experimental results from the LUNIKs obtained in 1959 by Gringauz *et al.* (1960a;b) were discussed in Chapter 1 and will not be repeated here. The next *in situ* cold plasma observations in the magnetosphere were collected in 1962 on board the MARS-1 spacecraft (Gringauz *et al.*, 1964). In 1964, *in situ* measurements of the ion number density in the plasmasphere were obtained from the ELECTRON-2 and ELECTRON-4 satellites (Bezrukikh and Gringauz, 1965; Bezrukikh, 1970). The first measurements of the plasmasphere ion composition were carried out by Taylor, Brinton and Smith (1965a;b) with a mass spectrometer on board OGO-1.

Unfortunately, the mass spectrometers on board OGO-1, -3 and -5 as well as the charged particle traps on ELECTRON-2 and 4 were used without variable retarding potentials, and there was therefore no means of determining the spacecraft electrostatic potential. Furthermore, the later PROGNOZ-2 and -4 had no variable potentials on the outer grids of the particle detectors. Therefore, in all these missions the data processing and analysis were weakened by the a priori assumption that the spacecraft potential was equal to zero! The errors on the calculated ion number density resulting from such an assumption appeared to be within a factor of 2 for PROGNOZ, PROGNOZ-2 and -4 measurements. This could be concluded from comparison of the results with later ones obtained with PROGNOZ-5, which was equipped with an RPA (Gringauz and Bezrukikh, 1977). According to Harris, Sharp and Chappell (1970), indirect estimates of the electrostatic potential of OGO-1 varied from -9 V in the ionosphere, to 0 V in the plasmasphere.

Figure 3.1 shows ion density distributions determined from the charged particle traps of ELECTRON-2 (Bezrukikh and Gringauz, 1965). This figure shows the region of steeply decreasing ion density corresponding to the plasmapause. The plasmapause gradient was also identified in both the proton and helium ion density distributions deduced from the Bennett radio-frequency mass spectrometer of OGO-1 (Taylor *et al.*, 1965b). Those OGO-1 observations confirmed another important feature of the ion density 'knee': the anti-correlation of its position (*L*-parameter) with indices of geomagnetic activity.

Near the time when Carpenter (1966) determined the local time distribution of the equatorial plasmapause position from whistler observations (see Fig. 2.10), the results of IMP-2's cold plasma measurements became available (Serbu and Maier, 1966; Binsack, 1967). These new data were obtained by two different groups using two different types of instruments: charged particle traps with retarding potentials in the case of Serbu and Maier, and traps of a modulation type in the case of Binsack. The data from the latter instrument indicated a pronounced plasmapause 'knee' in the observed plasma density distribution, while the data from the former failed to detect the steep gradient. The cause of this discrepancy is still not clear, but, as discussed in Section 1.3.19, there was reason to suspect that the effective collecting area of the IMP-2 trap had not been correctly estimated. Also, the IMP-2 data reported were from the local afternoon sector, and could well have been influenced by cases such as those discussed below, in which the plasmasphere radius is exceptionally large or there is no well-defined plasmapause effect.

The intercomparison of data from five independent experiments, including OGO-1, ELECTRON-2, IMP-2 and whistler data, showed that near solar activity minimum, the ion densities at a geocentric distance of $2R_E$ ranged between 2×10^3 cm^{-3} and 4×10^3 cm^{-3} (Brinton, Pickett and Taylor, 1968).

Figure 3.2 shows ion density profiles obtained in March 1968 with a mass spectrometer on board OGO-5, a satellite which was on a highly eccentric orbit, scanning a wide range of *L* values, up to $L = 9$ (Harris *et al.*, 1970). At $L = 2$, the H ion density was of the order of 10^3 cm^{-3}, but it occasionally reached values as high as 10^4 cm^{-3}. Except for profiles 1 and 4, of 7 and 9 March 1968, respectively, the plasmapause density gradient is clearly seen. In the case of 12 March 1968

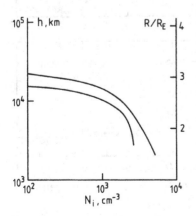

Figure 3.1 Distribution of ion density with height obtained by means of the charged particle trap on board the ELECTRON-2 satellite. The upper curve corresponds to the flight of 2 February 1964, the lower to that of 13 February 1964 (after Bezrukikh and Gringauz, 1965).

(profile 6), two steep density gradients were observed at $L = 4$ and $L = 5.8$, with a region of constant ion density in between. These direct spacecraft measurements confirmed that the plasmapause position can vary over a wide range of L-values, in these cases from $L = 3.5$ (profile 8) to $L = 6.5$ (profile 3).

Beside the ion-mass spectrometer there was a spherical Langmuir probe on board OGO-5. This probe, 6 cm in diameter, measured the flux of electrons in the plasmasphere (Freeman, Norman and Willmore, 1970). During the flight the probe was sometimes in the shadow of the solar cells. Under these conditions, photoemission from the probe surface did not distort the measurements, al-

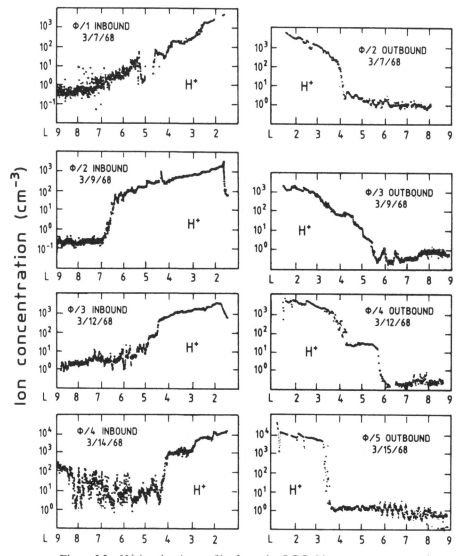

Figure 3.2 H^+ ion density profiles from the OGO-5 ion mass spectrometric data (after Harris *et al.*, 1970).

though some photoelectrons from other illuminated parts of the spacecraft
could have reached the detector. The aim of the experiment was to measure
electron densities larger than $10\,\text{cm}^{-3}$, i.e. above the level at which the Debye
length is comparable with the dimensions of the satellite $\sim 2.5\,\text{m}$. The photoemission from the satellite in the plasmasphere was assumed to be equal to its
value in the interplanetary medium. The corresponding current was then sub-
stracted from the current measured by the probe in the plasmasphere.

Figure 3.3, taken from Freeman *et al.* (1970), represents sample profiles of the
measured electron density obtained by this instrument. The arrows marked B
and M indicate the positions of the bow shock and the magnetopause, respectively; the letter P indicates the position of OGO-5 perigee. The electron densities
shown in Fig. 3.3 were obtained at the same time as the ion densities shown in
two of the panels of Fig. 3.2. The plasmapause crossings are clearly identifiable
by steep density gradients. This was the first time that a Langmuir probe
succeeded in directly detecting the 'knee' in the cold electron density distribution.

The plasmapause positions given in geocentric distances in Fig. 3.3 correspond rather well with the L-values of the plasmapause in Fig. 3.2, since the
inclination of the orbit of OGO-5 was small and L and R are not too different.

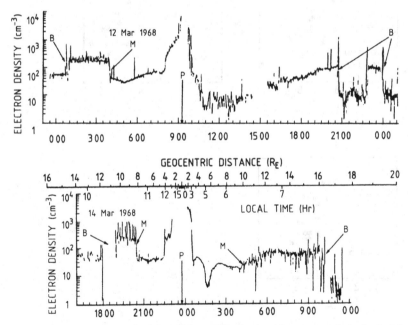

Figure 3.3 Samples of $N_e(R)$ profiles obtained by means of a spherical Langmuir probe on board the OGO-5 satellite, 12 and 14 March 1968. The arrows
marked B and M indicate the positions of the bow shock and the magnetopause, respectively; the P marks the perigee position (after Freeman *et al.*,
1970).

Differences between the two figures in density levels outside the plasmasphere can be understood in terms of the problems of interpretation noted earlier.

Figure 3.4 shows the equatorial position of the plasmapause, L_{pp}, as a function of the maximum value of K_p during the 24 hours before the time of the measurement. The ELECTRON-2 and -4 results obtained by Bezrukikh (1970) are shown in Fig. 3.4a, and those obtained in 1966 from OGO-3 by Taylor, Brinton and Pharo (1968a) in Fig. 3.4b. Those obtained in 1968 from OGO-5 by Chappell, Harris and Sharp (1970) are illustrated in Fig. 3.4c. These figures confirm that the size of the plasmasphere decreases when the level of geomagnetic activity increases.

A more detailed analysis of OGO-5 mass spectrometer observations by Chappell, Harris and Sharp (1970a;b, 1971a;b) and summarized by Chappell (1972) indicates that the relationship between the equatorial position of the plasmapause and K_p is not a simple one, but depends on the local time angle. For instance, the post-midnight plasmapause position was found to be fairly

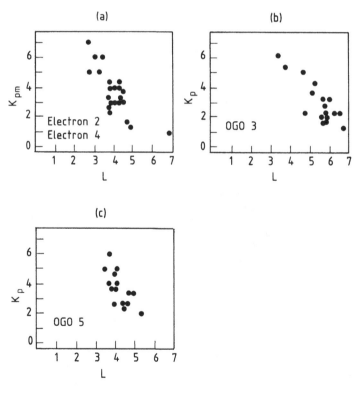

Figure 3.4   The relation between the L coordinate of the plasmapause (determined from ion density measurements) and K_p-index averaged over the preceeding 24 h: (a) from ELECTRON-2 and ELECTRON-4 data (after Bezrukikh, 1970); (b) from OGO-3 data (after Taylor *et al.*, 1968a); (c) from OGO-5 data (after Chappell *et al.*, 1970).

well correlated with the K_p-index averaged over the 2 to 6 hours before the measurement, whereas the plasmapause radius on the dayside appeared to be determined by the level of magnetic activity at the time when the plasma region under observation had been rotating with the earth through the midnight sector.

Furthermore, OGO-5 mass spectrometer observations also indicated that the density gradient not only shifted to lower L-values when K_p increased, but that it became steeper when the geomagnetic activity was enhanced. These features are shown in Fig. 3.5, taken from Chappell *et al.* (1970a). The hydrogen ion concentrations versus L correspond to the local time region from midnight to 0400. The values of K_p referred to are averages over the period from 2 to 6 hours prior to the measurements. The error bars shown correspond to the maximum spread in each group for the density level of $10\,\mathrm{cm}^{-3}$. Note that this spread decreased when the value of K_p increased. It was also noted by Chappell *et al.* (1970a) that the levels of concentration inside the plasmasphere and outside in the trough remained about the same as the plasmapause changed position. But, as indicated above, absolute values of cold ion densities obtained from direct measurements are subject to large errors under conditions of uncontrolled spacecraft charging. In Fig. 3.5, the density jumps at the plasmapause are at least

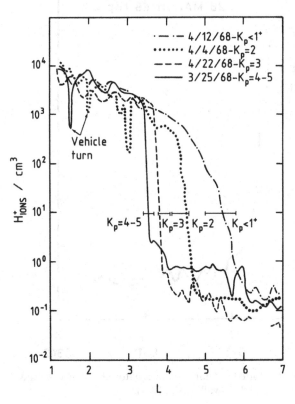

Figure 3.5 A composite of several typical plasmapause crossings representing different levels of magnetic activity. These plots of hydrogen ion concentrations versus L represent outbound passes in the local time region from midnight to 0400 (after Chappell *et al.*, 1970a).

an order of magnitude larger than those indicated later by radio measurements
from ISEE (see Fig. 3.28); values in Fig. 3.5 inside the plasmasphere were
apparently overestimated and those outside underestimated.

At other local times the observed ion density profiles appeared quite different,
less steep and often unchanged immediately after the onset of a substorm when a
pronounced disturbance in the midnight sector had occurred. From this study of
plasmapause densities in different local time sectors it was inferred that a new
and steeper plasmapause density gradient was formed near midnight local time
at the onset of a substorm. The new 'knee' in the equatorial ion density
distribution then propagated to the noon LT sector with the corotation velocity.
In the dayside time sector, the inward shift of the plasmapause position was
observed with a time delay corresponding to the time of corotation between
midnight LT and the LT corresponding to the point of observation (Chappell,
1972).

The ion density profiles deduced from OGO-5 direct measurements displayed a great variety of forms, depending on local time and geomagnetic

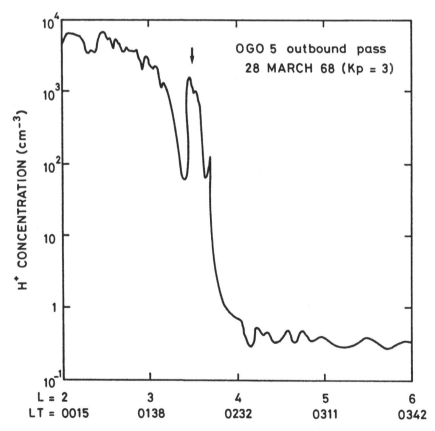

Figure 3.6   H$^+$ ion density plotted as a function of equatorial distance (L) and
local time (LT) (after Lemaire and Kowalkowski, 1981).

activity. Often large amplitude fluctuations or dips were observed in the plasmaspheric density distributions, as illustrated in Fig. 3.6 (Kowalkowski and Lemaire, 1979) and Fig. 3.7 (Chappell *et al.*, 1970a) . These large depressions appeared not only in the dayside local time sector (curve c in Fig. 3.7 obtained around 1000 LT) but also in the nightside density profiles (Fig. 3.6).

It can also be seen from Figs 3.6 and 3.7 that the ion concentration in the dayside trough region sometimes exceeded $10 \, cm^{-3}$ (curve c), which was not usually the case near midnight where densities tended to be lower (Figs 3.6 and 3.7, curve b). This effect has been explained as a consequence of polar wind type flow in flux tubes which have been depleted near midnight during an earlier substorm and which subsequently corotate and gradually refill. All these aspects have been reviewed by Rycroft (1975).

3.2.2 Temperature measurements

Cold plasma particles of less than 1 eV energy coexist in the plasmasphere with other components of magnetospheric plasma covering a wide range of energy up to hundreds of MeV, i.e. the radiation belt population. These different particle populations have different characteristic 'temperatures', depending on the en-

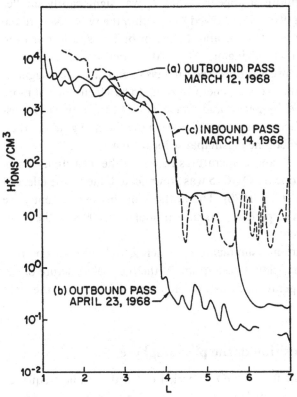

Figure 3.7 Examples of plasmapause crossings as shown by plots of the concentration of hydrogen ions versus L. (a) an example of a case of multiple plasmapauses in which the plasmapause position is not uniquely identified; (b) a case of a well-defined plasmapause position; (c) a case showing high concentration in the trough region outside the plasmapause (after Chappell *et al.*, 1970a).

ergy range considered and the method used to determine the parameter identified as the 'temperature' from a measured energy spectrum. One has often questioned the usefulness of defining a temperature in the case of collisionless plasmas where the velocity distribution is usually neither isotropic nor Maxwellian nor close to detailed balance thermodynamic equilibrium. But the notion of temperature is nevertheless widely adopted in space plasma physics. It can be useful in characterising either the slope of an energy spectrum or the dispersion or second order moments of a velocity distribution function. However, one should be rather careful when comparing temperatures based on these alternative working definitions and which are obtained with instruments having different experimental characteristics and energy ranges. Indeed, these different 'temperatures' are equal only when the velocity distribution is a Maxwellian.

The first evaluation of the ion temperature (T_i) in the plasmasphere was based on observations of the LUNIK's ion traps. Values of T_i smaller than 10 000 K were deduced by Gringauz et al. (1960a, b). Serbu and Maier (1966, 1970) evaluated electron temperatures (T_e) and ion temperatures (T_i) from charged particle traps with retarding potentials on IMP-2 and OGO-5. The electrons were found to have a Maxwellian distribution at energies below 2 eV and a warmer component at higher energy. Electron temperatures reached values of $\sim 2 \times 10^4$ K at $5R_e$. T_e was found to increase in proportion to the square of the radial distance R. Furthermore, it appeared that T_e was significantly smaller than T_i. Serbu and Maier (1970) also showed that within the plasmasphere the day to night ion temperature ratio was about a factor of 2; this ratio was even higher in PROGNOZ data sets (Gringauz, 1983). At the plasmapause a sudden increase in the ion temperature was observed on days of enhanced geomagnetic activity ($\Sigma K_p > 15$). This increase was generally found to be a factor of 5 or more within $0.8R_E$ and was used by Serbu and Maier (1970) to identify the plasmapause location. A satisfactory agreement between the thermal gradient data and the whistler average quiet time plasmapause location was found.

The determination of electron temperatures based on the measurements of the spherical Langmuir probe on OGO-5 was complicated due to the effect of photoemissions (Freeman et al., 1970). The authors of this experiment gave values ranging from 3000 K at perigee (300 km altitude) to 35 000 K in the outer regions of the magnetosphere.

These initial temperature measurements were neither reliable nor comprehensive enough to provide a global description of the thermal structure of the plasmasphere. Better temperature measurements were to come in the next decade.

3.2.3　Chemical composition of the plasmasphere

According to measurements from OGO-1 carried out with a radio-frequency mass spectrometer, the density of He$^+$ ions in the plasmasphere was $\sim 1\%$ of that of the H^+ ions (Taylor et al., 1965a;b, 1970). The plasmapause, as identified by the steep ion density gradients, was found to be almost at the same geocentric distance for both the He^+ and H^+ ions (Taylor et al., 1965a;b). Figure 3.8a shows an example of H^+ and He^+ density profiles versus L obtained by Taylor et al. (1968a) from OGO-3 observations. Figure 3.8b shows the H^+, He^+ and O^+ ions density distributions versus L from the mass spectrometer on OGO-5 (Harris et al., 1970). These results as well as those of OGO-6 showed that the He^+ number density was about 10^2 times smaller than the H^+ ion density.

3.3   Experiments and results from the decade 1970–80

3.3.1   Measurements with the PROGNOZ series of satellites

Orbits and measurement techniques

The PROGNOZ series of satellites (up to PROGNOZ-6) were launched into similar orbits with an apogee of 200 000 km, perigee of 940 km, inclination of 65°, and an orbital period of about four days; the satellite spin axis was directed toward the Sun. These satellites traversed the plasmasphere in orbits similar to that illustrated in Fig. 3.9, which enabled them to collect data over a wide range of L-shells in a time of about three hours. Observations were collected for a decade beginning in 1972 with instrumentation that was similar on all satellites of the series.

Two ion traps were installed aboard PROGNOZ and PROGNOZ-2 with hemispherical outer grids electrically connected to the satellite surface and the flat collectors. One of the traps was located on the illuminated side of the spacecraft and oriented toward the sun, while the other one was located on the shaded side. The ion temperature, T_i, was determined from the ratio of the currents measured by both ion traps. This ratio depends on the anisotropy of the ion velocity distribution function in the frame of reference of the moving spacecraft and on the characteristic ion temperature (Bezrukikh and Gringauz, 1976). The velocity distribution was assumed to be a displaced Maxwellian, and within the plasmasphere the contribution of the heavier ions was considered negligible. The electric potential of the satellite surface was assumed to be equal to zero. The relative error is estimated to be less than 20%.

The plasmasphere ion trap collector current is

$$I_i = N_i G(V, T_i, \beta) \tag{3.1}$$

where N_i is the ion density, G is a function describing the dependence of the collector current on the satellite velocity V and on β, the angle of the detector with respect to the ram direction. The ratio of currents collected by two traps at different orientation is

$$I_{i1}/I_{i2} = G(V, T_i, \beta_1)/G(V, T_i, \beta_2) \tag{3.2}$$

I_{i1} and I_{i2} are measured nearly simultaneously. If V, β_1 and β_2 are known, the only unknown quantity in (3.2) is T_i, the temperature of the ions.

On the PROGNOZ satellite $\beta_2 = 180° - \beta_1$; V and β values were determined from the trajectory data. The measured values of the collector current of the trap

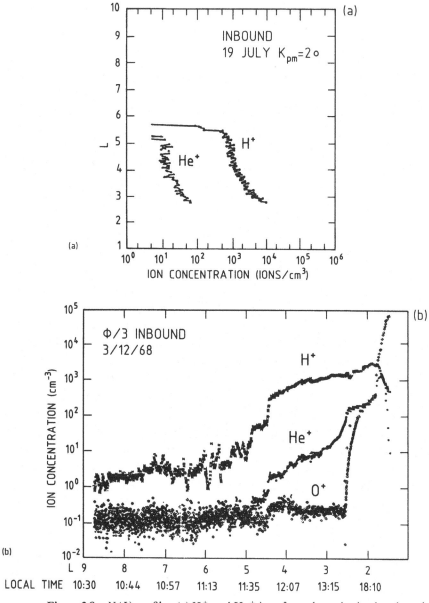

Figure 3.8　$N_i(L)$ profiles. (a) H^+ and He^+ ions from data obtained on board the OGO-3 satellite (after Taylor *et al.*, 1968a); (b) H^+, He^+ and O^+ ions from data obtained on board the OGO-5 satellite (after Harris *et al.*, 1970).

illuminated by the sun were corrected for the collector photocurrent. The photocurrent value was determined from comparison of hemispherical trap readings and those of the modulation trap (Faraday cup), which was also oriented toward the sun. In Fig. 3.10a the relation between I_{i1}/I_{i2} and T_i for different angles β_1 are shown for a satellite velocity $V = 8$ km/s, and in Fig. 3.10b for $V = 5$ km/s. One can see from Fig. 3.10b that the accuracy of determining T_i from the known ratio I_{i1}/I_{i2} decreases with increasing values of T_i and β_1. In the case of PROGNOZ satellite, the values of T_i were determined only for $T_i <$ 25 000 K within the range of angles $\beta_1 = 0$–$35°$ (i.e. $\beta_2 = 180°$–$145°$).

These limits for β allowed the authors (Bezrukikh and Gringauz, 1976) to determine T_i only in the dayside plasmasphere, and the indicated upper limit on T_i made it possible to determine only the lower limit of possible values of T_i in some regions of the plasmasphere. The curves shown in Fig. 3.10 were calculated with the assumption of zero satellite potential.

Beginning with the PROGNOZ-5 satellite, density and temperature measurements were performed with the retarding potential analyzer whose position on board the spacecraft is shown by an asterisk in Fig. 3.11 (Gringauz, 1985). The analyzer was located on the shaded part of the satellite; its aperture was perpendicular to the sunward axis of the satellite rotation. A sketch of the analyzer is shown in Fig. 3.12. The outer grid is at the potential of the satellite surface, while the middle analyzing grid has a potential varying from 0 to 25 V

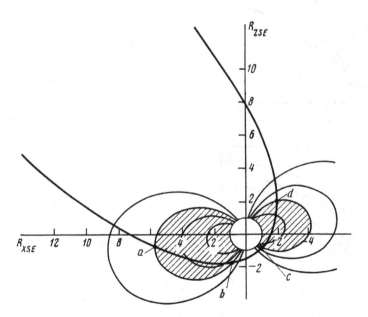

Figure 3.9 Projection of the near-earth parts of PROGNOZ, PROGNOZ-2, and PROGNOZ-5 orbits onto the X, Z-plane in solar-ecliptic coordinates (after Gringauz, 1985).

with respect to the satellite and the inner supressor grid is at a constant potential of -70 V to prevent photoelectrons from reaching the analyzer collector.

Processing of the retardation curves (ion probe characteristics) was performed as follows (see Fig. 3.13) (Gringauz, 1985). The current intensity measured in the 25 V step was first subtracted from the other measured currents (i.e. the effect of the ion fluxes with $E_e > 25$ eV and of electrons with $E > 70$ eV, corresponding to the voltage -70 V on the suppressor grid of the analyzer, were eliminated). The slope of the retardation curve, when plotted on a semi-logarithmic scale, gives T_i, with smaller slopes corresponding to greater T_i, according to Langmuir theory. The presence of two linear parts of the retardation curve was interpreted as being due to H^+ ions with different T_i. The data processing was performed assuming Maxwellian ion velocity distributions. When a retardation curve showed a single linear part, all the ions were taken to be protons and the least-squares method was used to select an analytical expression including the known satellite velocity, the ion number density N_i, their temperature T_i and the satellite potential ϕ_{sc} such that the theoretical and the experimental data were

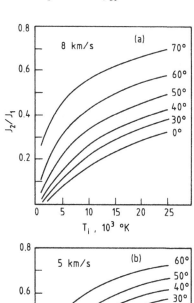

Figure 3.10 The ratio of ion trap collector currents I_{i1}/I_{i2} as a function of T_i at different angles of the normal to the collector with respect to the satellite velocity vector. (a) Case of $V_s = 8$ km/s and (b) $V_s = 5$ km/s (after Bezrukikh and Gringauz, 1976).

best fitted. Plasma drift can only shift the retardation curve but does not change its slope, and the ions with $E_i > 25\,eV$ contribute to the collector current in a way which does not vary with the retarding potential and, hence, does not affect the determination of T_i. A similar procedure was used in the RPA data processing of the DE/RIMS retardation curves (Comfort, Waite and Chappell, 1985) with the only advantage that the other species could be separated from the protons by means of the RIMS mass spectrometer. The above procedure for the

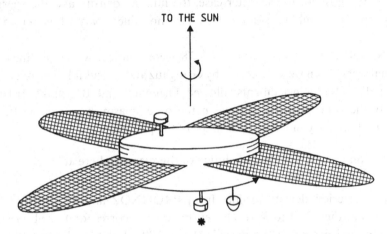

Figure 3.11   The location (*) of the ion retarding potential analyzer on board the PROGNOZ-5 spacecraft (after Gringauz, 1985).

Figure 3.12   A sketch of ion-retarding potential analyzer on board PROGNOZ-5 (after Gringauz, 1985).

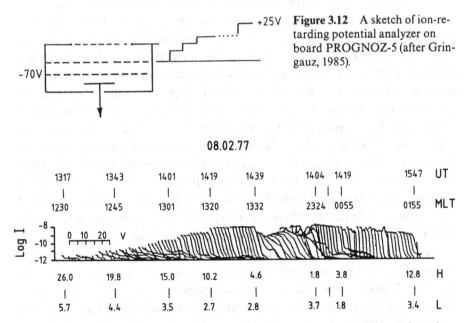

Figure 3.13   Composite of the ion retardation curves for the near-earth section of a PROGNOZ-5 orbit (after Gringauz, 1985).

PROGNOZ/RPA data processing precluded the possibility of confusing the effect of ring current ions with those of ions $E_i \leq 25\,\text{eV}$ in the determination of ion temperatures T_i.

The acceptance cone of the analyzer was wide (aperture angle cone of $40°$) and did not allow precise measurements of ion pitch angles. However, one could roughly estimate the number density and energy of ions with energy $E_i < 25\,\text{eV}$ from changes in the shape of the retardation curves and from the non-linearity in some parts of them. In the latter case, the number density and the energy of anisotropic ions could be estimated from the non-linear parts of the retardation curves.

PROGNOZ-5 and -6, flown in 1977–78, were equipped with retarding potential analyzers which were described by Gringauz and Bezrukikh (1976). The data collected by these instruments allowed Gringauz and Bezrukikh (1976) to identify the zone of warm plasma in the outer plasmasphere as well as the noon–midnight asymmetry in the shape of the plasmasphere.

Ion density profiles; the bulge of the plasmasphere at noon local time

Plasmasphere ion density profiles from PROGNOZ and PROGNOZ-2 are presented in Figs. 3.14 to 3.16. The left-hand sides correspond to observations from inbound passes on the dayside (LT = 1000–1400), while the right-hand side plots correspond to outbound nightside passes (LT = 2200–0400). The K_p value corresponding to the time when the satellite was in plasmasphere, as well as the sum of K_p over the 24 hours prior to the measurements, are indicated for each pass. Three sets of passes are shown, corresponding respectively to quiet (Fig. 3.14), moderate (Fig. 3.15) and disturbed conditions (Fig. 3.16) (Gringauz and Bezrukikh, 1977). From Fig. 3.14 it can be deduced that the plasmasphere was more extended in the dayside (L_d) than in the post-midnight sector (L_n). By comparing Figs. 3.14 and 3.16 it can be seen that this noon–midnight asymmetry decreased with increasing geomagnetic activity, but that L_d-L_n always remained larger than 1 during quiet conditions. When geomagnetic activity increased from quiet to disturbed conditions, L_d-L_n decreased from a maximum of ~ 1.6 to zero. It can also be seen that the nightside profiles were usually steeper than the dayside profiles. Moreover, sometimes considerable fluctuations were observed outside the dayside plasmasphere, where $N_i \leq 10^2\,\text{cm}^{-3}$, whereas on the nightside no such large density fluctuations were generally observed.

Figure 3.17 illustrates schematically the equatorial plasmapause radius as deduced from PROGNOZ observations over almost one cycle of solar activity. The dotted curve corresponds to quiet geomagnetic conditions (Gringauz and Bezrukikh, 1977), the dashed line to disturbed conditions (Remizov, Gringauz and Bassolo, 1990).

The double-probe measurements of EXPLORER-45 (Maynard and

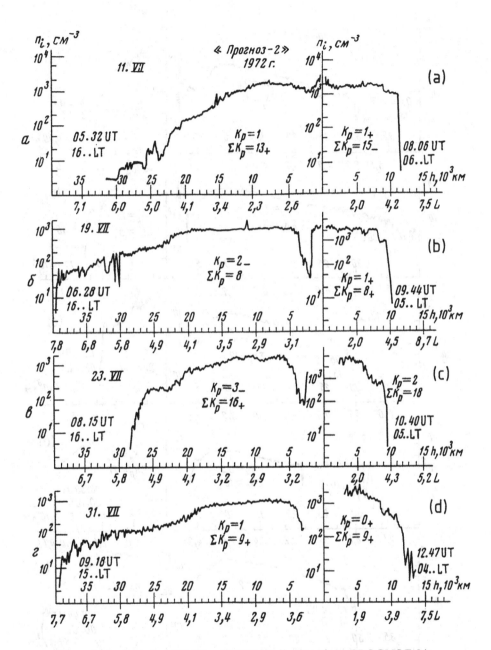

Figure 3.14 Samples of $N_i(L)$ and $T_i(L)$ distributions from PROGNOZ-2 in 1972, illustrating the dawn–dusk asymmetry. The altitudes are also shown; the data for altitudes below $(5–6) \times 1000$ km at $L > 3$ are omitted due to uncertainty in the L-values. The UT-values are given for the beginning and the end of every satellite pass through the plasmasphere; the LT-values are also given for the beginning and the end of each curve; ΣK_p is the sum of K_p-indices for the day prior to the observations (after Bezrukikh and Gringauz, 1976).

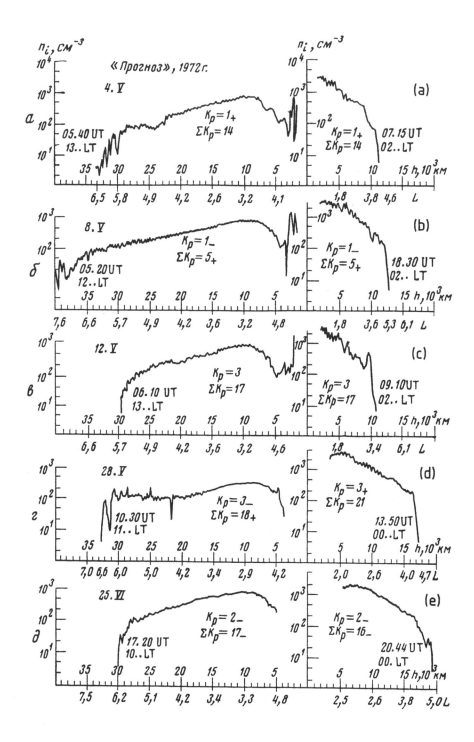

Figure 3.15 The same as in Fig. 3.14 but for PROGNOZ data, again showing the noon–midnight asymmetry of the plasmapause position.

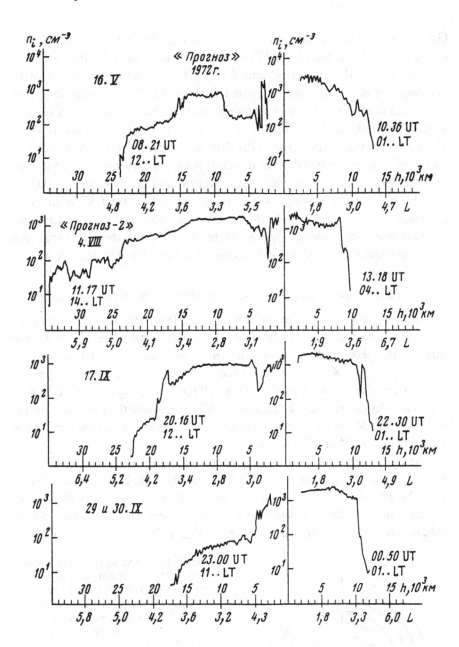

Figure 3.16 The same as in Fig. 3.15 but for different cases: (a) a case in which the plasmapause is almost symmetric; (b) a case of very strong noon–midnight asymmetry (after Bezrukikh and Gringauz, 1976).

Grebowsky, 1977), the data from the mutual impedance wave experiment on board GEOS-1 (Décréau *et al.*, 1978a) and those from the relaxation sounding experiment on GEOS-2 (Higel and Wu, 1984) have also confirmed that the plasmapause does not have a symmetric shape but has a bulge near dusk. McComas *et al.* (1993) found that the plasmapause has a highly variable shape showing a duskside bulge during active geomagnetic conditions, i.e. when $K_p > 2$. From their multipoint MPA (Magnetospheric Plasma Analyzers) observations they confirmed that the duskside plasmaspheric bulge seems to move to earlier local times with increasing geomagnetic activity. They also found cases of stable bulges (non-moving), radial expansions and contractions of the whole plasmapause, as well as complicated behavior indicative of a highly structured plasmapause and detached plasma blobs. For $K_p < 2$ the equatorial plasmapause could be located beyond geosynchronous orbit at all local times.

Plasmasphere temperatures; warm zone in the outer plasmasphere

The proton temperature distribution deduced from PROGNOZ-5 measurements is illustrated in Fig. 3.18. The ion density distribution for two passes through the plasmasphere is also shown by solid dots. The temperature increased from 3000 K to $\sim 10^5$ K between $L = 3$ and the plasmapause region. The observations from PROGNOZ satellites led Gringauz and Bezrukikh (1976) to infer the existence of an inner 'cold' zone, where the ion temperature is less than 8000 K, and a 'warm' outer zone at $L > 3$ where the ion temperature ranges between 10^4 and 2×10^5 K.

The existence of this warm outer zone, discovered in 1976, was later supported by experimental data from GEOS-1 and 2, ISEE-1 and 2, and DE-1 and 2, to be presented below. Note that in some instances PROGNOZ-5 and GEOS-1 did not detect a warm outer zone; in these cases the whole plasmasphere was cold up to the plasmapause, i.e. $T_i \leq 1$ eV.

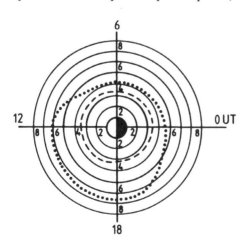

Figure 3.17 Average plasmapause positions in the equatorial plane from PROGNOZ data. The dotted curve corresponds to geomagnetically quiet periods (after Gringauz and Bezrukikh, 1976), the dashed curve corresponds to disturbed conditions (after Remizov *et al.*, 1990).

3.3.2   GEOS multi-experiment measurements

Orbits and scientific objectives

GEOS-1 was launched into an elliptical orbit with an apogee of 38 000 km, a perigee of 2050 km, an inclination of 26.5°, and an orbital period of 12.06 hours; data were collected from April 1977 to June 1978. The planned orbit for GEOS-1 was a geostationary one like that of GEOS-2, which was launched in 1979. The unexpected eccentric orbit of GEOS-1 allowed it to sample the plasmasphere from $L = 3$ to 8 and over a wide range of geomagnetic latitudes. In retrospect, it is for that reason that in terms of plasmaspheric studies, the GEOS-1 mission was more productive than the GEOS-2 mission in geostationary orbit.

Three groups of experiments were carried out on GEOS-1 : (1) mass spectrometric measurements of ions and electrons; (2) two types of active wave experiments : (i) mutual impedance measurements, (ii) a relaxation sounding experiment, and (3) electric field measurements by a floating potential probe. The main characteristics of these experiments are presented in Table 3.1; the locations of the instruments on the spacecraft are shown in Fig. 3.19 (Décréau *et al.*, 1978a).

Ion composition experiment (ICE)

The scientific objectives of the Ion Composition Experiment were to study the energy per charge distribution from thermal energies up to 16 keV/e. The ICE instrument is a mass spectrometer designed to determine the elemental abundance and velocity distribution function of the ion component. A detailed technical description of the instrument is given by Balsiger, Eberhardt and Geiss (1976) and it is schematically illustrated in Fig. 3.20, taken from Geiss, Balsiger and Eberhardt (1978). After passing through a retarding potential analyzer (RPA), ions of charge Q and energy E are accelerated by a fixed potential difference $U_a = 2.7$ kV to the energy $E + QU_a$. They then enter a cylindrical electrostatic analyzer (EA). About 10% of the ions are intercepted by a channeltron which gives a 32-step energy per charge spectrum of the ion flux. The major part of the ion beam enters the mass analyzer (MA) where a combination of cylindrical electric field and perpendicular magnetic field provides mass separation. By varying the MA electric field, a mass is registered for the energy per charge window selected by the EA (Geiss *et al.*, 1978). The thermal mode of the E/Q analysis covers the zero to 110 eV/q range with 32 RPA voltage steps. In the simplest operating program, 64 steps of mass per charge are scanned for each of the 32 energy steps. This energy–mass scanning is particularly suitable for the low-energy ion component, which is usually isotropic in the plasmasphere. In the low-energy mode of operation, the RPA is fixed at $+25$ V, and the EA covers the range from 25 to 16 000 eV/q in 32 logarithmically spaced steps of $E + QU_0$.

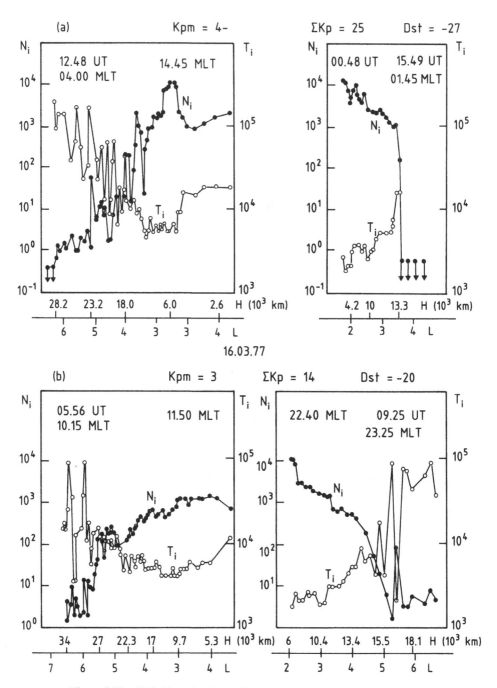

Figure 3.18 $N_i(L,H)$ and $T_i(L,H)$ distributions derived from PROGNOZ-5 RPA data. The UT values are given for the beginning and the end of each satellite pass through the plasmasphere. MLT times are also given for the beginning and the end of each curve. K_{pm} is the maximum K_p-index of the preceding day and ΣK_p is the sum of K_p-indices for the day prior to the observations. The D_{st}-index corresponds to the time of measurement (after Gringauz, 1983).

Table 3.1 *GEOS experiments and techniques used in density study*

Experiment designation	Measurement	Technique	Parameters measured	
			Directly	Derived
S-300 DC	DC electric field	2 × 20 m radial booms	$V_s - V_p$	V_s, N_e
S-301	Relaxation sounding of plasma resonances 10–77 kHz	VLF transmitter and receiver	f_p, f_{uhr}, nf_e	$N_e, |B|$
S-304	Mutual and self impedance 0–77 kHz	VLF transmitter and receiver	$Z/Z_0(f)$	N_e, T_e, λ_D
S-302	Electron and ion energy spectra (0.5–505 eV)	two electrostatic analyzers, ⊥ and ∥ to spin axis on 1.8 m boom	J_e, J_+	$N_e, T_e, N_+, V_+, T_+, V_s$
S-302	Ion composition (M/Q = 1 to >138) and energy spectra (0–16 keV/e)	Ion mass spectrometer body mounted ⊥ to spin axis	$J_+(M/Q = 1,2,4 \ldots)$	N_+, V_+, T_+, V_s

M = ion mass
Q = ion charge
V_s = satellite potential
V_p = probe potential
V_{pl} = plasma potential

f_p = plasma frequency
f_{uhr} = upper hybrid frequency
f_c = electron gyro frequency
Z/Z_0 = normalized plasma impedance
$I_e \, I_+$ = electron/ion differential energy fluxes

N_e, N_+ = electron/ion density
T_e, T_+ = electron/ion temperature
λ_D = Debye length
eV = electron volt
$|B|$ = magnetic field magnitude

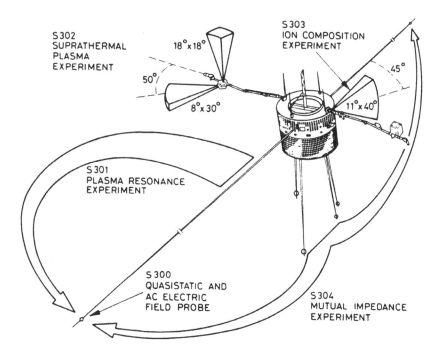

Figure 3.19 Schematic diagram of the GEOS satellite showing the location of instruments listed in Table 3.1. The viewing angles of the particle instruments are given (after Décréau *et al.*, 1978a).

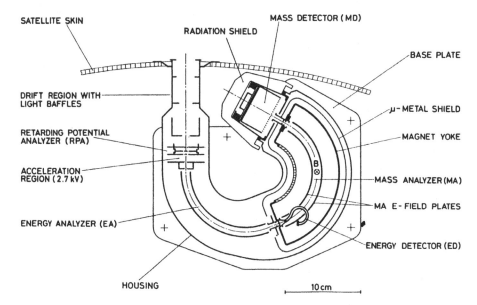

Figure 3.20 Schematic view of the ICE ion mass spectrometer (after Balsiger *et al.*, 1976).

GEOS/ICE results

Six different thermal ion species have been identified by Geiss *et al.* (1978) with the ICE experiment: H^+, He^+, O^+, D^+, He^{2+} and O^{2+}. The last three were not detected in the plasmasphere prior to GEOS-1 measurements. They found also that O^{2+} and He^{2+} can be produced locally in the magnetosphere and do not necessarily come from the interplanetary medium. The high densities of O^{2+} ions at high altitudes are the result of production in the ionosphere combined with upward transport by thermal diffusion.

Diffusive transport along the magnetic field lines from the lower altitude up into the plasmasphere occurs over time scales of 1–3 days. At altitudes greater than 2000 km O^+ ions are lost by charge exchange with hydrogen atoms at an appreciable rate. This charge exchange process leads to increases of the relative abundance of O^{2+} with respect to the O^+ ions. The GEOS/ICE data, from June 1977 to May 1978, were used by Farrugia, Geiss, Young and Balsiger (1988, 1989) to study the density, temperature, and composition in the plasmasphere under rather quiet geomagnetic conditions: $\Sigma K_p < 16$ and $K_p(\max) < 3$. They found that He^+ was the second major ionic component of the plasmasphere after H^+ over the entire range $L = 3$–6. Its density relative to H^+ was highly variable, ranging from 1% to over 100% on some occasion, with the most frequent values in the range 2–6%. In the evening sector (1700–2000 LT), H^+ ion densities followed closely an inverse fourth power variation with L. On strings of quiet days when the same region was sampled at daily intervals, the constant of proportionality for both ions increased with day number. These observations indicated that the flux tubes were continuously refilling; the adherence of the equatorial densities to a fixed power law is indicative of a uniform rate of filling.

The average He^+ densities decreased slightly more rapidly with increasing L than did H^+ ions. He^{2+} was generally the major contributor to the mass/charge = 2 ions, but sometimes D^+ prevailed. Near $L = 3$, the densities of O^+ and O^{2+} ions were comparable.

The temperature was found to be between 4×10^3 K and 1.5×10^4 K in the quiet plasmasphere, slowly increasing with L. Usually, the main ionic components were in thermal equilibrium. However, deviations were sometimes observed, in which $T(He^+)$ was nearly 4 times larger than $T(H^+)$. On occasion, negative temperature gradients were observed over some L range. The variations in time usually dominated the spatial (L) variations. Thus the differences in temperatures seen on any two passes well separated in time were generally more pronounced than those within the same pass at different L-values. The temperature profile for each ion on one pass tended to be repeated on the next day, suggesting the approach of a steady thermal structure with increasing duration of magnetic quiet. The density and temperature of the major ion

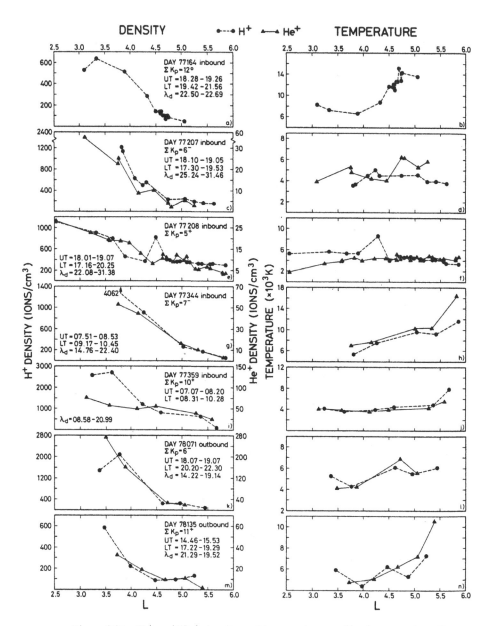

Figure 3.21 H$^+$ and He$^+$ density and temperature profiles for a number of GEOS 1 plasmasphere crossings. The data points have been joined by straight lines. Note the difference in H$^+$ and He$^+$ density scales. He$^+$ data from day 164 of 1977 are not shown since these data were acquired from a mass channel next to the main one (after Farrugia *et al.*, 1988).

species H^+ and He^+ are shown in Fig. 3.21 versus L for 7 days. Note the difference in the H^+ and He^+ density scales.

The mass/charge spectrum is shown in Fig. 3.22 as a function of time on day 164 of 1977. A histogram of the observed values of the ratio $N(He^+)/N(H^+)$ are shown in Fig. 3.23, for all the data obtained by ICE between June 1977 and May 1978 and between $L = 3$ and 6 (Farrugia *et al.*, 1988, 1989).

GEOS wave experiments

GEOS-1 was also equipped with wave experiments : S-301 was the identifier of an instrument performing relaxation sounding of plasma resonances (0–77 kHz). The electron density (N_e) and/or magnetic field intensity (B) could be determined from these measurements. The mutual and self-impedance experiment (S-304), also operating in the 0–77 kHz frequency range, allowed determination of the electron density (N_e), electron temperature (T_e), and Debye length; finally, the DC electric field probe allowed determination of the floating potential (S-300). The locations of the antennae on the GEOS-1 spacecraft are shown in Fig. 3.19 from Décréau *et al.* (1978a). In addition to these wave experiments, two electrostatic analyzers were mounted on a short radial boom to measure the electron and ion densities, temperatures and bulk velocities (S-302).

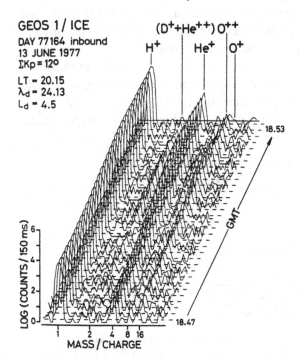

Figure 3.22 GEOS-1/ICE data showing evidence of six ion species. Shown is a three-dimensional plot of raw detector count rates (vertical axis) against mass per charge (horizontal axis), with 0 to 1 count per sample (150 ms) plotted as zero; L_d, λ_d and LT give the spacecraft location at the time of observation, L_d and λ_d being the equatorial distance of the magnetic shell and the magnetic latitude, respectively; LT is the local time. A 6-min angular scan of thermal mode ($E <$ 110 eV) high-resolution data on the inbound pass on 13 June 1977 (Day 164) is shown. The retarding potential analyzer was set at 1 V during the period. The analysis is based exclusively on a high-resolution angular scan, where one spin is subdivided into ∼35 angular bins in azimuthal angle (after Farrugia *et al.*, 1989).

The mutual impedance experiment (S-304)

The mutual impedance $Z_m(f)$ is measured between two double-sphere dipoles as a function of frequency in the vicinity of plasma resonance frequencies. The shape of the impedance curve is a function of the velocity distribution of the electrons in the ambient plasma. With the assumption that the plasma is homogeneous, Maxwellian and isotropic, the mutual impedance $Z_m(f)$ is a well determined function of the plasma frequency (f_p) and the Debye length (λ_D). A two-parameter fit between the measured impedance $Z_m(f)$ and the theoretical curves makes it possible to determine f_p and λ_D and hence the electron density (N_e) and temperature (T_e), provided that the *a priori* assumptions concerning the electron velocity distribution are satisfied at the time of the measurement. For nominal operation of the instrument the ambient Debye length should not exceed 6 m nor be smaller than 0.5 m due to the dimensions and geometry of the antennae system. Furthermore, the plasma frequency had to be higher than the electron gyrofrequency, but lower than the maximum of the sweep frequency (i.e. 77 kHz). For these reasons the density measurements were limited to the range from 0.3 to 70 cm^{-3}. Since the electron velocity distributon is often non-Maxwellian and non-isotropic in the magnetosphere, only a limited fraction of the data collected could be interpreted.

Three main regions have clearly been identified with this GEOS-1 mutual impedance probe: (i) the plasmasphere, where the temperature is of the order of 10^4 K or less; (ii) an intermediate region located immediately beyond the inner-most steep density gradient, where T_e is very sensitive to the level of geomagnetic activity, with an average of 2×10^4 K, and electron densies range from 2 to 20 cm^{-3}. This intermediate zone is generally absent in the nightside sector except during recovery phases; (iii) a third region, observed in the nightside and

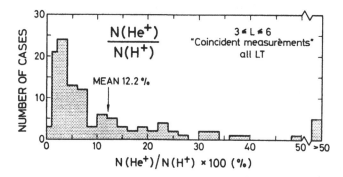

Figure 3.23   Histogram of He$^+$/H$^+$ density ratios obtained on GEOS-1 between June 1977 and May 1978. 116 'coincident' angular spectra of He$^+$ and H$^+$, a particular pair of spectra being separated by an increment in the L coordinate not exceeding 0.2, make up the database for the histogram (after Farrugia *et al.*, 1989).

morning local time sectors, characterized by very low electron densities and a non-Maxwellian velocity distribution under disturbed conditions.

Figure 3.24 presents three typical profiles of electron density and temperature obtained with the mutual impedance technique. Note the profile measured on 4 September 1977, in which a hole in the temperature distribution was observed at $L \approx 7$ where the density changed rapidly (instead of the increase usually seen near the plasmapause). Such an effect was seldom seen, but this was not a unique example (Décréau, Beghin and Parrot, 1978b).

Figure 3.25 from Décréau, Beghin and Parrot (1982) shows an example of electron density and temperature distributions versus L obtained by this technique in the bulge region on 3 June 1978. It can be seen that the electron density

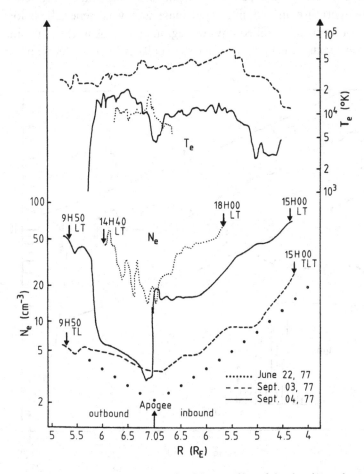

Figure 3.24 Illustrative GEOS-1 profiles of the density and temperature for three different days versus geocentric distance R in Earth radii. The magnetic index ΣK_p for the preceding 24 h had the following values : 12+ for 22 June, 16+ for 3 September and 9+ for 4 September. The curve with the small filled circles corresponds to an R^{-4} profile (after Décréau et al., 1978b).

decreased almost as L^{-4} (see the slope of dot dashed lines P_1 and P_2). The apparent position of the plasmapause is indicated by PP near $L = 7.6$. The electron temperature increased with L as found with the PROGNOZ-5 data and with the GEOS-1/ICE experiment described above.

The plasma density distributions obtained have confirmed most of the earlier findings. The electron temperatures measured were in the range of 10 000–15 000 K. and thus were a factor 3 to 5 larger than those found at low altitudes in the ionosphere. They were qualitatively consistent with those determined in the outer region of the plasmasphere by other techniques, as for instance the ion temperatures reported by Serbu and Maier (1970), Bezrukikh and Gringauz (1976), and Lennartson and Reasoner (1978). Warmer plasmas tended to be present in the outer region of the plasmasphere and during active periods. The temperature in the inner plasmasphere was generally below 15 000 K. In the outer freshly refilled dayside region as well as in the nightside during the recovery phase, the temperature was usually close to 20 000 K, while

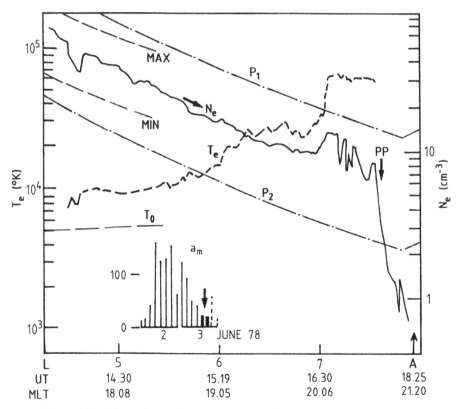

Figure 3.25   Example of $N_e(L)$ and $T_e(L)$ distributions measured by the GEOS-1 'mutual impedance probe' method. P_1 and P_2 are L^{-4} profiles. MIN and MAX are the density profile models proposed by Chiu et al. (1979); T_0 is their model temperature (after Décréau et al., 1982).

in the plasmatrough region the plasma was mostly non-Maxwellian or could not be completely characterized from the GEOS 1 mutual impedance probe observations. In the night and morning sectors thermal plasma populations of high temperatures were sometimes found under magnetically disturbed conditions. The origin of these temperature enhancements is not yet understood.

The temperature gradients were generally not coincident with density gradients; they could be separated from each other by more than 1000 km and were not always in the same direction.

Relaxation sounder experiment (S-301)

The relaxation sounder experiment is based on the observation of echo signals of waves emitted in the vicinity of the resonance frequencies of the ambient plasma. It is a method similar to that used in topside sounding of the ionosphere except that, in the plasmasphere, the plasma densities are $\sim 10^3$ times lower than in the ionospheric F-region. Short radio wave pulse of 3.3 ms duration at frequencies in the range 0.3–77 kHz are transmitted through the long radial boom into the ambient plasma. The sounder then monitors the echo signal received by the two spheres during 83 ms. The resonance echoes are observed near the electron gyrofrequency and its harmonics nf_c, at the plasma frequency f_p, and above the upper hybrid resonance frequency, f_{UHR}. The identification of the plasma frequency makes it possible to calculate the electron density:

$$N_e(\text{cm}^{-3}) = f_p^2(\text{kHz})/81$$

This method is effective for electron densities smaller than $70\,\text{cm}^{-3}$, as in the case of the mutual impedance experiment. The electron densities determined by Higel and Wu (1984) from this technique on GEOS 2 are consistent with those deduced from the other experiments. The great advantage of this type of *in situ* experiment is that they are independent of the spacecraft potential, and could be used to check or readjust other plasma measurements on board GEOS satellites. Another major advantage of these wave experiments is their ability to measure very low densities (down to $1\,\text{cm}^{-3}$), such as those existing in the plasmapause region.

Figure 2.16 shows an example of electron density profile obtained by this technique along the geostationary orbit of GEOS-2 (Higel and Wu, 1984). The figure shows that N_e was significantly larger in the noon and afternoon sectors, and smoother on the dayside than on the nightside. The much more irregular density distribution in the midnight sector confirms that this is the region where the plasmasphere is most sensitive to the level of geomagnetic activity. This is also the local time sector where the plasmapause is forming according to OGO-5 observations (Chappell, 1972).

3.3.3   ISEE: General feactures of the experiments and main results

Plasma composition experiment (PCE)

The ISEE-1 satellite was launched in October 1977 into a highly elliptic orbit with an apogee of $22.5R_E$, a perigee at about $2R_E$, an inclination of $20°$ and an orbital period of about 2.5 days. Plasma measurements were obtained with the Plasma Composition Experiment (PCE) operated in the 'thermal plasma' mode. Details concerning the operation of this plasma probe have been reported by Shelley *et al.* (1978). The procedure for data analysis is described in Comfort, Baugher and Chappell (1982). In the thermal plasma mode the PCE operates as a retarding potential analyzer over the energy range 0–100 eV. The effective time resolution of the instrument in this operational mode was 2 to 3 minutes. The ion temperatures were then derived from the energy spectrum deduced from the flux of particles, assuming that the total flux was composed of H^+ ions. According to the OGO-5 results of Chappell *et al.* (1971b), it was reasonable to assume that protons were the dominant ions in the plasmasphere.

Figure 3.26 illustrates ion temperature distributions versus L in the predawn, noon and dusk local time sectors. These observations correspond to very quiet geomagnetic conditions. Although in all cases the temperature increased with L in the warm outer zone, it can be seen in the second panel (day 340) that a negative temperature gradient was sometimes observed at $L < 4$. This peculiar feature was not an exceptional case in ISEE/PCE observations (Comfort, 1986); it has also been observed with GEOS-1/ICE (Farrugia *et al.*, 1988) (see above, Fig. 3.21b, l, n).

Figure 3.27 (Comfort, 1986) shows statistical distributions of the mean temperature and of its standard deviations in the (a) dayside and (b) nightside plasmasphere. The negative temperature gradient is again clearly evident in the dayside for $L < 3$. This effect is still unexplained. The second important feature illustrated in this figure is that beyond $L \sim 3$–4 the nightside temperatures are much higher than the dayside ones. It should be pointed out that PROGNOZ-5 and PROGNOZ-6 had observed similarly low ion temperatures on the dayside (Gringauz, 1985). It has been suggested that the low dayside plasmasphere temperatures are related to the low level of solar activity when these observations were made. On the contrary, the much higher nightside temperatures indicate the existence of an ion heating mechanism in this local time sector of the outer plasmasphere, even during low solar activity conditions.

The Sweep Frequency Receiver (SFR)

Active and passive wave experiments were also operated on board ISEE-1 and -2. The active wave experiments used both a relaxation sounder and a mutual impedance probe similar to those of GEOS-1 and -2 and described above

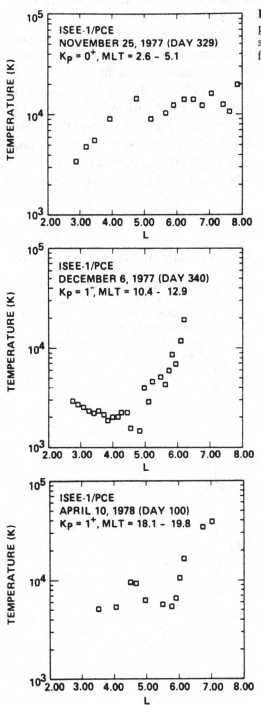

Figure 3.26 Examples of plasmaspheric H$^+$ temperatures profiles observed by ISEE-1/PCE (after Comfort, 1986).

(Etcheto and Block, 1978). Therefore, these ISEE active wave experiments also had the same technical limitations as did those on GEOS.

The passive wave experiment developed by the University of Iowa measures the natural emissions in the ambient magnetospheric plasma at frequencies near the upper hybrid resonance frequency and near the cutoffs at the electron plasma frequency (Gurnett and Shaw, 1973). Electron densities are measured in a wide range, from less than $1\,\mathrm{cm}^{-3}$ to $\sim 2000\,\mathrm{cm}^{-3}$ with this experiment. The Sweep Frequency Receiver (SFR) of the Plasma Wave Experiment (PWE) had 128 channels in the range 100 Hz–400 kHz which were scanned every 32 seconds in the manner described by Gurnett *et al.* (1979). The most intense emissions within the dense plasmasphere are observed near the upper hybrid resonance frequency

$$f^2_{\mathrm{UHR}} = f^2_{\mathrm{p}} + f^2_{\mathrm{c}}$$

where f_{p} is the electron plasma frequency and f_{c} is the electron cyclotron

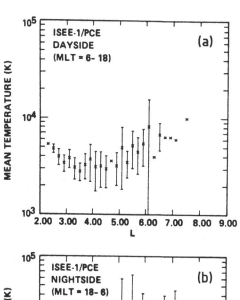

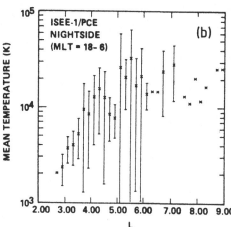

Figure 3.27 Mean H$^+$ temperature profiles with respect to *L*-shell, averaged over intervals of $\Delta L = 0.2$ from a limited number of ISEE-1/PCE observations (a) dayside; (b) nightside (after Comfort, 1986).

frequency. These emissions were observed on almost every pass through the plasmasphere, although sometimes their intensity was moderate; hence it was possible to determine the electron number density throughout most of the plasmasphere beyond $L \sim 2$ (Anderson, 1987). The electron density N_e was derived either directly from the value of the measured plasma frequency f_p or from the measured value of f_{UHR} combined with the value of f_c from the magnetic field measurements of ISEE. Indeed, using the equation above, the plasma frequency can be determined when f_{UHR} and f_c are both known.

Figure 3.28 shows ISEE profiles representing three different levels of geomagnetic agitation during the nine hours preceding the observations (courtesy of R. R. Anderson). Profiles of this kind were used in a number of studies reported in Chapter 2.

3.4   Plasmaspheric measurements during the decade 1980–90

3.4.1   The Dynamic Explorer (DE) mission

General characteristics and scientific objectives

The scientific aim of the DE dual-spacecraft mission was to study the different plasma components of the magnetosphere and their interactions. DE-1 and DE-2 were launched simultaneously in August 1981 into near-polar orbits ($i \simeq 90°$). The apogee of DE-1, the high altitude spacecraft, was $4.56 R_E$; its initial perigee was 570 km and its orbital period 7.5 hours. The altitudes of DE-2 ranged between 300 km and 1000 km; DE-2 made measurements only until March 1983. Sometimes the apogee of DE-1 was near the equator, and at such times the orbit was rather suitable for study of the plasma distribution along a given L-shell.

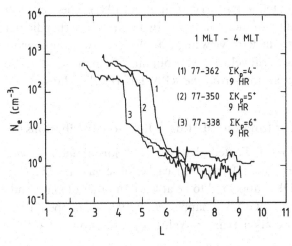

Figure 3.28 Nightside profiles of electron number density versus L from ISEE-1 representing various levels of magnetic activity. The nine-hour sums of K_p up to the observing times were 4-, 5- and 6+, respectively (courtesy R. R. Anderson, 1993).

The Retarding Ion Mass Spectrometer (RIMS) of DE-1 was especially designed to perform sensitive measurements of cold H^+, He^+, O^+, He^{2+} and O^{2+} in the density range from 0.1 to $10^6 \, cm^{-3}$ with an accuracy of 10%, and the corresponding ion temperatures in the range 0 to 45 eV with an accuracy of 5%. The spacecraft potential would be controlled with an accuracy of 0.1 V in the range from several volts positive down to 45 V negative. High resolution energy/ pitch angle distributions could be obtained for a selected pair of ion species. At altitudes lower than 1500 km, the range of operation of the mass spectrometer was 1–32 a.m.u. A detailed description of the RIMS instrument operation can be found in Chappell et al. (1981). A cut-away view of the RIMS sensor is presented in Fig. 3.29a.

In addition to RIMS ion measurements, the Plasma Wave Instrument (PWI) of DE-1 was designed to determine electron densities and temperatures by measuring the natural wave emissions along the DE-1 orbit as well as the ambient plasma response to the PWI radio signal. In parallel with these measurements, the PWI instrument was aimed at studying the wave activity in the plasmasphere. For a detailed description of this instrument see Shawhan et al. (1981).

Figure 3.29b illustrates the orbits of DE-1 and DE-2, as well as some of the important physical processes to be studied with these two sophisticated spacecraft (Brace, Chappell and Chandler, 1988; Chappell et al., 1981). The orbit of the DE-2 spacecraft was located at low altitude in the upper ionosphere so as to provide rendezvous opportunities for DE-1 – DE-2 studies of plasmasphere–ionosphere coupling. Indeed, DE-2 was equipped with instrumentation which was complementary to that of DE-1. The DE-LANG instrument of DE-2 provided measurements of the low altitude electron temperature (T_e), electron and ion densities (N_e and N_i), as well as the spacecraft potential (Krehbiel, Brace and Theis, 1981). The Retarding Potential Analyzer (RPA) measured the bulk ion velocity in the direction of the spacecraft motion and the ion composition and temperature along the satellite path (Hanson et al., 1981). The Ion Drift Meter (IDM) measured the flux of thermal ions in two mutually perpendicular directions, both orthogonal to the RPA view angle. In this way, two components of the ambient thermal plasma drift velocity were obtained from the IDM, while the third component was determined from the RPA data (Heelis, Hanson and Lippincott, 1981a).

Main scientific results from the DE-1 data; light ion distributions

The DE-1/RIMS observations indicated that the He^+ ion abundance was considerably higher than the values inferred from earlier observations. The He^+ densities measured aboard DE-1 appeared to be at least an order of magnitude higher than those measured by OGO-5 (Chappell et al., 1970a; Chappell, 1972) and on the average 2–3 times larger than the values deduced from the GEOS

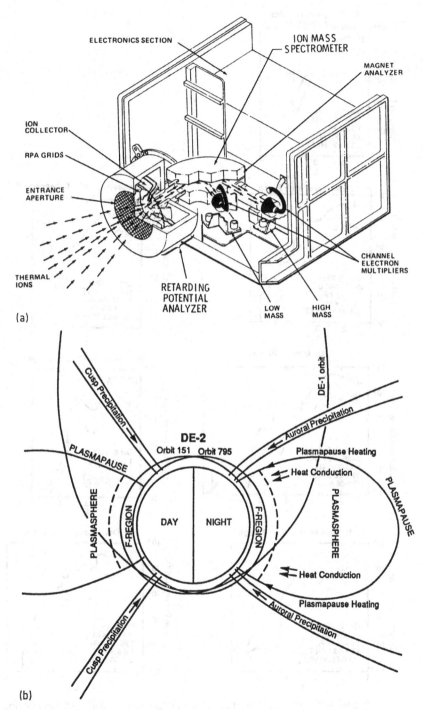

Figure 3.29   (a) A cut-away view of the RIMS sensor head used on DE. Arrows
show the path of thermal ions through the analyzer (from Brace *et al.*, 1988).
(b) Schematic view of plasmasphere–ionosphere system with typical DE-1 and
-2 orbits during the latter part of 1981.

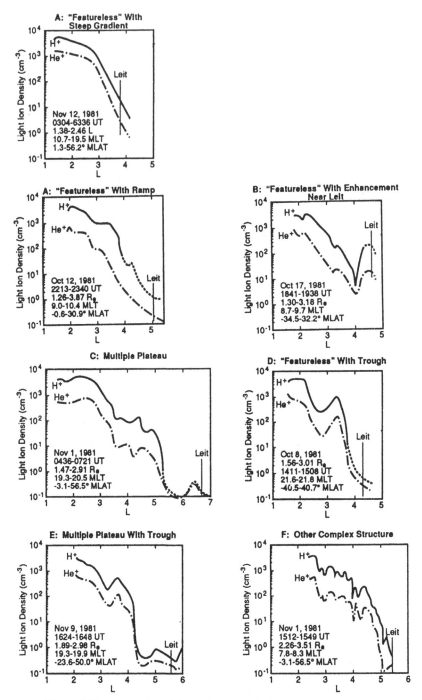

Figure 3.30   Examples of the various categories of density profiles obtained on DE-1. The solid (for H⁺) and dash–dot (for He⁺) segments were obtained by the full RPA analysis techniques described by Comfort *et al.* (1985), while the dotted (for H⁺) segments were based on the rammed flux approximation (after Horwitz *et al.*, 1990).

observations by Geiss *et al.* (1978), Farrugia *et al.* (1988, 1989) or by the ISEE observations of Horowitz, Comfort and Chappell (1986b) and Farrugia *et al.* (1988). The average $N(He^+)/N(H^+)$ ratio measured on DE-1 was 0.2–0.3 and was fairly constant for *L*-values ranging from $L = 2$ up to the plasmapause (Horwitz *et al.*, 1986a); sometimes the helium ion density exceeded that of H^+. This usually occurred at low altitude for *L* close to the plasmapause (Newberry, Comfort, Richards and Chappell, 1989).

Figure 3.30 shows examples of H^+ and He^+ density profiles compiled by Horwitz, Comfort and Chappell (1990). It can be seen that, indeed, $N(He^+)$ sometimes became equal to $N(H^+)$ in the plasmapause region. Newberry *et al.* (1989) further found that the ratio of He^+ and H^+ concentrations tends to increase toward the plasmapause, especially in the evening sector. Based on model calculations by Richards and Torr (1988), Newberry *et al.* (1989) explained the large observed values of $N(He^+)/N(H^+)$ as being due to an unusually high level of solar EUV flux associated with the high level of solar activity during the DE observations considered. Thus, according to the DE/RIMS observations, He^+ ions must be considered to be a major component of the plasmasphere.

Heavy ion 'torus'

Just inside the plasmasphere inner density gradient ($L = 3$–4), heavy ion (O^+, O^{2+} and N^+) densities have been found to be enhanced by a factor 10 or more, while there was no such variation in the light ions (H^+ and He^+) (Chappell, 1982). The location of this 'torus' or 'shell' of enhanced heavy ions is often correlated with ionospheric temperature peaks, suggesting a relationship with SAR arcs and an interaction between hot ring current particles and cold plasmaspheric plasma (Horowitz *et al.*, 1986a; Olsen, 1981).

Minor N^+ and N^{++} ions

Singly and doubly ionized nitrogen were detected for the first time by the DE-1/RIMS instrument. The measured N^+ fluxes were about equal to 5–10% of the flux of O^+ ions. The N^{2+} fluxes were equal to 1–5% of the N^+ fluxes (Chappell *et al.*, 1982).

Redistribution of different ion species in the plasmasphere following a geomagnetic storm

Figure 3.31 shows a sequence of plasmaspheric ion composition profiles versus *L* following a geomagnetic storm which occurred on 12 November 1981. K_p decreased gradually from 7 at 03–06 UT on 12 November to $K_p = 0$–1 at 1800 UT on 13 November. It can be seen from Fig. 3.31 that during the period of quieting, the region outside $L \sim 4$ extending to $L \sim 7$ began to fill. The He^+/H^+ density ratio remained nearly constant (0.2–0.3) in the whole plasmasphere despite the two orders of magnitude decrease of the light ion densities from $L = 2$ to $L \geq 5$. The remarkable feature is the dramatic enhancement of the O^+ ion density which reaches values comparable to the H^+ density in the region $L = 3$–4. Such an oxygen-rich dense plasma had never been observed before and remains unexplained (Singh and Horwitz, 1992).

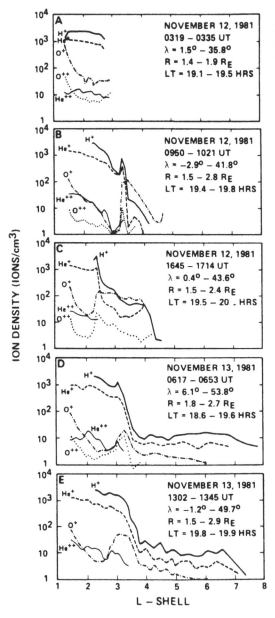

Figure 3.31   A sequence of plasmaspheric ion composition profiles measured by DE-1 during and following a geomagnetic storm (after Horwitz *et al.*, 1984).

Heated light ions trapped in the equatorial plasmapause region

Another interesting feature which was discovered with the DE-1/RIMS instrument is the minimum in the electron number density in the equatorial region for latitudes from $\pm 5°$ to $\pm 20°$ and from 2 to $5R_E$ in altitude. Density depletions of 10–70% are found in regions where the off-equator density ranges from 10 to 1000 cm^{-3}. This feature is associated with trapping and heating of light ions in the equatorial region; their energy is enhanced beyond 50 eV near the plasmapause (it is interesting to note that such ion heating in the direction perpendicular to the magnetic field was not observed for the heavier ions). There is a rough pressure balance provided by the warm ions forming the tail of the velocity distribution. These warm ions contribute only a few percent to the density, but 1–2 orders of magnitude more than the cold ions to the temperature. It should, however, be cautioned at this point that the field aligned density/temperature structures referred to here are not the norm for the plasmapause region, although they do appear to be reasonably common (Olsen, 1992). A detailed study of this feature has been presented by Olsen, Chappell and Gallagher (1987) and Olsen (1992).

Heating of thermal ions (~ 1–10 eV) by the low-frequency electromagnetic emissions between the cyclotron frequency and lower hybrid frequency that have frequently been observed near the earth's magnetic equator (Russell, Holzer and Smith, 1970; Gurnett, 1976) could explain the high correlation that has been observed between the occurrence of these waves and the occurrence of trapped equatorial ions (~ 50–100 eV) near the outer edge of the plasmapause (Olsen, Chappell and Gallagher, 1987). Often these magnetosonic or whistler waves are observed to have a funnel-shaped appearance on a frequency time spectrogram when a spacecraft like DE-1 passes latitudinally through the magnetic equator. Boardsen *et al.* (1992) demonstrated by ray tracing calculations that the funnel-shaped structure is largely a wave propagation effect in which the lower frequency waves are more strongly confined to the magnetic equator. A proton ring distribution in velocity space was detected by the energetic ion composition spectrometer (EICS) in the same region where equatorial wave emissions were observed and was considered to be responsible for the growth of the emissions. Such ion ring distributions can form as a result of the loss of low-energy ions, thereby causing a 'hole' to form in velocity space. These 'holes' were first observed by McIlwain (1972).

Thermal structure of the plasmasphere

The thermal structure of the plasmasphere is based on data collected in the evening and midnight–morning sectors. Despite this local restriction the following general features were identified:

(1)   Suprathermal light ions (10 eV) are observed in the outer plasmasphere
      in addition to the cold ions. This is illustrated in Fig. 3.32 where the H^+
      ion density is given versus L, along with the temperatures of the 'cold'
      and 'warm' components of the H^+ ion populations (Comfort, 1986). Note
      that these warmer ions are observed close to the plasmapause region
      where ring current particles interact with the plasmasphere.

(2)   These observations confirm that the ion temperature increases from
      4000 K to 10 000 K in the dayside outer plasmasphere while it is ranging
      from 2000 to 10 000 K in the evening sector.

(3)   The plasmasphere responds differently on the evening and morning sides

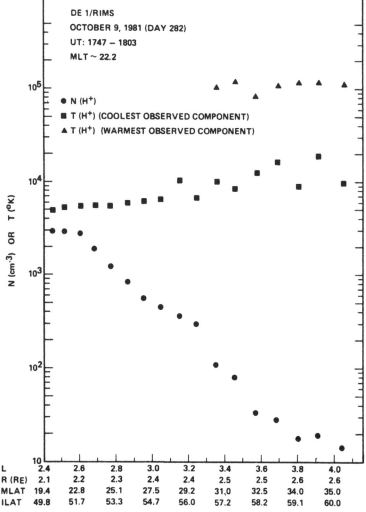

Figure 3.32   Warm and cold H^+ temperature components in the outer plasmasphere observed by DE-1/RIMS (after Comfort *et al.*, 1985).

to magnetic activity. On the morning side the inner zone temperature is independent of K_p up to $L \simeq 3$. Beyond this L value the ion temperature is enhanced when K_p increases. On the evening side T_i is independent of K_p only up to $L \simeq 2$. Between $L = 2$ and 3 the ion temperatures are usually depressed when K_p is larger than 6 (Comfort, 1986).

(4)   Field-aligned temperature gradients have been inferred from DE-1/RIMS observations throughout the whole plasmasphere in the evening sector. On the morning side field-aligned temperature gradients were mainly observed in the outer zone beyond $L = 3$. This is illustrated in Fig. 3.33 where low and high altitude mean temperatures are given versus L in the morning and evening local time sectors (Comfort, 1986).

(5)   It has been found that the He^+ and H^+ ions are in thermal equilibrium in most case studies (Comfort et al., 1985) as well as from a statistical point of view. This is illustrated in Fig. 3.34, where it is shown that the ratio $T(He^+)/T(H^+) \simeq 1$ in the whole plasmasphere, with the largest standard deviation in the evening sector.

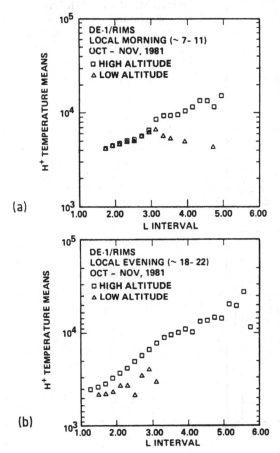

(a)

(b)

Figure 3.33   Mean H^+ plasmaspheric temperature profiles with respect to L, with corresponding observations from low altitude segments of the DE-1 plasmasphere transits for (a) local morning and (b) local evening (after Comfort, 1986).

(6)   The average plasmaspheric ion temperature is nearly three times larger
      than the ionospheric temperature.

Density distributions and the location of various plasma boundaries

Based on DE-1/RIMS data Horwitz *et al.* (1984;1986b;1990) and Comfort *et al.*
(1985) identified different characteristic types of ion density profiles versus L and
proposed a schematic classification for them; this is illustrated in Fig. 3.30. In
these figures the acronym LEIT stands for Low Energy Ion Transition. It
corresponds to the boundary in or beyond the plasmasphere where the cold ions
change from an isotropic pitch angle distribution to a field-aligned one. The
LEIT is generally co-located with a steep plasma density gradient. In other cases
it may mark a location where some of the conditions for plasmapause formation
currently exist, but at which a substantial density jump is not present, as
discussed in connection with Fig. 2.1.

   Figure 3.35 shows the equatorial distances of observed LEIT locations in
different local time sectors for different ranges of K_p. As in the case of the

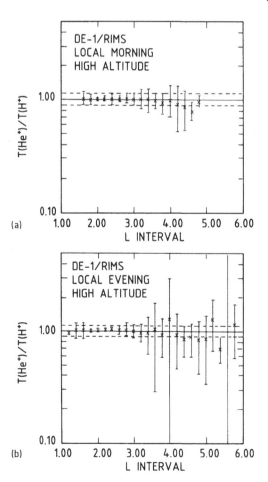

Figure 3.34   Ratios of He$^+$ and
H$^+$ temperatures versus L from
DE-1 in the (a) local morning and
(b) local evening (after Comfort,
1986).

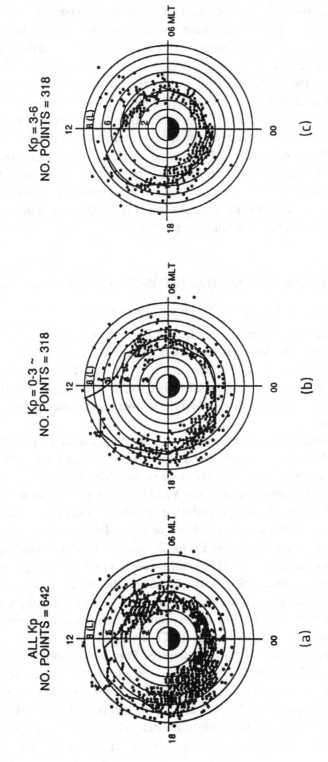

Figure 3.35 Distribution of LEIT (low-energy ion transition) locations with *L* and MLT for all K_p conditions (a) and for low (b) and high (c) K_p ranges (after Horwitz *et al.*, 1990).

plasmapause, the positions of the LEIT are characterized by a day–night asymmetry and an afternoon bulge, with a maximum asymmetry in the noon–midnight direction for periods of quiet geomagnetic conditions: $K_p = 0$–3 (Horwitz et al., 1990).

Figure 3.30 includes density profiles corresponding to 'multiple plateaux' or steplike radial distributions of cold plasma densities. The innermost steep density gradients are then usually located between $L = 3$ and 4. As in whistler studies, these inner 'knees' are interpreted by Horwitz et al. (1990) as the vestiges of earlier 'plasmapauses' formed during geomagnetic disturbances prior to the density measurements. These authors also found that 'multiple plateaux' profiles occurred predominantly in the afternoon and evening sectors, while 'featureless' profiles with a steep density gradient (see Fig. 3.30) were statistically most common in the midnight and morning sectors.

3.4.2 The Exospheric Satellite (EXOS-D) or AKEBONO Mission

The orbit, the payload and preliminary results

EXOS-D, the 12th Japanese satellite, was launched on February 1989 into an eccentric orbit with apogee at 10 470 km, perigee at 272 km, inclination of 75° and orbital period of about 3 hours. The orbit of EXOS-D is shown in Fig. 3.36. The payload contains instruments designed to study the density distribution of cold plasma in the outer plasmasphere, its temperature, composition and charge distributions.

Suprathermal ions are measured with the Suprathermal Ion Mass Spectrometer (SMS) in the energy/charge range 0–25 eV/q and mass/charge range 0.8–60 amu/q. Preliminary results concerning SMS observations have been published by Watanabe, Wallen and Yau (1992). These concern depletion and refilling of the outer plasmasphere during and after magnetic disturbances. They indicated that, on the equatorward side of a high-latitude boundary, cold ions were found to corotate with the Earth. These authors also showed that in this region the field-aligned ion bulk velocities were low (< 1 km/s), implying that the cold ions were of plasmaspheric origin. The densities measured near the boundary at 5000–10 000 km altitude were 10–100 cm^{-3}, indicating that the boundary can be located in the trough region, beyond the conventional plasmapause, as is also on occasion the case for the LEIT discussed above. Poleward of this boundary the convection pattern deviates from corotation and large field-aligned flows of ionospheric ions were observed. These outward flows of ions were consistent with those predicted by polar wind models. The observed field-aligned velocities of H$^+$, He$^+$ and O$^+$ ions appeared to be 12 km/s, 7 km/s and 3 km/s, respectively. The correlation study of these plasma parameters with the K_p geomagnetic activity index suggests that it takes about 2–3 days between the time when K_p increases and the establishment of new, large scale steady conditions in the ionosphere.

The mid-altitude ion composition measured with EXOS-D in the plasmasphere changed from light-ion-dominated plasma during quiet times to a heavy-ion-dominated one during magnetic disturbances (Watanabe *et al.*, 1992). However, as emphasized by the authors, these are preliminary results obtained from a limited sample of EXOS-D data.

3.5   The latest results

The number of scientific satellite missions devoted to study of the plasmasphere and the magnetosphere has shrunk dramatically over the last ten years in favor of different types of spaceborne activities. This is mainly due to budgetary restrictions.

Fortunately, a few missions have escaped these severe 'cut backs'. The mother–daughter pair of space probes of the ACTIVNY, APEX and INTER-BALL missions are among these exceptions and have been successfully launched by Russia. In the USA the CRRES satellite has made observations in the plasmasphere but only for a limited period of time of 15 months. In order to update this chapter we have added a few of the results obtained by these most recent missions to the plasmasphere.

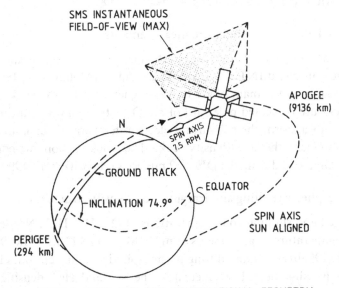

EXOS-D ORBIT AND SMS OBSERVATIONAL GEOMETRY

Figure 3.36   Sketch of the EXOS-D orbit and the SMS's field of view on the orbit (after Sawaga *et al.*, 1987).

3.5.1 The ACTIVNY (MAGNION-2) and APEX missions

The orbits of the pair of satellites forming the ACTIVNY mission are nearly polar ($i = 82.5°$). Their perigees and apogees are about 500 and 2500 km, respectively. A cylindrical Langmuir probe was mounted on the MAGNION-2 subsatellite for *in situ* diagnostics of cold plasma, while a Bennett type ion mass spectrometer (HAM-5), as well as electric and magnetic antennae were flown on the mother spacecraft INTERCOSMOS-24. A similar instrument was operating on INTERCOSMOS-25 which is part of the APEX mission.

Observations of thermal O^{++}

Smilauer *et al.* (1996) published O^{++} density observations obtained in the outer ionosphere on near polar orbits by the HAM-5 spectrometers of both ACTIVNY and APEX projects. They report characteristic enhancements in the latitudinal profiles of the O^{++} ion density. These enhancements are often observed at middle latitudes. The value of the *L*-parameter for these O^{++} density peaks varies in a non-linear way with the ring current activity index D_{st}: when the D_{st} index changes from $+20$ to -90 nT the locations of the density peaks change from $L = 5$ to 2. The positions of the O^{++} density peaks are found to be located predominantly within the plasmasphere. Their distance from the plasmapause decreases, however, when the level of geomagnetic activity during the preceding 24 hours increases.

Similar regions of high O^{++} density were observed earlier in the outer plasmasphere with the DE-1 satellite (Roberts *et al.*, 1987).

Thermal electron temperature measurements

Simultaneous measurements of the electron temperature and ion composition have been made on board INTERCOSMOS-24 and 25 (Afonin *et al.*, 1994). It was found that during the main phase of geomagnetic storms the peak of the auroral temperature is located poleward of the O^{++} trough, while during the recovery phase of a storm the subauroral electron temperature enhancement (SETE) corresponds to the O^{++} density peaks. These observations are consistent with those obtained during the DE-1&2 missions (Horwitz *et al.*, 1990).

Plasmaspheric refilling and irregular ion density structures

Using combined and simultaneous observations of O^{+}, H^{+}, He^{+}, N^{+} density and electron temperature measurements from HAM-5 on INTERCOSMOS-24, Boskova *et al.* (1993b) found that during plasmaspheric refilling episodes localized increases (small-scale enhancements) of the H^{+} and He^{+} densities are detected in narrow *L*-intervals within the plasmasphere. Subauroral electron temperature enhancements and VLF noise plasmaspheric emissions (see Section

2.6.3) are frequently observed in this same ionospheric region. The authors interpret these density inhomogeneities as evidence that the upward ionization flow refilling plasmaspheric flux tubes is not a uniform function of latitude; i.e. that the refilling can be significantly larger than the average in these narrow regions where the light ion density peaks and where electron temperature enhancements are observed. They also found that in the nightside sector at low ionospheric altitudes, below 600 km, the He^+ ions are more sensitive indicators of the refilling process than the H^+ ions; indeed, at those altitudes the H^+ ions are strongly bound to the O^+ ions by the charge-exchange reaction.

Dynamics of the plasmasphere

Irregular ionospheric structures observed for different levels of magnetic activity have been studied by Förster et al. (1992) using the MAGION-2 electron and temperature measurements. Maximum entropy spectral analysis of these irregular structures showed that the occurrence of some characteristic wavelengths in the upper ionosphere depends on local time and on the level of geomagnetic disturbances. As expected, the fluctuations observed in the electron density and temperatures are more prominent during disturbed conditions then during quiet times. The characteristic scale length of these electron density and temperature irregularities increases when the magnetic activity increases.

Using electron temperature and ion density measurements from the AC-TIVNY mission, Jiricek et al. (1996) have studied the dynamics of the plasmasphere and plasmapause position during the magnetic storm of 20 March 1990. The successive passages of the satellite in the dusk and dawn sectors showed how the latitude of the high altitude plasmapause and the low altitude ion and electron troughs changed during this magnetic storm. The ACTIVNY results nicely confirm results obtained during previous missions. They offer in addition the possibility to determine whether an inner (fossil) plasmapause exists in the dusk sector and whether a plasmaspheric bulge has formed or not in this local time sector (see also Förster et al., 1992).

The plasmaspheric community is looking forward to more such results and to confirmation of earlier findings. Indeed, only detailed and comprehensive observations will tell us what the plasmasphere really looks like; simultaneous, multipoint observations like those recently reported by McComas et al. (1993) and Moldwin et al. (1995) are needed in the future as well as a study of the wealth of data on total electron density now available in the records of satellites such as ISEE 1, CRRES, ACTIVNY, etc. Through such work it should be possible to verify theoretical models proposed to describe the distribution of plasma and electric fields in the plasmasphere and at the plasmapause.

3.5.2   The INTERBALL mission and the ISTP and GGS programs

Russia has launched the INTERBALL mission which consists of two sets of space probes: (1) the TAIL PROBE with an orbital period of 48 h, inclination of 65 degrees, apogee of 200 000 km and perigee of 500 km; (2) the AURORAL PROBE with an orbital period of 6 h, inclination of 65 degrees, apogee of 20 000 km and perigee of 700 km. The latter is most suitable to study the dynamics of the plasmasphere during magnetic storms. Indeed, due to its much shorter orbital period it traverses the plasmasphere four times per day.

The ALPHA-3 cold plasma experiment on board both satellites will offer the possibility of separating time variations from spatial ones. It is designed to measure fluxes of ions with energies ranging between 0 and 25 eV; i.e. in the range of energies corresponding to the thermal particles forming the bulk of the plasmasphere (Bezrukikh and Gringauz, 1995).

These new magnetospheric missions will form, with GEOTAIL, WIND, SOHO, CLUSTER, ULYSSES and YOHKOH, major components of the International Solar Terrestrial Physics (ISTP) program that seeks to develop a new level of understanding of the near earth environment, or geospace.

All these space missions fit into the Global Geospace Science (GGS) program of correlated space-based and ground-based campaigns which are coordinated by the Inter-Agency Consultative Group (IACG). All these grand programs open a new era of solar terrestrial research. We hope that the chapters of these books will contribute, at least a little, to consolidate the foundations on which these ambitious international programs will be elaborated during the coming decade.

Chapter 4

A global description of the plasmasphere

4.1 Introduction

In the previous chapters, we have described different methods of observations which have been used over three decades to study the plasmasphere, its shape, ionic composition, dynamics and deformations during geomagnetic substorms. In this chapter we wish to put together all pieces with the hope that an up-to-date global picture of the plasmasphere will emerge. We briefly mention theories and models proposed to explain various features of the plasmasphere, empirical models being presented where relevant. Discussion of theoretical aspects will be found in Chapter 5.

4.2 The ionosphere as a source and sink for plasmaspheric particles

The ionosphere of the earth has been divided into different layers, the D-, E-, and F-regions. Above $\sim 300\,\mathrm{km}$ altitude where the maximum ionization occurs in the F-region, the thermal ion and electron densities steadily decrease with altitude. This region, which is the base of the topside ionosphere, extends deep into the magnetosphere. It forms the plasmasphere at low and middle latitudes, the plasmatrough at high and mid-latitudes, and the polar wind at high latitudes (see Fig. 4.1).

Figure 4.2a illustrates the daytime ionospheric and atmospheric composition, based on early IQSY mass spectrometer measurements. It shows that, below

1000 km altitude, ions remain minor constituents of the earth's atmosphere. Below 500 km O^+ is generally the dominant ionic constituent. Above 600 km, there is a transition level where H^+ become the dominant ions. Under exceptional geophysical conditions the He^+ ion density can exceed the density of O^+ ions and the density of H^+ ions in an intermediate altitude range (He^+ belt). Regions of dominance of He^+ ions have been observed at altitudes of 900–950 km with DE-2. A neutral helium belt has also been predicted by Nicolet (1961) based on theoretical considerations and drag measurements of artificial satellites (see also Kockarts and Nicolet, 1962).

The distribution of ions and electrons in the ionosphere up to an altitude of 1000 km can be determined from an empirical model called the International Reference Ionosphere (IRI). The earliest version of this model is described by Rawer *et al.* (1978). The latest version of IRI is now available as a software package which can be obtained from the NSSDC Request Coordination Office, NASA/Goddard Space Flight Center. The general description of this model can be found in Rawer and Bilitza (1985, 1989) or Bilitza *et al.* (1993) (see also Bilitza (1990) for a technical description of IRI). Another comprehensive model of the electron and ion temperatures has been developed by Köhnlein (1986) based on several sets of satellite data and incoherent scatter radar observations.

The O^+, H^+, He^+, O_2^+ and NO^+ ion densities, as well as the ion and electron temperatures (T_i and T_e) predicted by the IRI model have been compared with the measurements of the AE-C satellite between 140 km and 2000 km altitude. These AE-C data are part of the Goddard Comprehensive Ionosphere Database (GCID) built and described by Grebowsky, Hoegy and Chen (1990, 1993). The comparison of the AE-C observations and IRI model predictions over all altitudes and latitudes sampled by the AE-C satellite indicates rather satisfactory agreement (Hoegy and Grebowsky, 1994). The H^+ composition agrees best among the ion species considered: $<n_i(\text{AE-C})/n_i(\text{IRI})> = 1.05$ for H^+ ions; $= 0.62$ for He^+ ions; $= 2.49$ for O^+ ions; $= 3.1$ for O_2^+ ions; $= 1.6$ for NO^+ ions. The ratio of the electron temperatures determined from AE-C observations and

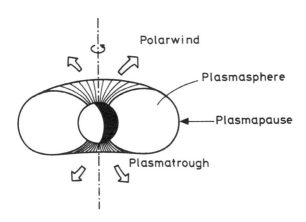

Figure 4.1 Illustration of plasmasphere, plasmatrough and polar wind as high-altitude extensions, respectively, of the low- and middle-latitude, high- and mid-latitude and high-latitude ionosphere.

the IRI model values is equal to 0.99. For the ions this temperature ratio is equal to 1.17. This confirms that the IRI model is probably reliable on the average to a factor 2 for the major O^+ ion density and even better for the other ions. Despite the large scatter of individual AE-C temperature measurements the average electron and ion temperatures are close to those determined from the IRI model.

Figure 4.2b shows the statistical distributions of O^+, H^+ and He^+ ion densities, and Fig. 4.2c the ion temperature (T_i) and the electron temperature (T_e) versus altitude in the topside ionosphere up to 1200 km altitude, for invariant latitudes ranging between 40° and 50° in both hemispheres. These distributions correspond to measurements made in the noon local time sector (9–15 MLT) during the summer season and for minimum solar activity conditions. Although the solar UV and X-radiation heating and ionizing the earth's atmosphere is absorbed mainly at altitudes below 200 km, it can be seen from Fig. 4c that the electron and ion temperatures increase with altitude up to 1000 km.

It can also be seen from Fig. 4.2b that at mid-day local time, O^+ ions remain the dominant ions in the plasmasphere at least up to 1000 km altitude. However, this is not the case at midnight, as illustrated in Fig. 4.2d, where the O^+ and H^+ average ion density distributions determined from the GCID have been plotted versus altitude. The standard deviation of the ion densities are also shown. The thick curves corespond to the diffusive equilibrium profiles computed by Hoegy and Grebowsky (1994), using the measured average plasma temperatures in 40 km altitude bins. The graph 4.2d corresponds to solar minimum conditions. From a comparison of 4.2d and 4.2b it is evident that the transition height where the O^+ ions and H^+ ions are equally abundant decreases from above 1000 km altitude at noon to 600 km at midnight. Furthermore, due to the reduction of ion and electron temperatures during the night, the O^+ density decreases faster with altitude and is considerably reduced at 1000 km while the H^+ ion density is enhanced at this altitude compared to the daytime profile shown in Fig. 4.2b.

Using these profiles of light ion averaged densities, Hoegy and Grebowsky (1994) showed that there is a dramatic departure from diffusive equilibrium near midnight local time above 400 km altitude along magnetic lines outside the plasmasphere ($\Lambda = 60$–$70°$). However, at lower invariant latitudes within the nightside plasmasphere, diffusive equilibrium is approximately satisfied up to 1000 km altitude. But, within the dayside plasmasphere at all latitudes poleward of $40°\Lambda$, the statistical results indicate that the H^+ and He^+ ions are accelerated upward as in the polar wind, at altitudes above 600 km during solar minimum, and above 1000 km altitude during solar maximum.

What is most important to note is that the upward flow is not only observed in the polar region, but also on closed magnetic field lines within the dayside plasmasphere. What is even more unexpected is that this upward acceleration of H^+ and He^+ ions leads to higher field-aligned bulk velocities at lower altitudes,

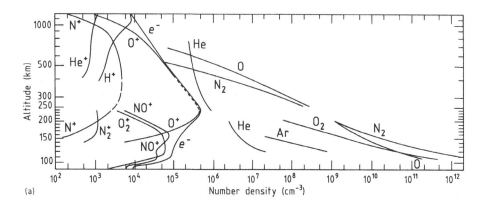

(a)

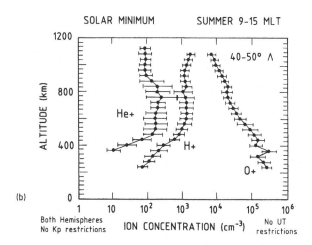

(b)

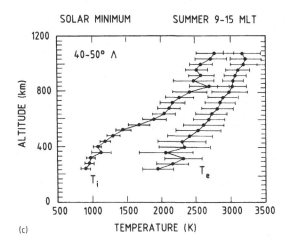

(c)

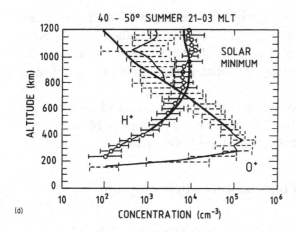

(d)

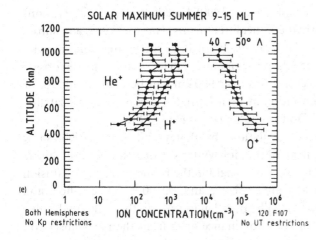

(e)

Figure 4.2 (a) The International Quiet Sun Year (IQSY, 1964) daytime ionospheric composition, based on mass spectrometer measurements. Ion and neutral distributions below 250 km are from two daytime rocket measurements above White Sands, New Mexico (32°N, 106°W). The helium distribution is from a nighttime measurement. Distributions above 250 km are from the Elektron II satellite results and Explorer XVII results (after Rishbeth and Gariott, 1969). (b) Summer solar minimum average density distributions of H^+, He^+ and O^+ ions in the noon magnetic local time sector (9–15 MLT). Standard deviations of the values measured by the AE-C satellite are also shown every 40 km altitude, for invariant latitudes of 40–50 °Λ: i.e. inside the plasmasphere. (c) The averages and standard deviations of the ion and electron temperatures (T_i and T_e) deduced from AE-C observations. These observations correspond to the same local time sector, the same season, and solar activity conditions as in panel (b). (d) Averages and standard deviations of H^+ and O^+ densities as in panel (b), but for the midnight magnetic local time sector (21–03 MLT). The thick lines correspond to diffusive equilibrium density profiles, using average plasma temperature determined from AE-C observations. (e) Same as panel (b) but for solar maximum conditions (after Grebowsky et al., 1993).

when solar activity is minimum, i.e. when the heat input to the thermosphere and ionosphere by solar UV and X-rays is minimum. Furthermore, the polar wind like flow is faster at higher invariant latitudes, where the ionospheric temperature is smaller than at lower latitudes, and where heating by solar radiation is weaker.

These are challenging observations which are emphasized here to set the stage for chapter 4, and to introduce the following description of the plasmasphere and of its low-altitude frontier: the topside ionosphere.

4.2.1 The distribution of the major ions in the topside ionosphere

The distribution of He^+ at middle and equatorial latitudes has been examined recently by Heelis, Hanson and Bailey (1990) based on DE-2 observations during solar maximum. Early theoretical treatments (Hanson, 1962) and subsequent satellite observations (Taylor et al., 1970), have shown that solar activity can significantly influence the relative abundance of He^+ and H^+ in the topside ionosphere. That He^+ concentrations near 900 km can be comparable to, or even exceed, the O^+ concentration at high latitudes is confirmed by satellite observations (Hoffman and Dodson, 1980; Heelis et al., 1981b).

Recent ionospheric model calculations by Bailey and Sellek (1990) show regions of He^+ dominance near 900 km for winter solstice and equinox during solar maximum. Such regions, appear at night in their time-dependent diffusion model at about 20:00 LT at winter solstice and, 02:00 LT at equinox, and persist until shortly after sunrise as illustrated in Fig. 4.3. In the Bailey and Sellek numerical models, regions of He^+ dominance arise from the post-sunset collapse of O^+ densities combined with the nighttime maintenance of He^+ and H^+. These results suggest that He^+ dominance in the topside ionosphere (at $L = 3$) is indeed a solar maximum phenomenon and most pronounced in winter.

Figure 4.3 shows calculated local time variations of O^+ (solid lines), H^+ (dashed) and He^+ (dotted) at different altitudes between 500 km and 2500 km along field lines at $L = 3$, for solar maximum conditions and different seasons. The density of O^+ is the dominant one during daytime at all altitudes below 2500 km. At sunset, the decrease in the production of O^+ from photoionization of O coupled with the loss of O^+ from recombination with N_2 and O_2 gives a large and rapid decrease in the O^+ concentration at lower altitudes. This decrease in O^+ concentration, combined with the decrease of O^+ density scale height arising from the lower temperatures, leads to much lower O^+ concentrations at most altitudes in the topside ionosphere during nighttime. This effect is so large during winter that a region is formed at 1000–1500 km altitudes in the post-midnight local time sector where He^+ ions are dominant (see Fig. 4.3). Note that at a fixed altitude the nighttime H^+ and He^+ concentrations are higher than their daytime values (see Bailey and Sellek, 1990; compare Figs. 4.2b and 4.2d).

At high latitudes both He^+ and H^+ are significantly reduced from their low-latitude and mid-latitude values (Taylor, 1972b). This phenomenon is known as the light ion trough (LIT) and is illustrated in Fig. 3.8a obtained from measurements made with the OGO-3 satellite by Taylor et al. (1968b). Taylor (1971) pointed out from latitudinal profiles of ion density measurements made in the topside ionosphere that the equatorward edge of the LIT is located on magnetic field lines which are within the plasmasphere.

The relationship between the light ion trough in the latitudinal profiles of He^+ and H^+ concentrations and the high-altitude plasmapause density gradient has already been mentioned in Section 2.4.3. Taylor (1971) related the invariant latitude of the LIT to a parameter α, defined as the angle between the Earth's magnetic equator and the Earth–Sun line. Subsequent work confirmed that this solar-geomagnetic-seasonal control of the topside ionosphere is indeed

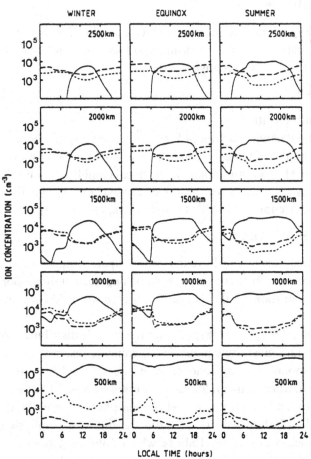

Figure 4.3 Calculated diurnal variation of O^+ (solid lines), H^+ (dashed) and He^+ (dotted) concentrations during solar maximum at different seasons and different altitudes along convecting magnetic flux tubes at $L = 3$ (after Bailey and Sellek, 1990).

determined by a complex universal time dependence resulting from the tilt of the Earth's magnetic dipole with respect to the Earth–Sun direction (see Sojka, 1989).

Hydrogen and helium ions are subject to chemical processes which can appreciably modify their distributions in the upper ionosphere. Helium ions are destroyed below 600 km by charge-exchange reactions with O_2 and N_2 (Bauer, 1966). Raitt, Schunk and Banks (1978) calculated the He^+ density profiles for a variety of geophysical conditions. They found that He^+ has a characteristic peak density in the 500–700 km altitude range whose value varies by more than a factor of 100 with geophyscial and solar activity conditions. The photoionization frequencies and chemical rate coefficients are given in the reviews by Schunk and Nagy (1980) and Torr and Torr (1982). See also Bailey and Sellek (1990) as well as Schunk (1988b) for more recent comprehensive reviews.

4.2.2 Production and transport of O^+ and H^+ ions at the bottom of the plasmasphere

As mentioned above oxygen ions are mainly produced by photoionization of atomic oxygen in the ionosphere. The O^+ ions undergo a series of reactions with N_2, O_2 and H of which the charge exchange reaction

$$O^+ + H \rightarrow O + H^+ \tag{4.1}$$

is the most important at high altitudes.

The resonant charge exchange reaction (4.1) proceeds very rapidly because the first ionization potential of atomic oxygen is nearly equal to that of atomic hydrogen. Consequently, H^+ ions are rapidly in chemical equilibrium with the other major constituents (O^+, O, H) of the upper atmosphere. Under these conditions Hanson and Ortenburger (1961) showed that n_{H^+} is given by

$$n(H^+) = (9/8)n(H)n(O^+)/n(O) \tag{4.2}$$

when the ionic and neutral constituents have the same temperatures. As a result of this reaction the concentration of H^+ ions should be approximately proportional to the O^+ concentration in the altitude range between 500 km and 1000 km.

Rishbeth and Garriott (1969, pp. 157–8) note that the altitude distribution of H^+ in chemical equilibrium given by equation (4.2) coincides exactly and by chance with that corresponding to an ionosphere where H^+ and O^+ are in diffusive equilibrium. But, when non-equilibrium processes such as field-aligned ionization transport between the ionosphere and plasmasphere are present, the chemical equilibrium distribution will depart from the diffusive equilibrium distribution. There is then a critical level above which diffusion equilibrium controls $n(H^+)$, while below that level it is mainly determined by the chemical process (4.1). This critical level is located at 700 km or higher, i.e. much above the F2-region. As also indicated in Figs 4.2b, 4.2d and 4.3 the changeover from O^+ to H^+ dominance occurs at altitudes ranging between 600 km and 3000 km depending on magnetic latitudes, local times, geographic longitudes, seasons, solar and magnetic activity (Titheridge, 1976a; Miyazaki, 1979; Greenspan et al., 1994).

Throughout the F-region there are sufficient collisions to maintain H^+ in chemical equilibrium. In the topside ionosphere where the O^+ density falls below $5 \times 10^4 \, cm^{-3}$, H^+ ions are able to diffuse upward to refill empty plasmaspheric flux tubes. For an H^+ ion created by reaction (4.1) to reach the plasmasphere, it must first diffuse upward until it can escape relatively freely into the plasmasphere. However, there is a limit to the rate at which the plasmasphere can be replenished by this upward flux of H^+ ions (Hanson and Patterson, 1964).

Because of the large collision cross-section of the H^+–O^+ Coulomb interaction, which is orders of magnitude larger than the H^+–O ion–neutral collision cross-section, the diffusion of protons through O^+ ions is a rather slow physical process, even in the diffusion region above 700 km. Therefore, the ionospheric region between 700 and 1000 km, where O^+ remain the dominant ions, constitues a kind of 'diffusive barrier' separating the ionosphere and the plasmasphere. It behaves like a porous layer which limits the flux of H^+ ions able to diffuse upward into the plasmasphere. The limiting flux depends on the temperature and on the concentration of neutral hydrogen. Hanson and Patterson (1963) estimated this flux to be of the order of $10^7 \, ions \, cm^{-2} s^{-1}$. However, actual observations of flux tube refilling indicate that this limiting flux must be one order of magnitude larger: $1–3 \times 10^8 \, cm^{-2} s^{-1}$ (Park, 1974a). Geisler (1967) found that the maximum flux occurs during daytime and is proportional to the neutral hydrogen density.

More recent studies have confirmed that the flux of hydrogen ions into the plasmasphere is predominantly limited by the rate at which H^+ is produced by the reaction (4.1) rather than by other factors. Richards and Torr (1985) have derived an approximate analytical expression for the maximum possible H^+ flux supplying the plasmasphere. A value of $1–1.5 \times 10^8 \, cm^{-2} s^{-1}$ for the field-aligned proton flux into the plasmasphere was determined by Krinberg and Tashchilin (1980, 1982), based on their ionosphere–plasmasphere interhemispheric diffusion model for O^+ and H^+ ions.

Considering that the maximum H^+ ion flux is correctly given by the Richards and Torr expression and following an approach similar to that of Li, Sojka and Raitt (1983), Rasmussen, Guiter and Thomas (1993) deduced a simple expression for the equilibrium plasmaspheric H^+ density.

Field-aligned transport is also driven by plasma pressure gradients and $\mathbf{E} \times \mathbf{B}/B^2$ drift. During the day, downward pressure gradients and upward transport are generated by the enhanced production of ions in regions that have larger neutral densities and receive more sunlight. At night, recombination at low altitude creates reduced pressure gradients and downward transport.

Greenspan *et al.* (1993) recently showed that DMSP observations at 840 km altitude are consistent with all these physical mechanisms. They confirmed that the most significant effect of magnetic disturbances is to raise the transition where H^+ and O^+ ion concentration become equal. Comprehensive time-dependent numerical models of the ionosphere, like those developed by Bailey (1983), Bailey, Simmons and Moffett (1987), Schunk (1988a) and co-workers, and by Bailey and Sellek (1990) include most of the electrodynamic, physical and chemical effects outlined above.

4.2.3 The effects of photoelectrons and wave–particle interactions

Photoionization of oxygen atoms produces copious amounts of electrons. The resulting photoelectrons with energies of tens of eV heat the ionospheric plasma by elastic and inelastic collisions. This transfers heat from the suprathermal photoelectrons to the bath of colder ions and to the thermalized ionospheric electrons, whose temperatures are therefore increased.

Among the first, Hanson (1964) pointed out that photoelectrons traveling along geomagnetic field lines can easily escape from their production region and enter into the plasmasphere. Geisler and Bowhill (1965) were the first to evaluate the fluxes of escaping photoelectrons and the associated heating rate of the plasmasphere. Sanatani and Hanson (1970) pointed out that photoelectrons may become trapped in the geomagnetic field, and that they contribute there a population of electrons with suprathermal energy. As a consequence, the electron temperature increases at high altitudes.

The transfer of energy from photoelectrons to thermal electrons and eventually to the ions and neutral atoms by collisions has been formulated in the Chapman–Enskog approximation by Schunk and Watkins (1979), in the Grad 13-moment approximation (Schunk, 1975, 1977) and in the kinetic approximation by Khazanov, Gombosi, Nagy and Koen (1992), Khazanov and Liemohn (1995) and Khazanov, Neubert and Gefan (1994) who modeled the formation of superthermal electron fluxes under non-stationary conditions.

Photoelectrons easily escape along magnetic field lines into the plasmasphere. Indeed, the Coulomb collision cross-sections for momentum and energy transfer are rapidly decreasing function of the energy. Therefore, the photoelectron exobase level, i.e. the altitude where the mean free path of a 10 eV photoelectron becomes larger than the plasma density scale height, is located around 600–800 km, i.e. nearly at the same altitude as the exobase for neutral atoms where the mean free path of H atoms become equal to the atmospheric density scale height. However, for thermalized (0.25 eV) ionospheric electrons the exobase level is at much higher altitudes (above 2000–3000 km) due to the larger Coulomb collision cross-section at lower energies. As a consequence the suprathermal electrons forming the tail of the velocity distribution escape more easily and they contribute a significant fraction to the upward particle flux and energy flux out the ionosphere and into the plasmasphere. In this way, photo-electrons increase the electron energy density in the plasmasphere. This is likely the reason why the electron temperature during the day is significantly larger in the plasmasphere than in the topside ionosphere as clearly illustrated by the solid line in Fig. 4.4.

The dashed lines in Fig. 4.4 show the diurnal variation of the electron temperature at 1000 km (a) and at the equator (b) along a magnetic flux tube, $L = 5$, when additional heating of the electrons by wave–particle interactions is switched on (Gorbachev, Gombosi and Nagy, 1988). In these model calculations waves are assumed to be generated by the ring current particles as described by Kennel and Petschek (1966). These calculations show that the presence of MHD waves in the plasmasphere can substantially enhance the electron temperature, and hence their escape flux and density distribution in the plasmasphere (see also Section 4.3.1).

4.2.4 Plasmasphere refilling fluxes

The experimental evidence of post-storm refilling of the plasmasphere has been presented in Sections 2.3.3 and 3.5.1. Empty plasmaspheric flux tubes, in the

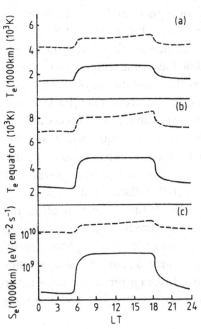

Figure 4.4 Diurnal variation of electron temperature for $L = 5$ at 1000 km altitude (a) and at the equator (b). Panel (c) shows the electron heat flux at 1000 km altitude. The solid and dashed lines correspond to model calculations respectively without and with 'wave–particle' interactions (after Gorbachev *et al.*, 1992).

early stage of their refilling tend to become first populated with suprathermal electrons and ions streaming along magnetic field lines in a nearly collisionless manner as described by Lemaire (1989). This scenario is confirmed by Barakat and Lemaire (1990), Barghouti *et al.* (1990), Wilson (1992) and Miller *et al.* (1993) using direct Monte Carlo simulation methods (see Chapter 5). These simulations demonstrate that in the exobase transition region where the mean free path becomes equal to the background density scale height, the velocity distribution of the outstreaming particles experiences large departures from an isotropic Maxwellian function (Fahr and Shizgal, 1983).

The existence of high speed polar wind flows along 'open' magnetotail field lines and their extension along closed geomagnetic field lines within the plasmasphere were confirmed by Hoffman and Dodson (1980) from the observations made by the ISIS-2 satellite in a polar orbit at 1400 km altitude.

According to GEOS-2 observations analysed by Sojka and Wrenn (1985), the equatorial plasma density increases at a rate of $30\text{--}50\,\text{cm}^{-3}\,\text{day}^{-1}$ during the initial 48 hours after a plasmaspheric erosion event. During the early refilling phase the plasmaspheric ion populations remain field aligned until the equatorial ion density reaches the limit of $\sim 20\,\text{ions/cm}^3$; they become isotropic only when the plasmaspheric density gradually exceeds this threshold. Thereafter, the refilling rate decreases. Furthermore, it appears from the GEOS-2 observations that the plasma flow becomes predominantly corotational only after this density threshold has been reached (Wrenn *et al.*, 1984).

Theoretical model calculations, numerical simulation and experimental observations by whistlers, radio beacon or *in situ* satellite observations indicate that maximum refilling flux, in the topside ionosphere at 1000 km altitude is $1 - 3 \times 10^8\,\text{cm}^{-2}\,\text{s}^{-1}$ (Park, 1970; Farrugia *et al.*, 1989; Higel and Wu, 1984; Song *et al.*, 1988a; Décréau *et al.*, 1986a). The net upward flux during daytime is larger than the downward flux in the nighttime local time sector; this implies a net daily average transfer of ionization from the mid-latitude $(L > 2.5)$ ionosphere to the plasmasphere, even after prolonged periods of quiet (Park, 1974a; Tarcsai, 1985). After eight days of uninterrupted refilling one expects all flux tubes at $L < 4$ to be 'saturated', i.e. to be in hydrostatic equilibrium with the underlying ionosphere. The time required to reach saturation depends of course strongly on the value of L and ranges from $\sim$ two days at $L = 2$ to $\sim$ eight days at $L = 4$ (Park, 1974a) (see Section 2.3.3 for a detailed description).

Since the average interval between magnetic disturbances that affect the plasmasphere is less than eight days, the plasmasphere beyond $L = 4$ rarely, if ever, reaches saturation. Thus in recovery periods the plasmasphere usually consists of two distinct regimes: an inner, or 'saturated', part that is in equilibrium with the underlying ionosphere in a diurnal average sense, and an outer or 'unsaturated' part that is still filling up from below (see Fig. 2.9). The transition between the two regimes typically lies between $L = 3$ and $L = 4$ and shifts to larger L-shell as a function of time (Park, 1974a). Large-scale density irregularities can exist in the unsaturated plasmasphere (Park and Carpenter, 1970) but they tend to smooth out as densities approach the saturation level.

Many competing refilling scenarios have been proposed since the first ones suggested by Banks, Nagy and Axford (1971), Schulz and Koons (1972) and others. According to the early hydrodynamical models flux tube refilling should proceed from top to bottom (i.e. from the equatorial plane to the ionosphere), with hydrodynamical shocks propagating in both hemispheres downward to the ionosphere during an initial stage. Subsequently, a subsonic refilling stage would begin when the shocks would have reached the topside conjugate ionospheres.

The modeling community is divided into two schools. The first group uses hydrodynamic fluid or moment equations with various degrees of sophistication (Banks and Holzer, 1968; Marubashi, 1970; Bailey *et al.*, 1977, 1987, 1990; Barakat and Schunk, 1982; Khazanov *et al.*, 1984; Mitchell and Palmadesso, 1984; Ganguli and Palmadesso, 1987; Ganguli, Mitchell and Palmadesso, 1987; Palmadesso, Ganguli and Mitchell, 1988; Singh, 1988; Rasmussen and Schunk, 1988; Guiter and Gombosi, 1990; Singh and Torr, 1990; Gombosi and Rasmussen, 1991; Singh and Chan, 1992; Demars and Schunk, 1986, 1987a, 1987b, 1991; Cordier, 1994); the second group uses kinetic approaches (Lemaire, 1985, 1989; Yasseen *et al.*, 1989; Barghouti *et al.*, 1990, 1993; Barakat *et al.*, 1990; Wilson, Horwitz and Lin, 1992; Lin *et al.*, 1992; Mitchell, Ganguli and Palmadesso, 1992; Tam, Yasseen and Chang, 1995) which are based on solutions of the Boltzmann equation. These different types of refilling models have been reviewed and compared by Singh and Horwitz (1992), and Singh, Wilson and Horwitz (1994). They will be discussed briefly in Chapter 5.

4.3   Thermal structure of the plasmasphere

As indicated in Chapter 2, Section 2.5.3 whistler observations are not suitable to determine the plasma temperatures in the magnetosphere. *in situ* measurements with electron and ion spectrometers are needed directly to determine the velocity dispersion of particles or their kinetic pressure. From these measurements parallel as well as perpendicular temperatures can be deduced for each particle species in the magnetosphere. The topic of temperature anisotropies in the terrestrial ionosphere and plasmasphere has been comprehensively reviewed by Demars and Schunk (1987b).

We will first consider temperatures of electrons and ions at the lower boundary of the plasmasphere. In the next Subsection (4.3.2) we will review our current understanding of high altitude ion and electron temperatures. The rich variety of pitch angle distributions which have been observed in the plasmasphere are described in Section 4.4.

4.3.1   Temperatures at low altitudes

Using ISIS-1 observations Brace and Theis (1974) found that the electron temperatures in the upper ionosphere peaks within the steep electron density gradient of the middle latitude trough on the nightside (see Section 2.4.3). They showed that this subauroral electron temperature enhancement (SETE) corresponds to the field aligned projection of the high-altitude plasmapause. They suggested also that this temperature peak which is clearly identifiable at all local times down to at least 600 km might be used to follow the movements of the plasmapause under variable geomagnetic conditions. A similar peak in the plasma temperature was found by Titheridge (1976b) in the topside ionosphere from a statistical analysis of the ALOUETTE 1 observations.

Green et al. (1986) and Horwitz et al. (1986a) found that the subauroral electron temperature enhancement in the post dusk sector occurs on field lines which thread the equatorward edge of a steep density gradient inside the plasmasphere but which is located closer to the Earth than the 'whistler knee' in this LT sector. To distinguish both steep plasma density gradients the former has been called the 'inner or vestigial plasmapause' while the latter is called the 'outer plasmapause'.

Brace et al. (1988) have confirmed with the DE-1&2 observations that the SETE temperature enhancements are statistically associated with magnetic field lines colocated with the inner plasmapause density gradient.

When there is a multiple plateau, or multiple plasmapause structure of the plasmaspheric densities as illustrated in Figs. 3.2f or 4.14c and e, the low altitude T_e signatures are magnetically connected to the inner 'vestigial plasmapause knee'.

In addition to these broad temperature peaks, narrow peaks have been observed in the electron and ion temperatures along the plasmapause projection where stable auroral red arcs (SAR) are seen during the recovery phase of large geomagnetic storms (Hoch and Smith, 1971) (see also Section 2.4.2).

Kozyra et al. (1986) have shown that Coulomb collisions between ring current O^+ and plasmaspheric electrons can be an effective heat source for these ionospheric stable auroral red arcs. The importance of Coulomb collisions, even for particles of higher energy, has recently been re-emphasized by Kozyra et al. (1987, 1990), Chandler et al. (1988) and Fok et al. (1993, 1995). They confirm that Coulomb collisions between plasmaspheric electrons and ring current O^+ ions is likely to be the dominant energy source for the elevated electron temperatures (T_e) observed in subauroral F-region, in the vicinity of plasmapause projection.

Ring current particle energy is transferred to cold plasmaspheric electrons via Coulomb interactions. This energy is then transported down along field lines into the ionosphere by heat conduction, as originally proposed by Cole (1965, 1970a, b, 1975) (see also Fok *et al.*, 1993).

At the location of the subauroral electron temperature peak, enhanced densities of thermal electrons exist in the tail of the velocity distribution. They excite atomic oxygen to the ^{1}D state during a collision ($E > 1.96$ eV). The unquenched ^{1}D state emits in about 110 s a red photon at 630 nm which characterizes stable auroral red arcs. Significant emission only results at altitudes where deexcitation collisions with N_2 molecules are infrequent. The intensity of 630 nm emission is dependent on the neutral atmospheric densities of O and N_2, the electron density in the F_2 peak region, the height of the F_2 peak and the electron temperature. Kozyra *et al.* (1990) have recently modeled these processes and studied the seasonal and solar cycle variations of the resulting SAR arcs.

4.3.2 Temperatures at high altitudes

High altitude ion temperatures

As mentioned in Chapter 1, the first determination of the ion temperature T_i in the plasmasphere was based on data from the LUNIK spacecraft in 1960. This temperature was found to be of the order of 10^4 K (Gringauz *et al.*, 1960a). Serbu and Maier (1966, 1967, 1970) used ion and electron analysers with retarding potentials on IMP-2 and OGO-5 to determine T_i and T_e (see Section 3.2.2).

In the first half of the 1970s, the data from PROGNOZ showed the existence of two adjacent zones in the plasmasphere differing by their thermal regimes. The inner region, at $L \leq 3$, was called the 'cold zone' where the ion temperature is rather steady and less than 8×10^3 K. The second zone, initially called the 'hot zone' ('warm zone' is used more often now) is characterized by ion temperatures changing rapidly with time and increasing with L. This discovery has been documented in Section 3.3.1. Sometimes values as large as 10^5 K were observed near the plasmapause as already noted by Serbu and Maier (1966). During prolonged quiet time periods the width of this warm zone can reach $1 R_E$ in the equatorial plane. However, during magnetic quiet conditions the ion temperature can be cold ($< 10^4$ K) in the whole plasmasphere. The electron temperature T_e, increases also with L (see Fig. 3.25, and Décréau, Beghin and Parrot, 1982).

GEOS-1, ISEE and DE-1 data confirmed the existence of two zones: the cold and warm ones. According to DE-1 measurements, T_i in the plasmasphere is about three times higher than in the ionosphere (Comfort, 1986). It is generally argued that interactions of the cold ionospheric plasma with waves generated by an unstable pitch angle distribution of ring current particles contribute to the observed temperature enhancement in the plasmasphere. This approach is justified with reference to a theory of the interaction between waves and particles proposed by Kennel and Petschek (1966). A consequence of this theory is that increasing the density of cold plasma should increase the rate of precipitation of energetic ring current particles.

Part of the observed ion temperature enhancement at the plasmapause can be expected as well from a combination of plasma kinetic transport processes: i.e. (1) the gravitational escape of the 'warmest' ions out of the gravitational potential well; (2) the upward acceleration of the lightest ions by the ambipolar charge separation electric field; and (3) the velocity filtration effect proposed by Scudder (1992a, b) for the 'heating' of the solar corona. The ion temperature at high altitude in the plasmasphere could be increased above the standard exospheric one, when a second ion population with a higher exobase temperature is assumed in addition to the ionospheric Maxwellian background. The former superthermal ions, from being a minor population at low altitude, become more prominent at high altitude because of their larger density scale height. Pierrard and Lemaire (1996) have shown that a non-Maxwellian ion velocity distribution which varies as a power-law instead of an exponential function for superthermal speeds, gives ion temperatures which are increasing with altitude instead of being isothermal in the plasmasphere. If this interpretation proves to be correct, the warm zone of outer plasmasphere can then be compared, in some respect, with the solar corona where a hot plasma of 10^6 K sits also above the much colder photospheric-chromospheric plasma (Scudder, 1992a; b).

Recent studies of ion temperatures observed at geosynchronous orbits with the Los Alamos National Laboratory's magnetospheric plasma analyzer show evidence that the warmer population (4–10 eV) of plasmasphere ions is usually observed in regions of the plasmasphere containing low plasma densities (1–$10 \, \text{cm}^{-3}$). Consequently, cold ion temperatures (1–4 eV) are most of the time associated with high plasma densities ($> 10 \, \text{cm}^{-3}$). These high-density, cold plasma elements are next to, but often interspersed with, less dense, warm trough-like plasma. The cold and warm plasma populations occasionally co-exist (see Fig. 3.32), but are most often separate (Gringauz, 1983; Comfort et al., 1985; Moldwin et al., 1995).

Regions with plasma densities larger than 10 ions/cm^3 have mean energies smaller than 1 eV with little temperature variability, while the low-density and warmer is much more variable or structured.

The anticorrelation between the ion number density and their temperature is attributed by Gary et al. (1994) to the action of the ion-cyclotron instability driven by anisotropies in the hot ion velocity distribution.

The velocity distribution functions are generally postulated to be a Maxwellian or a superposition or two Maxwellians. It is, however, difficult to check this working hypothesis usually adopted by experimentalists to interpret their ion flux measurements and to infer densities, bulk velocities and temperatures. At this state of the art other non-Maxwellian velocity distributions with enhanced suprathermal tails cannot be excluded. On the contrary, for simple theoretical reasons, they should be more often the rule than the exception (see Fahr and Shizgal, 1983).

Comparison of high- and low-altitude ion temperature profiles from DE/ RIMS showed the existence of 'field-aligned' temperature gradients (see Fig. 3.33). Temperature gradients are observed at high altitudes for all $L < 5$, both in the morning and evening sector. But they are not observed in the morning sector for $L > 3$. Comfort (1986) infers that these observations provide evidence that the evening plasmasphere cools down in the darkened ionosphere. Some model calculations by Gorbachev et al. (1992) illustrated in Fig. 4.4 seem to support these conclusions.

Other numerical simulations made with the Field Aligned Interchemispheric Plasma (FLIP) hydrodynamical code developed by Richards et al. (1994) indicate that additional sources of heating, other than the conventional photo-electron heating (Section 4.2.3), are needed in order to produce the high ion temperatures observed in the outer plasmasphere (Jorjio et al., 1985; Newberry et al., 1989). Comfort et al. (1996) argue that within the framework of the FLIP model, indirect heating of the plasmaspheric ions via the photoelectrons trapped at high altitude do contribute, but cannot account for the largest ion temperatures observed. They show that direct heating of the ions either by Coulomb interaction with ring current ions (Kozyra et al., 1987; Fok et al., 1993, 1995) or by wave–particle interaction (Barakat and Barghouthi, 1994a,b; Barghouthi et al., 1994) is required to produce ion temperatures as high as those observed near the plasmapause. Of course, other alternative mechanisms cannot be excluded at this stage.

All measurements of ISEE/PCE, GEOS/ICE and DE/RIMS indicate that H^+ and He^+ ions are in close thermal equilibrium ($T_{H^+}/T_{HE^+} \sim 1$) all over the plasmasphere (Comfort, 1986, 1996; Farrugia et al., 1988).

The latest and most comprehensive survey of ion temperatures in the plasmasphere has been published by Comfort (1996). This statistical study is based on the RIMS observations made with the DE-1 satellite. Both in the morning and evening local time sectors, the average H^+ temperature at high altitude increases with L through the plasmasphere and beyond. Larger statistical deviations in the observed values of T_{H^+} are found in the range $3 \leq L \leq 4$, indicating the uncertainty in the location of the plasmapause and the increased temperature gradient which typically is associated with the plasmapause. In the outer region of the plasmasphere the temperatures tend to level off near 10 000 K for both morning and evening. At lower L ($L < 2$) a pronounced diurnal variation is observed with $T_{H^+} < 3000$ K in the dusk sector and $T_{H^+} > 4000$ K at dawn.

From these observations Comfort (1996) confirmed also that within the plasmasphere, thermal equilibrium generally prevails among the H^+, He^+ and O^+ ions. Comfort's comprehensive study also shows a clear tendency for parallel ion temperature gradients to increase beyond $L = 2.5$ with typical values ranging from 0.05 K/km to 1.0 K/km. The field-aligned gradients of the perpendicular temperature for $L > 3$ is about an order of magnitude higher than the parallel temperature gradients.

Equatorial heating of thermal ions outside and near the plasmapause is supported by DE-1 observations (Olsen, 1982; Olsen *et al.*, 1987). The observations of heated equatorially trapped ions, predominantly H^+ ions, are often associated with observations of equatorial electrostatic noise: i.e. equatorially confined broadband, relatively low-frequency, electrostatic waves. Low-frequency electromagnetic emissions between the proton frequency and lower hybrid frequency are also frequently observed near the Earth's magnetic equator (Russell, Holzer and Smith, 1970; Gurnett, 1976). Heating of 1–10 eV thermal ions by these waves has been proposed and described by Curtis (1985).

Often the waves are observed to have a funnel-shaped appearance on a frequency–time spectrogram, particularly when the spacecraft passes latitudinally through the magnetic equator. Boardsen *et al.* (1992) have shown that the funnel-shaped emissions are consistent with a generation by protons with ring-type velocity space distributions.

High altitude electron temperature

Besides the heating of plasmaspheric electrons by Coulomb interactions with ring current O^+ ions, collisionless damping of Alfvén waves and right-hand polarization fast magnetosonic waves can also contribute to the heating of thermal electrons in the plasmasphere. This latter heating mechanism, by wave–particle interactions, has been re-evaluated by Konikov *et al.* (1989). Assuming a simplified spectral distribution for the resonating ion-cyclotron waves, they calculated the effect of these waves on the electron velocity distribution. Their results were then compared with the effects of Coulomb interactions based on the formulation of the 13-moment approximation (Schunk, 1975, 1977).

For wave intensities comparable with those observed in the outer plasmasphere, Konikov *et al.* (1989) estimate that the main contribution to the heating of the thermal plasmaspheric electrons is by Alfvén waves. They calculated that this heating rate due to ion cyclotron (IC) wave resonant interaction exceeds in general that resulting from superthermal electrons interacting with the ambient thermal electrons (Krinberg and Tashchilin, 1984).

The downward heat flow from the plasmasphere into the topside ionosphere has also been evaluated. The IC-wave particle interaction seems to be the largest contributor to this downward conduction flux ($\sim 10^{11}$ eV cm^{-2} s^{-1}) (Konikov *et al.*, 1989). Chernov, Khazanov and Tanygin (1990) recalculated this flux by including the effect of wave refraction and found that the heating by ion-cyclotron waves is most intense away from the geomagnetic equator. They argue that this can lead to a significant rise of the electron temperatures in the plasmasphere and ionosphere.

There is little information concerning possible effects of wave–particle interactions on low-energy electrons in the near-equatorial outer plasmasphere. However, Olsen (1981) demonstrated the existence of perpendicular heating of electrons in this region. Indeed, the pitch angle distributions of these electrons have a peak at 90°, like the ions; their temperature range is between 100 and 200 eV. See also Wrenn *et al.* (1984). Trapped populations for both electrons and ions were observed in association with equatorial electrostatic noise in the frequency range from 20 to 200 Hz.

Norris *et al.* (1983) have observed increases in low-energy field-aligned electron fluxes in the equatorial region which they associated with the presence of ion cyclotron waves with frequencies just above the local helium cyclotron frequency. This brings us to the topic of pitch angle distributions in the plasmasphere (see also Abe *et al.*, 1990, 1993).

4.4   Pitch angle distributions

When Coulomb collisions are frequent enough in the plasmasphere and plasmatrough, the pitch angle distribution of low-energy ions is expected to be isotropic. On the other hand, when these collisions and wave–particle interactions are not important, the pitch angle distribution can be anisotropic, either field-aligned or on the contrary, peaked at 90° pitch angle.

Figure 4.5a shows retarding potential analyzer (RPA) spectrograms for 90° pitch angle He^+ ions observed with the retarding ion mass spectrometer (RIMS) on DE-1 satellite for energies below 50 eV. Figure 4.5b displays the corresponding spin angle count rates distribution versus Universal Time and versus location in the magnetosphere. The white lines near the center and edges indicate, respectively, the maximum and minimum magnetic pitch angles corresponding to the closest aperture approach to the parallel and antiparallel direction of the magnetic field. Between ~ 1955 UT and 2045 UT DE-1 encountered a region of dense but structured plasma as the satellite moved toward the equator. Figure 4.5c shows the electron density profile from this region, as derived from the sweep frequency receiver (SFR) spectrograms for the same orbit.

From Fig. 4.5b, Carpenter *et al.* (1993) inferred that the plasma in this region has a rammed distribution (i.e. ordered by spin phase instead of the magnetic field direction) which implies that the pitch angle distribution is primarily isotropic. However, as indicated in the energy spectrograms of Fig. 4.5a, the temperature of the He^+ ions varied widely within this plasma density enhancement. It can be seen that the pitch angle distributions and energy spectrograms exhibit more variation than the value of the electron density, n_e, shown in Fig. 4.5c. This enhancement has been identified by Carpenter *et al.* (1993) as either a detached plasma element or a tail like structure beyond the main plasmasphere in the afternoon local time sector.

As DE reached $L = 4.7$ at 2130 UT, much higher count rates were recorded, extending above 10 eV. A peak in the count rate was then observed at 90° in the spin-time plot of Fig. 4.5b. The maxima at $+90°$ and 90° RAM angles identify warm anisotropic plasma trapped near the magnetic equator.

Between 2230 and 2330 UT when the spacecraft was moving along the magnetic field line $L = 4.7$, a bidirectional field aligned pitch angle distribution is observed for the He$^+$ ions (see Fig. 4.5b). However, at the same time a trapped pitch angle distribution was detected for the H$^+$ ions (not shown here). This indicates that the physical processes which determine the ion pitch angles can

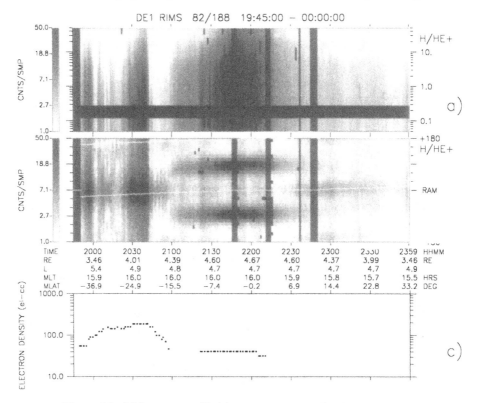

Figure 4.5 (a) Low-energy He$^+$ ion energy spectra (0–50 eV) versus time as observed by RIMS instrument along an orbit of DE-1 (2000–2400 UT on 7 July 1982); (b) RIMS data of pitch angle distribution versus time on the same orbit; (c) plot of $\log n_e$ versus time for the same orbits inferred from the observations of the upper hybrid resonance (UHR) in the data of the sweep frequency receiver (SFR) of the University of Iowa (Courtesy, B. L. Giles, 1993).

clearly be charge and mass dependent, and lead to a complex intermingling of the different ion populations. Often conical superthermal ion beams, with a maximum flux at a pitch angle different from $0°$ and $90°$, are detected in the plasmatrough during magnetically disturbed times (Baugher et al., 1980).

4.4.1 Field–aligned pitch angle distributions

Monte Carlo simulations by Barghouti et al. (1990, 1993) and Wilson, Horwitz and Lin (1992) confirm that Coulomb collisions should not a priori be overlooked and neglected in so-called collisionless magnetospheric plasmas (see also Horwitz et al., 1993). The usual tendency is to emphasize pitch angle scattering and heating of ions by wave–particle interactions only. But, for this latter mechanism to be adequate the wave amplitudes must be large enough and the wave spectrum wide enough to isotropize and Maxwellize over the whole range of particle energies. These conditions for wave–particle interactions to be prevailing are not necessarily met in the plasmasphere all the time.

Let us now come back to observations of pitch angle distributions in the plasmasphere. Décréau et al. (1986a) found DE 1 observations that indicate a phase of refilling of flux tubes with a double field-aligned structure in the vicinity of the equator. These observations do not support the earliest refilling scenarios (i.e. refilling from top to bottom) since in this case the plasma would be nearly isotropic in the 'high density equatorial region' whereas it should be field-aligned at non-equatorial latitudes (i.e. on the low-altitude sides of the presumed downward propagating shocks models). The Décréau et al. observations are consistent, however, with Lemaire's (1989) kinetic refilling scenario. Indeed, according to this scenario, the plasma density is always smallest in the equatorial region. As a matter of consequence, it is there that Coulomb collisions are less efficient to destroy a cigar-like field-aligned pitch angle distribution. On the contrary, at lower altitudes where the density is larger, Coulomb collisions are more frequent and pitch angle distributions tend to be isotropic.

Olsen et al. (1985, 1987) have found both an isotropic background and a low-energy field-aligned flowing population of light ions; the latter field-aligned population is observed under eclipse conditions. The isotropic population is primarily H^+ ions, with 10% He^+ and 0.1–1% of O^+. The field-aligned streaming light ions have densities of a few cm^{-3}; their thermal energy is 0.25–1 eV and their bulk velocities 5–20 km/s. This corresponds to a Mach number of 1 for H^+, and > 1 for He^+. Watanabe, Whalen and Yau (1992) also found from EXOS-D measurements rather large bulk speeds for H^+ (> 12 km/s), for He^+ (7 km/s) and 3 km/s for O^+ ions.

Watanabe, Oyama and Abe (1989) observed anisotropic electron temperatures and non-Maxwellian features in the electron velocity distribution even at low altitudes in the mid-latitude ionospheric trough with $T_{\parallel}/T_{\perp}$ up to 2.

4.4.2   Pancake pitch angle distributions

Low-energy plasma observations at the equator in the vicinity of the plasmapause often reveal distributions with peaks at 90° pitch angle, i.e. pancake
or trapped distributions as illustrated in Fig. 4.5b (Wrenn *et al.*, 1979). The
characteristic energy of these equatorially trapped ions is several eV (Horwitz
and Chappell, 1979; Comfort and Horwitz, 1981; Olsen *et al.*, 1987; Sagawa *et
al.*, 1987 (see also Section 3.4.1)). Since these ions originate in the ionosphere
they must have been subjected to perpendicular heating and pitch angle scattering. It is generally assumed that both effects are due to their interactions
with waves. Although wave–particle interactions are invoked to explain these
anisotropic pitch angle distributions, it cannot be excluded that other acceleration mechanisms, like weak electrostatic double layers, might contribute to
produce the observed pitch angle deflection and energization of equatorially
trapped ions.

Sagawa *et al.* (1987) find the peak occurrence frequencies, for such pancake
pitch angle distributions for low-energy H^+ ions, to be located within 5° of the
magnetic equator and between $L = 3.5$ and 5. Olsen *et al.* (1987) find H^+ and
He^+ ions of $\sim 50\,\text{eV}$ confined within only 3° of the magnetic equator. Their
perpendicular temperature ($\sim 10\,\text{eV}$) is 10 times larger than their parallel temperatures ($T_{\parallel} \sim 0.5\text{--}1.0\,\text{eV}$). Figure 4.6b illustrates the enhanced suprathermal
tail in the perpendicular velocity distribution. Figure 4.6a indicates that
further away from the equatorial plane perpendicularly heated ions are almost
absent.

Pancake and field-aligned distributions have also been observed in the electron pitch angle distribution. Norris *et al.* (1983) found that the latter were
associated with the presence of ion cyclotron waves in the equatorial region.

4.4.3   Conical pitch angle distributions

Besides field-aligned or pancake pitch angle distributions, conical pitch angle
distributions have also been observed in the outer plasmasphere and middle
magnetosphere (Klumpar, 1979; Sagawa *et al.*, 1987; Peterson *et al.*, 1992).
Here again wave–particle interactions are generally invoked to explain their
formation; in this case the ions are assumed to have been heated at lower
altitude, along the magnetic field lines. Several theories of ion heating perpendicular to the geomagnetic field have been reviewed and assessed by André
and Chang (1992). They suggest that the majority of ion conics are generated
over extended altitudes above 3000–4000 km by local ion cyclotron resonance
heating. At lower altitudes, energization by the turbulence around the lower
hybrid frequency is an important heating mechanism, according to these
authors.

Electrostatic potential barriers of the order of 1–50 V can also energize, deflect or reflect ions flowing out of the ionosphere or toward the Earth. This alternative mechanism can produce the observed conics and field-aligned light ion flows without necessarily invoking wave turbulence. Inclined or field-aligned weak double-layers are formed at the edges of equatorially trapped plasma clouds. At the interface between two plasma regions weak double-layers can form as a result of the thermoelectric charge separation. This charge separation is the result of the different inertia, thermal speeds and pitch angle distributions of the electrons and ions (see Alfvén and Fälthammar, 1963; Persson, 1966; Lemaire and Scherer, 1978; Carlqvist, 1995; see also Section 5.4.5).

4.4.4   Isotropic pitch angle distributions

In the innermost flux tubes which are 'saturated' (i.e. filled up), Coulomb collisions are frequent enough to establish and to maintain nearly isotropic ion and electron velocity distributions at energies below 1–5 eV. The characteristic plasmaspheric temperature in these saturated flux tubes are 0.5–1 eV, which

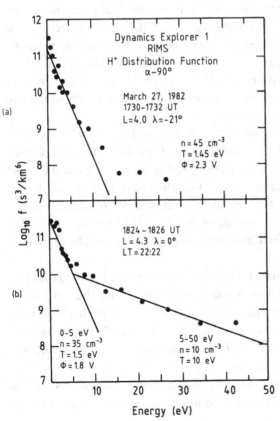

Figure 4.6   Energy distribution functions of H^+ ions measured perpendicular to the magnetic field (i.e. at 90° pitch angle) at 21° geomagnetic latitude i.e. well away from the equatorial plane (a); the energy distribution shown in panel (b) was observed by the RIMS instrument near the magnetic equator as DE-1 was moving nearly along a constant L-shell $L = 4.0$–4.3. The data are indicated by dots, and Maxwellian distribution fits with the indicated parameters for density, temperature and spacecraft potential are shown with straight lines. The equatorial distributions display a heated superthermal population ($T \sim 10$ eV) in addition to the cold (T ~ 1.5 eV) ion population seen at both locations (after Olsen *et al.*, 1987).

is, however, a factor 2–4 times larger than the underlying ionospheric temperature.

To conclude these two last Sections on plasmaspheric temperatures and pitch angle distributions, it can be said that the ions in the inner plasmasphere are in near thermal equilibrium as a result of local Coulomb collisions. However, flux tubes beyond $L = 3$–4, which are most of the time in a phase of refilling, in between two successive substorms events, are rarely in thermal and diffusive equilibrium. The low densities in these flux tubes imply slow pitch angle scattering of upward streaming light ions and even slower energy transfer between these ions and the background ones. Wave–particle interactions should contribute also to maintain a near-isotropic and Maxwellian velocity distribution function, but in many instances both dissipation mechanisms lack efficiency. As a consequence, the thermal ions and electrons remain anisotropic (field-aligned, pancake, conical) and non-Maxwellian in particular during the early phases of refilling process.

4.5 Ion composition

The most abundant ions in the plasmasphere are protons. This is why the region beyond the topside ionosphere has often been called the protonosphere, in the 1960s before the word plasmasphere was introduced. The second most important ion species in the plasmasphere is singly ionized helium. Although the He^+ concentration is usually of the order of 5–10%, cases are mentioned, in Sections 3.2.3 and 3.4.1 of Chapter 3, where the concentration of He^+ was almost equal to that of the protons (see below).

From the first mass spectrometer measurements performed by Taylor et al. (1965a,b, 1968a,b) onboard OGO-1 and OGO-3, it was found that the plasmasphere contains a few percent ($\sim 5\%$) of He^+ (see Fig. 3.8a). Subsequent OGO-5 observations enabled identification of O^+ ions (less than 1%) in the plasmasphere. The discovery of D^+, He^{++} and O^{++} ions in the plasmasphere was made by the GEOS/ICE experiment described by Balsiger et al. (1976). The results of these mass-spectrometric measurements were presented by Geiss et al. (1978).

A statistical study covering data for over a one year interval showed that He^+ is the second most abundant ion in the plasmasphere, and that its relative concentration varies in a wide range: $N(He^+)/N(H^+)$ varies from 1% to more than 50%, with the most frequent values between 2% and 6% (Farrugia et al., 1988, 1989). These comprehensive GEOS results are not inconsistent with earlier data from OGO-3 and OGO-5.

The most surprising result obtained during the DE-1 mission was that the concentration of He^+ ions sometimes becomes comparable to that of protons.

The average value of $N(He^+)/N(H^+)$ is about 0.2 (Newberry et al., 1989). Such high relative abundances of He^+ ions were first considered to be inconsistent with their relative concentrations at low altitude. They are also significantly higher than the values for $N(He^+)$ deduced from the earlier GEOS low-latitude measurements. The large relative abundances of He^+ ions are inconsistent with current hydrostatic and exospheric models: however, the FLIP model of Richards and Torr (1985) seems to be able to account for such large values as being due to the higher level of solar activity (solar maximum) in 1981 during the DE measurements. Indeed, during the years of solar maximum the solar UV radiation is enhanced and more photoelectrons are produced. Newberry et al. (1989) argue that high values for the He^+/H^+ ratio can be obtained by incorporating enhanced trapping of these photoelectrons in plasmaspheric flux tubes leading to enhanced plasmaspheric heating. For a complete description of the FLIP model see Richards et al. (1994).

Doubly charged He^{++} ions have been reported by Young et al. (1977). With the mass spectrometer of DE-1/RIMS the minor N^+ and N^{++} constituents were identified in the plasmasphere for the first time. The N^+ fluxes are 1000 times lower than the flux of O^+ ions; the N^{++} fluxes are 100 times lower than the O^{++} fluxes (Chappell, 1982). The ratio of $N(O^{++})/N(O^+)$ occasionally approaches unity in the plasmasphere, while it is of the order of 3×10^{-3} in the underlying mid-latitude topside ionosphere (Hoffman et al., 1974; Young et al., 1977; Geiss et al., 1978; Horwitz, 1981; Farrugia et al., 1989).

Different mechanisms have been proposed for the formation of the heavy ion enhancement in the outer plasmasphere. Beside the 'geomagnetic mass spectrometer' effect mainly invoked in polar regions and in the magnetotail (Lockwood et al., 1985; Waite et al., 1985), ionospheric electron heating has been suggested by Horwitz et al. (1986b). This latter scenario is illustrated in Fig. 4.7. The outer plasmasphere electrons interact with the more energetic ring current particles either through Coulomb collisions or through wave–particle interactions. The conduction of heat from the outer plasmasphere to the topside ionosphere is responsible for heating of ionospheric ions and electrons, resulting in increases in their scale heights. Raising the plasma scale heights results in larger densities at high altitudes, i.e. in the plasmasphere. Since heavy ions dominate the plasma composition of the ionosphere, the outer magnetic shells where wave–particle interaction with ring current particles is assumed to take place, should contain relatively higher densities of heavy ions than the magnetic shells at lower L. This could possibly explain the O^+ and O^{++} ion density peaks seen in Fig. 4.8 at large L-values, i.e. close to the plasmapause (Roberts et al., 1987; Horwitz, Comfort and Chappell, 1990).

The second mechanism for accelerating heavy ions from the ionosphere into the outer plasmasphere is through thermal diffusion, which results from the field-aligned temperature gradient established in a H^+ dominated plasma by an equatorial heat source. This mechanism was discussed by Geiss and Young (1981) in an attempt to explain that $N(O^{++})/N(O^{+})$ is much larger in the plasmasphere than in the topside ionosphere.

The results illustrated in Fig. 4.8 seem to support the former scenario: i.e. heating of the plasmaspheric and ionospheric plasma creates the plasmaspheric heavy ion density enhancements through expansion of the heavy ion scale heights. Indeed, there is a rather striking resemblance between the peak in the O^{+} temperature in the topside ionosphere (4.8b) and the plasmaspheric heavy (O^{+}, O^{++}) ion density peaks in the plasmasphere (4.8a).

Singh and Horwitz (1992) suggested that the buildup of O^{++} should be examined and compared with that of O^{+} ions during refilling events. Indeed, owing to thermal diffusion of O^{++} in a background of O^{+} ions, the former should buildup more rapidly in the plasmasphere than the latter (Young *et al.*, 1977; Geiss and Young, 1981). Chandler *et al.* (1987) contested this theoretical prediction.

Note also from Fig. 4.8: (i) the coincidence of location of the light ion density gradient and ion temperature peak in the ionosphere (4.8b) and plasmasphere (4.8a); (ii) the 'inversion' of the He^{+}/H^{+} density ratios between the ionosphere and plasmasphere; (iii) the plasmaspheric H^{+} ion temperature ($T_{H^{+}}$ in 4.8a) is a factor 3 larger than the ionospheric O^{+} temperature (4.8b), but both increase gradually with invariant latitudes between 47° and 60° invariant latitude.

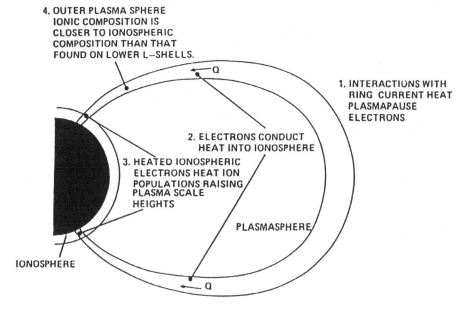

Figure 4.7   Schematic illustration of a chain of mechanisms resulting in the enhancement of O^{+} and O^{++} in the outer plasmasphere (after Horwitz *et al.*, 1986b).

The $O^{++}, O^+, H^+, He^+, N^+$ ion composition in the high-altitude ionosphere and its changes during magnetic storms has also been examined by Förster *et al.* (1992), Boskova *et al.* (1993a,b) and Smilauer *et al.* (1996) based on the near-polar ACTIVNY satellites.

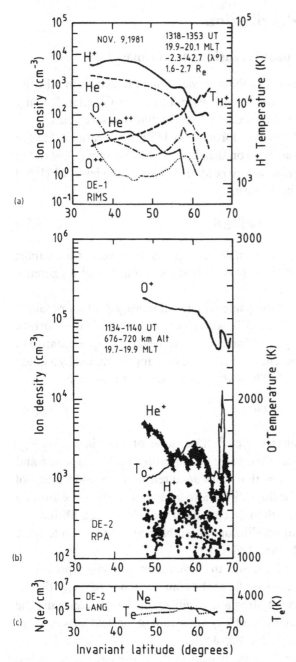

Figure 4.8 (a) Plasmaspheric ion densities together with the H^+ temperature versus invariant latitude in the dusk–evening sector. (b) O^+ temperature and ion densities at low altitude measured simultaneously in the same local time sector. (c) Electron temperature and density from the Langmuir probe on DE-2 (after Horwitz *et al.*, 1990).

The results quoted above indicate that despite the bits and pieces we have learned concerning the ion composition in the plasmasphere, we are far from a clear and comprehensive picture.

4.6   Plasma density distribution

4.6.1   Equatorial density profiles inside the plasmasphere

The equatorial electron density distribution was first obtained from whistler measurements (see Chapters 1 and 2). But as indicated in Chapter 3, many density profiles have also been obtained by *in situ* satellite measurements.

Berchem and Etcheto (1981) deduced from their ISEE observations that the radial distribution of equatorial electron density varies like a power law of the radial distance R, close to Earth. Moore *et al.* (1987) and Soloviev *et al.* (1989) have adopted this power law in the form

$$\log n_e/N_o = -[3.5 + 0.5 \sin \phi]\log R \tag{4.3a}$$

in their empirical models of the electron density n_e; ϕ is the magnetic local time; N_o is a reference density determined to fit boundary conditions at a reference level.

Figure 4.9 shows a serie of 11 dayside equatorial density profiles obtained with the sweep frequency receiver (SFR) along near-equatorial ISEE-1 orbits. These profiles selected by Carpenter and Anderson (1992) are representative of very quiet magnetic conditions. They appear to be well approximated by a linear relation between $\log_{10} N_{eq}$ and L. The least squares linear fit is given by

$$\log_{10} N_{eq} = -0.3145L + 3.9043 \tag{4.3b}$$

where N_{eq} is expressed in electrons/cm^3. The slope of this line, $d \log N_{eq}/dL = -0.3145$ is more gradual than values deduced by Park, Carpenter and Wiggin (1978) from whistler data that were acquired under a wide range of magnetospheric conditions. The slope differs also from that given by the formula (4.3a). It is also steeper at large L than that expected for diffusive equilibrium, i.e. when hydrostatic and thermal equilibrium is established with the underlying ionosphere (Angerami and Thomas, 1964).

The results shown in Fig. 4.9 confirm that even after prolonged periods of quiet, at 'saturation', a plasmaspheric flux tube contains fewer particles than it would in the case of isothermal hydrostatic equilibrium. This is especially the case when the ion temperatures T_{O^+} and T_{H^+} at the foot of the field lines increase with invariant latitude as indicated in Fig. 4.8.

Empirical models for the total plasma density in the plasmasphere have been developed by Rycroft and Alexander (1969), Rycroft and Jones (1985, 1987), Chiu *et al.* (1979), Gallagher and Craven (1988), Carpenter and Anderson (1992). The Chiu *et al.* (1979) model is a 3-D equilibrium model for e^-, H^+, O^+ and He^+ in the plasmasphere for $L \leq 5$. It is based on numerical integration of 'modified' hydrostatic equations along dipole magnetic field lines with given density and temperature distributions at an arbitrarily chosen reference altitude of 500 km. The temperature distribution of each constituent is specified to fit both ionospheric and high altitude measurements. Although their 'modified' hydrostatic equilibrium equations are not the standard ones, they obtain solutions which are in reasonable agreement with the observations. The 'modified' momentum balance equation which they employ in their model does not truly correspond to the standard equation of hydrostatic equilibrium, but it approximates hydrodynamic flow with an uniform bulk velocity along magnetic flux tubes. It also approximately simulates the field-aligned flow in flux tubes for which the volume is continuously expanding, as is expected in the case of a slowly expanding plasmaspheric wind (Lemaire and Schunk, 1992, 1994).

The Chiu *et al.* (1979) semi-empirical model of the (O^+, H^+, e^-) densities and

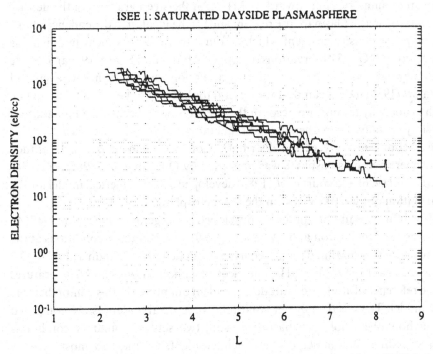

Figure 4.9 Equatorial density in the plasmasphere obtained from ISEE-1 observations during the dayside hours 09–15 MLT. In each of the 11 'saturated' density profiles illustrated, the plasmapause, if identiable, was beyond $L = 5$, in two cases it was beyond $L = 8$. Only profile portions interior to the plasmasphere are shown (after Carpenter and Anderson, 1992).

temperature is sufficiently flexible to fit a wide range of low- and high-altitude observations of ion and electron densities or temperatures in the plasmasphere and topside ionosphere. Unfortunately, it requires numerical integration to determine the field aligned density profiles. This makes it more difficult to implement in a computer code than, for instance, the Carpenter and Anderson (1992) analytical model. On the other hand Chiu *et al.* (1979) offer a 3-D model of the whole plasmasphere for $L < 5$, while Carpenter and Anderson (1992) give only a 2-D electron density profile for the equatorial region of the magnetosphere.

The Gallagher, Craven and Comfort (1988) model is an analytical 3-D model based on a statistical analysis of DE-1/RIMS H^+ ion observations. Grouping these observations according to K_p, MLT, geomagnetic latitude λ and L, they used empirical mathematical functions of L, λ, MLT to fit the values of log $N(H^+)$ observed in the plasmasphere and in the plasmapause region. Despite the limited coverage of the database used in this compilation Gallagher, Craven and Comfort (1988) have been able to model (1) the L-dependence of log $N_{eq}(H^+)$, which they found to be a linearly decreasing function of L; (2) the local time distribution of L_{pp}, the equatorial plasmapause position, which they found to have an asymmetric shift toward 21 MLT; (3) the steep gradient of the density distribution in the plasmapause region; and (4) the local time dependence of this plasmapause density gradient which is found to be steepest near midnight, as known from OGO 5 observations (Chappell, Harris and Sharp, 1970a). Despite its somewhat complicated analytical formulation, the Gallagher, Craven and Comfort (1988) model can be a very useful tool. It would be valuable to update such a model by taking into account the ISEE or more recent CRRES data, and possibly simplify the mathematical fit functions which have been used.

A theoretical study of the distribution of ionization in the high-latitude ionosphere and plasmasphere was presented by Quegan *et al.* (1982) and Millward *et al.* (1996). Delcourt *et al.* (1989) developed a 3-D numerical model for the distribution of 1.0–20 keV ions in the magnetosphere.

The diffusive equilibrium theory developed by Angerami and Thomas (1964), has been used by Rycroft and Alexander (1969) with boundary conditions at the reference level of 900 km determined from Alouette I and II, Injun 3 and OGO 2 observations. The O^+, He^+, H^+ and electron concentrations and temperatures at the reference altitude are latitude-dependent quantities in this comprehensive 3-D model. The electron density and relative ionic abundances are computed throughout the whole plasmasphere using two sets of boundary conditions corresponding to summer day and winter night, i.e. the two most extreme conditions. Unfortunately this model is not well known, since it was only distributed as a preprint of a paper presented at the COSPAR meeting in 1969. It has many interesting and realistic features which compare favorably with low-altitude as well as high-altitude measurements. Furthermore, since it is portable and simpler than the previous ones, it might be useful to consider updating it with all the additional data currently available at NSSDC (e.g. the Goddard Comprehensive Ionosphere Database (GCID)), and at other World Data Centers or laboratories around the world.

The Carpenter and Anderson (1992) model is based (1) on a large number of high-resolution measurements from the radio wave sweep frequency receiver (SFR) which were sampled along the near-equatorial orbits of ISEE-1, and on (2) published whistler measurements made over a 7.5-year period from 1957 through 1964 (Park, Carpenter and Wiggin 1978). The dayside equatorial density inside the plasmasphere in this model is given by equation (4.3b). A diurnal variation of N_{eq} in the range of ± 10–15% around the 24-hour average, with a minimum after midnight LT, was reported by Park, Carpenter and Wiggin (1978) for $L \simeq 3$. But, beyond $L = 3$, they could not identify a systematic diurnal variation because of the large day-to-day variations associated with magnetic disturbances (see Section 2.5.1).

Time-dependent theoretical models by Rasmussen and Schunk (1990) show that the H^+ density of the plasmasphere should indeed be relatively independent of local time for $L > 2$. ISEE-1/SFR data for $L > 2.5$ did not reveal notable departures from the average dayside profile given by equation (4.3b) and illustrated in Fig. 2.22 by the straight line for $L < 4$–4.5. Therefore, Carpenter and Anderson (1992) propose to use equation (4.3b) as a 'preliminary estimate' of the 'saturated' plasmasphere throughout the 00–15 MLT sector, with the expectation that it may be extrapolated between 15–24 MLT and improved when more statistical data have been assembled.

Although, diurnal variations are small inside the magnetosphere, significant annual, semi-annual, and solar cycle variations have been identified from whistler studies. Carpenter and Anderson (1992) found it convenient to incorporate these secular variations by adding new terms to the right-hand side of equation (4.3b):

$$
\begin{aligned}
\log N_{eq}(L,d,R) = {}& -0.3145L + 3.9043 \\
& + 0.15\cos[2\pi(d + 9)/365]\exp[-(L - 2)/1.5] \\
& - 0.5\cos[4\pi(d + 9)/365]\exp[-(L - 2)/1.5] \\
& + (0.00127R - 0.0635)\exp[-(L - 2)/1.5]
\end{aligned}
\tag{4.4}
$$

The third term of the right-hand side corresponds to the annual variation when d refers to the day number: the annual variation has been found to be largest within the inner plasmasphere, with a maximum to minimum ratio of 2 or more near $L = 2.5$. There are indications that the amplitude of this annual variation is longitude dependent with a maximum at 75°W (Park, Carpenter and Wiggin, 1978; Clilverd, Smith and Thomson, 1991). This longitude dependence has not been taken into account in equation (4.4). Furthermore, the December-to-June ratio at $L \sim 2.5$ has been taken to be 1.6.

Guiter *et al.* (1995) concluded from their model calculations that the most likely cause of the observed annual variations in the plasmasphere are variations of ionospheric O^+ ion density, brought about by changes in the composition of the neutral atmosphere. These model calculations did not indicate a similar correlation between ionospheric and plasmaspheric H^+ density.

The multiplicative exponential factor in equation (4.4) reflects the observed amplitude falloff with increasing L. The fourth term in equation (4.4) corresponds to the semi-annual variation with equinoctial maxima as noted by Carpenter (1962b) and reported by Bouriot, Tixier and Corcuff (1967). The fifth term represents the solar cycle variation of the equatorial electron density when R is the 13-month average sunspot number. Indeed these whistler studies suggest that, at $L \sim 2$, the ratio of the density at high sunspot number to that at solar minimum is ~ 1.5. In equation (4.4) a reduced amplitude of ~ 1.2–1.3 is postulated at $L = 3$ as suggested by the data of Park, Carpenter and Wiggin (1978). See also Tarcsai, Szemerédy and Hegymegi (1988).

4.6.2   Equatorial density profiles in the plasmatrough

Analytic expressions have also been given by Carpenter and Anderson (1992) for the density profile in the plasma trough. In all cases identified in the ISEE-1 data as plasma trough profiles, the density levels beyond $L = 3$ were a factor of 5 or more below the saturated plasmasphere levels on the dayside and a factor of 10 or more below on the nightside. In view of the evidence from whistlers that the flux tube electron total content beyond the dayside plasmapause tends to be roughly constant with L (Angerami and Carpenter, 1966) and in view of the roughly L^4 variation of flux tube volume, an L^{-4} variation is expected for $N_{eq}(L)$. But from the selected profile a slightly steeper logarithmic slope $(d \log N_{eq}/dL < -4)$ was found for $L < 6$. Beyond $L = 6$, the profiles tended to decay less steeply and to approach a constant value near 1 el/cm^3 beyond $L = 8$.

The equatorial density in the trough region exhibits day/night differences. The day/night differences in trough density are evident because of the depleted nature of the flux tube electron content. Two empirical linear variations in scale of the trough profiles with time were assumed, a slower one across the nightside to 06 MLT and a faster one for the local time sector between 06 MLT and 15 MLT:

$$N_{eq}(L) = (5800 + 300t)L^{-4.5}$$
$$+ \{1 - \exp[-(L-2)/10]\} \qquad \text{for } 00 < t < 06 \text{ MLT} \qquad (4.5a)$$

$$N_{eq}(L) = (-800 + 1400t)L^{-4.5}$$
$$+ \{1 - \exp[-(L-2)/10]\} \qquad \text{for } 06 < t < 15 \text{ MLT} \qquad (4.5b)$$

The second term in (4.5) is intended to approximate the decrease in decay rate and approach a constant value beyond $L = 6$. Evaluated at $L = 6.6$, these expressions are in good agreement with the data reported from the relaxation sounder on GEOS-2 by Higel and Wu (1984).

The two segments beyond $L = 4.2$ and 4.8 in Fig. 2.22 illustrate the equatorial density profiles in the plasma trough in the nightside at 02 MLT and in the dayside at 12 MLT. The amount of ISEE data thus far was insufficient for identifying possible secular changes in the plasma trough such as annual and solar cycle variations.

Moore et al. (1987) constructed a 3-D plasma density distribution with an exponential dropoff in the plasmatrough beyond a variable plasmapause radial distance. The exponential drop of density versus altitude continues until an asymptotic number density is reached which is representative of near-Earth plasmasheet values at night ($1.0 \, \text{cm}^{-3}$), with a somewhat higher value at noon.

4.7 The shape of the equatorial plasmapause

In Section 2.3.1 the K_p dependence of the plasmapause radius has already been discussed, based on whistler observations. From geosynchronous satellite observations, it has been confirmed by Higel and Wu (1984) and by McComas et al. (1993) that the position and extent of the plasmapause varies characteristically with geomagnetic activity. The plasmapause moves generally outward with decreasing geomagnetic activity, while the 'plasmaspheric bulge' moves sunward with increasing activity level. GEOS-2 and other satellite observations indicate that during quiet times, when $K_p < 2$, the geosynchronous orbit ($R = 6.6 R_E$) can lie entirely within the plasmasphere. During more active times only the afternoon to evening portions of these geostationary satellite orbits are typically within the plasmasphere.

4.7.1 The equatorial distance of the plasmapause

Among the 1977–1983 ISEE data, plasmapause profiles were selected according to the criterion that a drop by a factor of 5 or more occurs within $\Delta L \leq 0.5$. The L-value of the plasmapause, L_{ppi}, was determined by Carpenter and Anderson (1992) as the L-value of the last measured point prior to a steep plasmapause fall off.

As already pointed out in Section 2.3.1 the statistical relationship (2.1) deduced from whistler observations is virtually identical to that deduced by Carpenter and Anderson (1992) from the ISEE/SFR:

$$L_{ppi} = 5.6 - 0.46 K_{pmax} \tag{4.6}$$

where K_{pmax} is the maximum value of K_p in the preceeding 24 hours, although in the case of periods centered at 09, 12 and 15 MLT, respectively, the K_p values have been ignored for one, two, and three immediately preceding three-hour periods to account for observed delays in the dayside response to enhanced convection activity in the formative nightside region (Chappell, Harris and Sharp, 1971a; Décréau et al., 1982, 1986a).

It was found from ISEE/SFR data that the relationship (4.6) did not change significantly when the data were separated according to night, day and dawn sector values. This indicates that, irrespective of the nature of the process controlling the establishment of a plasmapause in the nightside profile, the equatorial plasmapause corotates nearly along a circular trajectory between the midnight sector where it is formed until at least 15 MLT in the afternoon sector. The ISEE/SFR also confirm OGO 5 observations showing that the nightside position tends to be communicated within 24 hours to much of the dayside. This is consistent with Lemaire's (1975, 1985) model of formation of the plasmapause. Indeed, according to this theory, the plasmapause is formed by detachment of cold plasma patches in the post-midnight sector. They drift away to the magnetopause like icebergs drift away from the ice-pack. The newly formed plasmapause corotates along the electric equipotential surfaces which are almost circular at $L < 3$–4.

The differences in L_{ppi} between day and night is less than $\sim 0.5R_E$ according to ISEE data. These data also seem to contradict the noon–midnight asymmetry in the plasmasphere radius reported from PROGNOZ and PROGNOZ 2 (Gringauz and Bezrukikh, 1976). However, it should be noted that Gringauz and Bezrukikh define the plasmapause as the outer edge of the plasmapause region (i.e. at the low density limit of the plasmasphere) while Carpenter and Anderson define L_{ppi} by the inner edge of the plasmapause region (i.e. at its high density limit). Since the width of the plasmapause is local time dependent and larger on the dayside than near midnight, a larger day–night asymmetry will naturally appear in Gringauz and Bezrukikh's plasmapause shape without contradicting Carpenter and Anderson's findings. Furthermore, Carpenter and Anderson (1992) have applied a rather stringent selection criterion, retaining only newly formed plasmapause gradients (i.e. very sharp gradients and very thin plasmapause regions), while no such selection criterion was applied to the much smaller number of density profiles considered in the Russian study. In the latter cases shown in Figs 3.14–16, rather wide plasmapause regions and irregular density profiles were not rejected as in the other statistical study. These different selection criteria may introduce an additional day-night asymmetry in studies like the Russian one.

Figure 2.17a is a plot in L versus MLT coordinates of all the scaled values of L_{ppi} in the ISEE/SFR data. All levels of magnetic activity are represented.

Despite the large dispersion, the distribution in MLT is relatively uniform. Figure 2.17b is an attempt to smooth the data of Fig. 2.17a by calculating the two-hour running mean of L_{ppi} centered at intervals spaced by 15 minutes. This figure provides an indication of typical plasmapause locations for periods when enough time has passed since the beginning of a plasmapause forming episode for a well defined plasmapause to appear locally, i.e. of the order of 10 hours for late afternoon. These results may appear surprising when compared with the first published shape of the equatorial plasmapause (see Fig. 2.10). Indeed these earlier results clearly showed a pronounced local time dependence with a bulge and a westward shoulder in the afternoon sector.

4.7.2 The paradigm based on the plasmapause bulge

It was the dawn–dusk asymmetry observed in the whistler plasmapause positions that led the early modellers to propose that the plasmapause coincides with the last closed equipotential of the magnetospheric convection electric field.

Although most space physicists have abandoned this belief, there is still some confusion about the relation of the bulge to the main plasmasphere which it is worthwhile discussing here for historical reasons. As mentioned above and in Chapter 2, it was convenient for early modeling purposes to assume that the bulge, illustrated in Figs. 2.10, 4.10a and b, is simply the elongated duskside part of a teardrop-shaped plasmasphere.

> However, the weight of the evidence suggests that while the teardrop model has been used with some success in dealing with statistics on plasmapause position in the dawn sector, most of what is observed in the bulge cannot be described in terms of such a model. From an observational point of view, the bulge and the main plasmasphere are essentially two separate entities. The main plasmasphere is an interior region of dense plasma that tends to assume a quasi-circular shape in the aftermath of an erosion event. The existence of a more nearly circular main plasmasphere explains why polar satellite measurements of a temperature peak, believed to be associated with the plasmapause, do not exhibit a significant bulge effect (i.e. a polarward excursion) in the dusk sector (Carpenter et al., 1993).

Carpenter (1966) already mentioned that the duskside bulge region seemed to be a 'region of new plasma' which is separated from the plasmasphere (see Fig. 4.10a). Furthermore, four years later, Carpenter (1970) insisted that 'the plasmapause radius at a given longitude is clamped at some time near dawn and then varies only slowly until mid-afternoon or later'. In retrospect, it is surprising that despite these early remarks the paradigm of the last closed equipotential (LCE) which had also been repeatedly challenged by Dungey (1967), Park (1970), Lemaire (1975, 1987), and perhaps others, has been durably implanted within the space physics community as is evident from a paper by Doe, Moldwin and Mendillo (1992) where the LCE concept was still used to determine the plasmapause position.

Rather than viewing the plasmapause as a last closed equipotential of a quasi-stationary convection electric field distribution, people tend now to define it in terms of the preceding history of exposure to the underlying ionosphere. But in none of these views proper account has been taken of the special flow pattern which develops at subauroral latitudes as a magnetic substorm builds up. Indeed, under these conditions fast eastward flows builds up in the post-midnight sector, and a sharp westward subauroral ion drift (SAID) often develops near midnight (Anderson *et al.*, 1991, 1993) . Although these flow patterns are most easily observed in the topside ionosphere, they must be driven by high-altitude substorm associated electric field penetrating into the nightside equatorial magnetosphere. The SAID structure as it develops during the substorm has not yet been included in any of the electric field models which will be outlined later in this chapter. In connection with this, it should be pointed out that Carpenter *et al.* (1993) have speculated that the plasmapause formation could indeed occur in the pre-midnight sector where SAID are often observed.

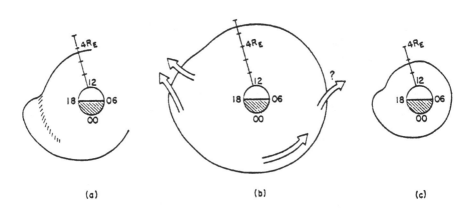

(a)   (b)   (c)

Figure 4.10 Equatorial sections of the magnetosphere illustrating several features of magnetospheric convection. At left (a), shading identifies a region where a local reduction or trough in electron density is sometimes observed. In the center (b), arrows indicate convection activity in the outer part of a large plasmasphere preceding or during the erosion phase of a magnetic storm. The arrow in the post-midnight sector indicates enhanced eastward convection at the onset of a substorm. An outward component of flow near dawn and dusk suggests the (at least) partial shielding of the dayside from perturbing convection electric fields. At right (c) is a small plasmasphere, reduced in magnetic flux content by the erosion process indicated in (b) (after Carpenter, 1970).

4.8   The plasmapause region

4.8.1   Density profile in the region of the plasmapause

The ISEE/SFR experiment had a high time resolution which allowed the density to be determined every 32 s. Hence it was particularly useful for studying the density profile at the plasmapause as well as irregularities with scale size greater than the ~ 150 km spatial sampling interval. From a series of 'smooth plasmapause' density profiles, Carpenter and Anderson (1992) determined the plasmapause thickness, ΔL_{pp}, the density drop, $\Delta \log N_{eq}$, and the scale width, $\Delta_{pp} = \Delta L_{pp}/\Delta \log N_{eq}$, over which the electron density varies by a factor of 10. The statistical distribution of Δ_{pp} is shown in Fig. 2.24.

It was found that the steepness of the profiles does not vary strongly with L_{ppi} and hence with the intensity of the associated convection activity. The relative number of smooth plasmapause density profiles (i.e. without irregularities) was largest on the nightside. The occurrence of fewer smooth profiles on the dayside is believed to be a real effect tending to indicate that when 'plasmapause knees' propagate from the night to the dayside they become wider and more irregular. Most of the values of Δ_{pp} range between 250 and 1250 km, i.e. less than $0.2 R_E$. Median values were $0.11 R_E$ on the nightside LT sector and $0.17 R_E$ during the day. These differences are consistent with results obtained from other satellite measurements in which the nightside plasmapause gradients were found to be steeper than those on the dayside (Gringauz and Bezrukikh, 1976; Horwitz, 1983; Nagai et al., 1985).

The local time distribution of $\Delta \log N_{eq}$ shows a density variation from a factor of roughly 15 after midnight to less than 10 in the afternoon sector. This local time change is attributed to dayside filling of the plasma trough region.

As a model for the smooth plasmapause density profile Carpenter and Anderson (1992) suggest the following empirical expression:

$$\log N_{eq}(t,L) = \log N_{eq}(L_{ppi}) - (L - L_{ppi})/\Delta_{pp} \tag{4.7}$$

where L_{ppi} corresponds to the inner edge of the plasmapause region (equation 4.6), and $N_{eq}(L_{ppi})$ is obtained from equation (4.4). To reflect the local time (t) variation of the plasmapause thickness these authors suggested that the day value of Δ_{pp} be taken to be equal to 0.2 and $\Delta_{pp} = 0.1$ for nightside local times. This plasmapause density profile corresponds to a straight line in Fig. 2.22, between $L = L_{ppi}$ and L_{ppo}. The trough segment of a given profile may now be defined as beginning at L_{ppo}, where L_{ppo} is determined by the simultaneous solution of (4.6) and (4.5) (see also Section 2.5.1).

Thus the experimental evidence presented above indicates that the thinnest and sharpest plasmapause density profiles are found in the nightside sector and less clearly at dusk. If it is agreed that the knee in the equatorial density profile is likely to be formed at the place where it has the sharpest gradient, one is led to conclude that the formation of the plasmapause occurs near midnight local time.

At this stage it should be emphasized that during a prolonged period of very quiet geomagnetic conditions the equatorial density has typically a gradual $10^{-0.31452L}$ profile with no evidence at all of a sharp knee like that drawn in Fig. 2.22. If the plasmapause were to form at the LCE boundary between closed and open equipotentials of a smooth and stationary convection electric field, one would expect to observe the sharpest plasmapause gradients at the end of the convection refilling phase, i.e. days after a substorm and not in the hour or so after the storm onset. On the contrary, after days of quiet magnetic conditions no 'knee' is observed in the equatorial density profiles (e.g. in Fig. 4.9).

The substorm associated earthward compression of the pre-storm nightside plasmasphere should steepen the pre-existing smooth density profile; but, a sharp density knee of less than $0.2R_E$ thickness is not likely to be formed in this way, i.e. as the result of an enhanced large-scale dawn–dusk electric field and ideal MHD convection in the sunward direction. Some other physical mechanism must exist in the nightside magnetosphere which leads to the prompt formation of the observed abrupt density drops when a substorm builds up. This same conclusion is reached also when considering that the steepness of the density profiles does not vary with L_{ppi} (i.e. the depth in the magnetosphere) and hence with the intensity of the dawn-dusk convection electric field. Indeed, according to the early ideal MHD scenarios, the steepness of the nightside plasmapause density gradient should be related to the intensity of the dawn-dusk electric field enhancement and, as a matter of consequence, dependent on L_{pp}. But this is not the case as indicated in Section 4.8.1.

In a revisited ideal MHD scenario proposed by Chen and Wolf (1972), the plasmapause is the boundary between plasma which has circulated for six days and that which has circulated for less than six days without reaching the magnetopause. According to this new definition the plasmapause becomes a very complex structure composed of a series of attached plasmatail extensions wrapped all around the plasmasphere. These stretched plasmatails could, in principal, twist up to 5 or 6 times around the plasmasphere! However, *in situ* satellite measurements of cold density profiles fail to show the multiple plasmatail structure which should, according to this theory, almost always be observed in all local time sectors. Similar models are also used by Kurita and Hayakawa (1985), Soloviev *et al.* (1989).

Due to the substorm associated dawn–dusk electric field enhancement producing an enhanced sunward drift of all plasma elements, the plasmasphere should be compressed inside the plasmapause surface on the nightside. As a matter of consequence the density earthward of the plasmapause should be increased. But quite the opposite effect is observed: after a substorm, the densities of the nightside plasmasphere are up to a factor three lower than before the storm or than the average density (see Section 2.5.1; Figs 2.1 and 1.11). This effect can hardly be explained by any ideal MHD theory for the formation of the plasmapause with the currently used magnetospheric convection electric field models.

It has been argued by Lemaire (1975, 1985) that the physical mechanism forming a new plasmapause in the post-midnight sector is plasma interchange motions driven by enhanced pseudo-centrifugal forces parallel to magnetic field lines. Large fragments of the nightside plasmasphere beyond the Zero Parallel Force surface are ripped off from the plasmasphere by interchange motion as a consequence of the enhancement of the eastward plasma drift velocity which occurs at the onset of a substorm. This is illustrated by the eastward directed arrow in Fig. 4.10b in the post-midnight sector. This implies that the erosion of the plasmasphere near midnight is determined by the radial distribution of the azimuthal convection velocity in this local time sector. This peeling off mechanism is discussed in Section 5.5.4 and illustrated in Fig. 5.23.

Satellite and whistler observations have also indicated that westward surging occurs in the pre-midnight and dusk sectors during substorms events as illustrated by the westward arrows in Fig. 4.10b near 1800 LT. Therefore, besides the peeling off process close to midnight, stretching and perhaps erosion of the plasmasphere can take place as well in the duskside at the onset of substorm events.

4.8.2 Relationship of the plasmapause to other plasma boundaries in the inner magnetosphere

Low energy ion transition

There has been a tendency to associate the plasmapause with other transitions or boundaries, such as for instance the LEIT (Low Energy Ion Transition). We recommend that this be avoided; a rather prudent strategy here would prevent later confusion and perhaps future controversies.

As pointed out in Section 3.4.1 (see Subsection: Density distribution and the location of various plasma boundaries) when the equatorial position of the sharp density drop, defining the 'classical' plasmapause, is compared with that of the LEIT (i.e. the transition where the ion pitch angle distribution changes from cold and isotropic to warm and field-aligned) one finds a rather good correspondence in the nightside region between 2200 and 0600 LT. However, on the dayside where the density profile is more gradual and smoother, the pitch angle transition is not necessarily associated with the equatorial plasmapause; it is generally observed where the density level is below $66 \, \text{cm}^{-3}$ (Nagai et al., 1985). Since the plasmapause determined by the density profile is not always observed on the dayside, whereas, according to DE/RIMS observations, the LEIT can be observed in all local time sectors, it was tempting to identify the LEIT with the 'plasmapause' (Nagai *et al.*, 1985; Horwitz *et al.*, 1986c).

Inner edge of the plasmasheet electrons

Horwitz *et al.* (1986c) have also compared the positions of the LEIT, the inner edge of the plasmasheet electrons at various energies and the equatorial boundary of the auroral oval electron precipitation region. Figure 4.11 shows the *L*-values of these three different boundaries in the evening sector as a function of time between 6 and 17 October 1981. The magnetic activity index K_p is also given at the bottom of this figure. All three boundaries shift inward and outward in *L* in response to changing values of K_p. For example, all three boundaries move outward from $L = 3$–4 to $L \sim 7$ during the decline of K_p from ~ 5 to $0+$. The equations (2.1) or (4.6) would have predicted a change from $L = 3.5$ to 5.4 for this range of change of K_p.

The inner edge of 100 eV plasmasheet electrons and the LEIT are generally within $0.3L$ of each other and similar in their inward and outward movements. The equatorward boundary of auroral electrons exhibits generally similar behavior. The close correspondence in the midnight sector of these three boundaries points to related physical origins which are not yet fully understood.

Substorm injection boundary

The correspondence between the substorm injection boundary described by McIlwain (1974; 1986), Mauk and McIlwain (1974) and Mauk and Meng (1983) and the plasmapause near midnight local time has never been examined, at least to our knowledge. Such a study might help to understand the complexity of substorm associated phenomena and their roles in the formation of the plasmapause.

Alfvén layers and shielding

It could be pointed out that Block (1966) was probably the first to identify the plasmapause with an equipotential of the magnetospheric electrostatic field distribution. He considered that the external (magnetospheric) electric field cannot penetrate inside the plasmasphere across this equipotential surface. This had set the stage for what has been called later on 'magnetospheric shielding' and 'Alfvén Layers' (Wolf, 1970). Pursuing this early suggestion, Karlson (1971) developed the idea that the plasmapause could be a surface (or a thin layer) across which the electric field changes intensity in a rather abrupt manner. Therefore, across such a charged layer, the convection velocity of cold plasma is expected to have a large shear: i.e. to jump from the corotation velocity, inside the plasmasphere, to a larger bulk speed in the subauroral region. Note that the theories of Alfvén layers by Block, Wolf and Karlson are based on the guiding center approximation of ion and electron drift paths. Finite gyroradius effects are not taken into account. Lemaire *et al.* (1996) point out, that the additional charge separation electric field resulting from the difference of the gyroradii of the plasmasheet electrons and ions may also contribute to the formation of Alfvén layers and SAIDs (see Section 4.10.3).

Using measurements of convection velocities made with COSMOS 184 at 630 km in the topside ionosphere, Galperin *et al.* (1975) found that during magnetically disturbed periods, ionospheric plasma drifts exhibit abrupt changes: i.e. large shears which they identified with the low altitude projection of 'Alfvén layers' as defined by Wolf (1970). Soloviev *et al.* (1989) identify this magnetospheric convection boundary with the actual plasmapause in the night-side sector.

4.8.3 Relationship of the equatorial plasmapause to the mid-latitude trough and light ion trough

As already discussed in Section 2.4.3, several ionospheric signatures of the equatorial plasmapause have been proposed, and it took many efforts and discussions to understand their connections. Rycroft and Thomas (1970) first reported that the minimum of the observed total density trough in the iono-sphere is associated with the statistical location of the high altitude plasmapause 'knee' (see also Rycroft and Burnell (1970)). However, the Alouette observations on which their study was based were more consistent with the mid-latitude electron density trough being 2–3 degrees equatorward of the plasmapause field line. Using a more extensive data set from the ESRO-4 satellite, Köhnlein and Raitt (1977) found that this trough minimum is observed at the statistical plasmapause latitude during very quiet conditions, but that it is located at lower latitudes than the plasmapause as K_p increases.

Brace and Theis (1974) observed latitudinal profiles of the electron density with ISIS-1. They found that the profiles across the plasmapause may be quite different. They note little diurnal variation of the invariant latitude of the ionospheric signature of the plasmapause, at altitudes just above the ionosphere (3000 ± 500 km). This led them to suggest that the plasmapause may not co-incide with a specific field line.

Grebowsky *et al.* (1976) compared the total density trough measured on Ariel-4 at 600 km with the high-altitude plasmapause determination of S^3. They found no consistent feature of the trough at the projection of plasmapause mapped down from S^3.

Using OGO-2 observations below 1000 km, Taylor and Walsh (1972) noted that the light ion trough (LIT) is a more consistently observable feature than the total density trough. They proposed the LIT as the low-altitude signature of the plasmapause. Direct comparisons of the LIT with the plasmapause as registered in VLF latitudinal cut off effects showed one to one agreement within 3 degrees invariant latitude (Taylor *et al.*, 1969; Carpenter *et al.*, 1969).

Comparing whistler observations and OGO-6 ion density measurements at 800 km Morgan *et al.* (1977) admitted the possibility that the H^+ trough may lie at a somewhat lower *L*-value.

Raitt and Dorling (1976) interpreted the global pattern of H^+ density in the topside ionosphere in terms of regions of light ion outflow or inflow from the plasmasphere. They found the region of continual daily outflow to be near the expected plasmapause position. But Titheridge (1976b), using Alouette observations, argued that the region of H^+ outflow extends within the plasmasphere, and is not a good signature of the equatorial plasmapause. These results indicating that the density troughs are not clear plasmapause signatures were consolidated by the intercomparison of ISIS-2 light ion trough (LIT) positions at 1400 km and positions of the whistler plasmapause by Foster *et al.* (1978). They confirmed that H^+ and He^+ densities (at 1400 km) attain their trough minimum at invariant latitudes which are smaller than those of the equatorial plasmapause observed at the same time and at the same longitude. Furthermore, they found that, unlike the equatorial plasmapause, the edge of the LIT does not exhibit a diurnal variation. But both the plasmapause and light ion trough move to higher latitudes as magnetic activity diminishes.

The observed positions of the plasmapause (L_{pp}) and the polarward edge of the ionospheric trough (L_{pe}) have been compared by Smith, Rodgers and Thomas (1987). They found that before midnight the trough edge was well poleward of the plasmapause. After local magnetic midnight the two features are roughly coincident in all of the five cases studied.

Foster *et al.* (1978) pointed out that there is, however, another low-altitude signature which constitutes a better indicator of the plasmapause field line. This ionospheric feature is the latitude where precipitated plasmasheet particles are identifiable at low altitude in the dawn and dusk local time sectors. This boundary between precipitated and trapped plasmasheet electrons determines indeed, the field line where trapped plasmasheet electrons are precipitated into the atmospheric loss cone, presumably due to wave–particle interactions. This is also the point of view developed by Sivtseva *et al.* (1984) in their article on 'coordinated investigations of processes in the subauroral upper ionosphere and the light ion concentration trough'. In this latter study the boundary between precipitated and trapped plasmasheet electrons is identified as the 'diffuse surge boundary' which corresponds to the ionospheric projection of the plasmapause. It results from this study that:

(1)   at altitudes below the level of the diffusion barrier (i.e. below
        500–700 km), the positions of the equatorial boundary of the H^+ ion

density trough (LIT) coincide almost with the equatorial boundary of the main ionospheric trough in the F-region during quiet geomagnetic conditions. However, during magnetic storms the LIT is located inside the plasmasphere at midnight;

(2)   during quiet conditions at altitudes of 1500–1800 km (i.e. much higher than the diffusion barrier) the equatorial boundary of the LIT is located approximately at the same invariant latitude as the equatorial boundary of the main ionospheric trough. During a substorm the 'polar wall' of the electron main trough (i.e. where the electron density increases sharply with latitude in the F-region) coincides with the 'diffuse surge boundary' and with the equatorial boundary of the light ion trough. All these features correspond then with the ionospheric projection of the plasmapause;

(3)   in the intermediate altitude range (500–1500 km) where the largest amount of experimental data has been accumulated with Ariel-4, OGO-2 and 6, ISIS-2 and AE-C, a wide variety of latitudinal profiles are obtained for the LIT and main ionospheric trough. Furthermore, their positions are not directly related with respect to the projection of the plasmapause in this range of intermediate altitudes. This lack of clear correspondence is probably a consequence of the competing roles (i) of charge exchange reaction between O^+ and H, (ii) of the increase of ion temperature versus latitude, and (iii) of field aligned flows of thermal ions (Sivtseva et al., 1984).

4.8.4   Relationship of the equatorial plasmapause to the subauroral electron temperature enhancement (SETE)

In the previous Section we reviewed in a historical perspective the expected relationship between the equatorial plasmapause and the location of troughs in the latitudinal profiles of the electron and ion densities observed at various altitudes between 400 km and 2000 km. In this Section we will outline its relationship with latitudinal profiles of the observed electron temperature in this same range of altitudes.

The subauroral electron temperature enhancement

It has already been pointed out in Section 2.4.3 that the electron temperature in the topside has a latitudinally narrow peak called SETE, for subauroral electron temperature enhancement. This SETE coincides with the outer edge of the plasmasphere. This important relationship between the high-altitude plasmapause and a well-defined feature observed in the subauroral ionosphere was first published by Brace and Theis (1974) (see also Section 4.3.1). This good correspondence is illustrated in Fig. 4.12 showing the electron temperature (T_e) and ion density (N_i) versus L, along two consecutive orbits of Cosmos-900, at an altitude of 480 km. The dashed and solid lines correspond to measurements made on 2 December 1977 along the orbits 3771 and 3772, when a large geomagnetic storm was in progress, and K_p increasing from 5 to 7. Both consecutive sets of measurements are made near dawn (05–06 LT) and separated by only 1.5 hour in universal time. It can be seen that the location of the subauroral electron temperature enhancement (SETE) coincides with the low ebb of the LIT. All these low altitude features moved equatorward to lower L-shells when $|D_{st}|$ and K_p increased between orbits 3771 and 3772.

The dotted line labeled N_p corresponds to the ion density collected at high altitude (3700–8400 km, see lower scale in Fig. 4.12) with Prognoz-6 in the nightside local time sector. The sharp gradient in this proton density profile corresponds to the high-altitude plasmapause. According to these observations, the high-altitude plasmapause is located at $L = 3.1$; i.e. between L = 3.2, the position of the SETE observed at 480 km less than an hour UT earlier with Cosmos-900, and $L = 2.9$, the position of the SETE observed on the following orbit, nearly one hour UT after the pass of Prognoz-6 in a nearby local time sector. This good correspondence between the L-parameter of the high-altitude plasmapause measured by Prognoz-6, and the L-parameters of the SETE at low altitude measured with Cosmos-900 just before and after the Prognoz-6 measurements, indicates the close relationship of both plasma features. But for various reasons these results obtained by V. Afonin in 1977, were not submitted for publication untill 1995.

Figure 4.12 illustrates also another general characterisitc of SETEs which was first pointed out by Brace *et al.* (1988): when $|D_{st}|$ is increasing, the electron temperature peaks become narrower and have a steep polar edge, while during quiet conditions the SETE has rather symmetric slopes on their poleward and

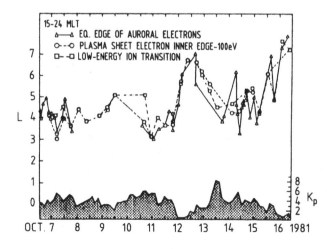

Figure 4.11 Variation of K_p and boundary L-values with time for the period 7–16 October, 1981. The boundaries displayed are for the 100 eV inner edge of the electron plasma sheet, the 100 eV precipitating auroral oval electron equatorward edge, and LEIT (after Horwitz *et al.*, 1986c).

equatorward sides. It is this characteristic signature that has been used in the Brace *et al.* (1988) statistical study to identify the ionospheric projection of the 'inner plasmapause'. This sharp signature of the SETE has been measured on Cosmos-900 with a spatial resolution of a tenth of a degree in latitude, i.e. 10 km.

Note that the maximum of T_e is located at a slightly lower L than this sharp poleward edge. The slope of the equatorward wing of the SETE is not steeper during disturbed conditions (orbit 3772) than it was during quiet times, i.e. during orbit 3771. The magnitude of the electron temperature peak increases

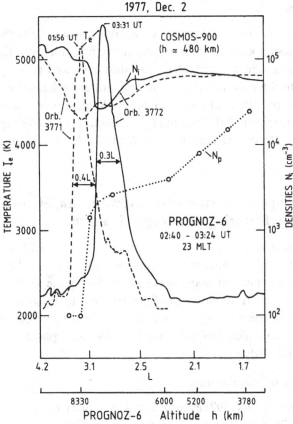

Figure 4.12 Coordinated observations of the electron temperature (T_e) and ion densities (N_i and N_p) with PROGNOZ-6 at high altitudes (3500–8500 km), and COSMOS-900 at low altitudes (480 km). The data are plotted as a function of the L-parameter. The dashed lines correspond to T_e and N_i observed along the orbit 3771 of COSMOS-900 nine hours after the beginning of the geomagnetic storm of 1–2 December 1977. The solid lines labeled (T_e) and (N_i) correspond, respectively, to the ionospheric electron temperature and ion density, on the orbit 3772, i.e. 1.5 hours later. The dotted line labeled (N_p) with the open circles correspond to ion density profile across the high-altitude plasmapause as observed in nearly the same local time sector (midnight to dawn) and at a universal time between the times corresponding to both COSMOS-900 passes 3771 and 3772 (after Afonin *et al.*, 1997).

with increasing values of the magnetic activity index $|D_{st}|$. This was first pointed out by Kozyra *et al.* (1986) and is confirmed by the Cosmos-900 observations shown in Fig. 4.12.

Figure 4.12 also shows that the width in L of the SETE peak is slightly narrower during disturbed conditions than during very quiet times. The reduction of the width of the SETE is consistent with the observed decrease of thickness of the equatorial plasmapause region when K_p increases (Chappell *et al.*, 1970a).

Stable auroral arcs

The SETE are thought to be the consequence of energy transfer from the equatorial region of the plasmapause down along magnetic field lines into the ionosphere, where observable signatures result during substorms and geomagnetic storms. In addition to the subauroral electron temperature enhancement and the light ion density trough, there are other low-altitude signatures related to the equatorial plasmapause; one of these remarkable signature is the 630 nm emission, called stable auroral red (SAR) arcs (Cole, 1965; 1970a,b; 1975; see also Sections 2.4.2 and 4.3.1).

4.8.5 Evolution and shape of the subauroral electron temperature enhancement (SETE) during a geomagnetic storm

Having confirmed with independent satellite observations that the SETE is a reliable low-altitude signature of the inner plasmapause, we will now describe in detail the evolution of the position and shape of the SETE during the development of a geomagnetic storm. It is based on Cosmos-900 high resolution measurements of the electron temperature (T_e) with a radio-frequency electron temperature probe, and measurements of the ion density (N_i) with planar and spherical retarding potential analyzers.

A unique sequence of observations

The data used in the Afonin *et al.* (1997) event study were collected along the post-midnight portion of 26 consecutive orbits (from orbit 3748 to 3776 of Cosmos-900) including the geomagnetic storm of 1–2 December 1977). The observations along nine of these orbits are displayed in Figs 4.13a–h. The values of K_p (not shown) were very low ($K_p \leq 1-$) from orbit 3757 to 3764, then K_p increased gradually up to a maximum value of 7- during orbits 3772 and 3773, and eventually decreased to reach 4+ during orbits 3774. During this sequence of orbits the D_{st} index dropped gradually from almost zero before orbit 3764, to $-150\,\mathrm{nT}$ during orbit 3773.

The density and temperature profiles shown by light solid lines in all eight

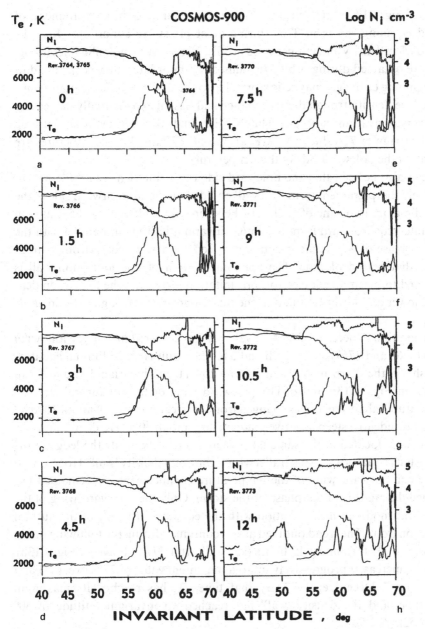

Figure 4.13 A series of latitudinal distributions of the ion density (N_i) and electron temperature (T_e) at an altitude of 480 km along successive orbits of COSMOS-900. Time $t = 0$ corresponds to the beginning of the geomagnetic storm of 1–2 December 1977. The observations are collected during the main phase of this geomagnetic storm at successive interval of 1.5 h which corresponds to the orbital period of COSMOS-900. Note the equatorward shift of the steep subauroral ionization front (wall) and of the associated sharp polar edge of the subauroral electron temperature enhancement (SETE), during the main phase of the geomagnetic storm (after Afonin *et al.*, 1997).

panels of Fig. 4.13 are reference profiles corresponding to the measurements of T_e and N_i along the orbit 3764 during quiet pre-storm conditions ($K_p = 1-$; $D_{st} \sim 0$). The density and temperature profiles shown by heavy solid lines in Fig. 4.13a are obtained during orbit 3765 almost 1.5 hour later; they correspond to the beginning of the geomagnetic storm, i.e. to the time $t = 0$. The equatorward wing of the density trough did not change and has an exponentially decreasing slope versus invariant latitudes. The SETE extends over $10°$ invariant latitude with a maximum T_e value of 6000 K at $\Lambda = 58$–$62°$. Small-scale structures are present on the poleward side of the temperature peak.

It can also be seen that the poleward side of the ion density profile has an extremely sharp gradient forming a wall or ledge on the equatorward side of the auroral region. In a time of 1.5 hours, between orbits 3464 and 3465, this wall has shifted equatorward from $64°$ to $63°$ invariant latitude in parallel with the slight increase of geomagnetic activity: indeed K_p has increased from 1- to 2- during that period of time. The apparent motion of this ionization wall is attributed to plasmasheet electron precipitation increasing the ionization density at lower and lower latitudes in the main ionospheric trough (Afonin et al., 1997).

An additional increase of K_p to a value of $3+$ occurred during the following orbits (3766 and 3767) at $t = 1.5$ h and 3 h. The result is an additional equatorward shift of the sharp auroral density wall down to $60°$ invariant latitude as can be seen in Figs. 4.13b and c. The poleward wing of the subauroral electron temperature enhancement has changed from an irregular slope of about 1000 K/degree latitude to a steep negative gradient larger than 40 000 K/degree latitude. This ledge is located at the same latitude as the density 'wall': the ledge of the SETE shifts equatorward in parallel with the auroral density 'wall'. However, so far the equatorward wing of the temperature enhancement has remained unchanged during this initial phase of the storm. Only the poleward wing of the SETE was modified in association with the equatorward shift of the auroral ionization front. The trend outlined above continues during the following orbits of Cosmos-900 for almost 12 hours (i.e. until orbit 3773, shown in Fig. 4.13h), while K_p increased progressively up to a maximum value of 7- between 03 and 06 UT on 2 December 1977. By that time, D_{st} has reached its maximum amplitude, and the density 'wall' has reached an invariant latitude of $50°$ ($L \simeq 2.52$).

Note that it is only 3 or 4 hours after the beginning of the storm that the distribution of temperature equatorward to the peak starts moving closer to the equator (see Fig. 4.13d corresponding to orbit 3768 at $t = 4.5$ h). The temperature and density distributions on the equatorward wing of the SETE did not change during the first 3 or 4 hours.

Erosion and motion of the plasmapause

The existence of subauroral electron temperature enhancements with such sharp poleward edges was also found in the observations of DE-2 (Brace *et al.*, 1988) which had a time/spatial resolution comparable to those of Cosmos-900. During very quiet conditions the sharp ledges are absent and rather irregular temperature profiles like that shown in Fig. 4.13a are observed. The irregular small scale structure in T_e profiles is attributed to irregular heating by ring current particle precipitation patterns.

The motion of the plasmapause, as identified by the equatorward shift of the SETE, fits well the relationships (4.6) or (2.1) giving the equatorial plasmapause position (L_{pp}) versus various geomagnetic activity indices. The erosion of the poleward wing of the SETE likely corresponds to the low altitude signature of the erosion of the equatorial plasmasphere. It illustrates probably with unprecedented resolution the formation of a new plasmapause.

With the unique sequence of observations shown in Fig. 4.13, Afonin *et al.* (1997) have captured an erosion event of the plasmasphere taking place in the post-midnight local time sector, where and when the equatorward edge of the auroral region is shifting closer to the equator in association with the building up of a magnetospheric storm. At high altitudes in the magnetosphere this is expected to be associated with the earthward surging of the edge of a cloud of plasmasheet particles. To our knowledge it is the first time that a sequence of observations has shown in detail the evolution of the ionospheric electron temperature and ion density profiles during the process of formation of a new plasmapause.

4.9 Plasma density irregularities outside and inside the plasmasphere

Whistler wave observations showing evidence of plasma density irregularities inside the plasmasphere have already been presented in Section 2.5.2. In Chapter 3, we reviewed satellite measurements of charged particle fluxes indicating also the presence of irregularities in the plasmasphere and plasmatrough. In this Section we summarize these observations and discuss the possible origin of these plasma density irregularities.

Different categories of equatorial density profiles

Figure 4.9 illustrates 'smooth' plasma density profiles. In their categorization, Horwitz *et al.* (1986b) would call these profiles 'featureless'. After prolonged time intervals of very quiet geomagnetic conditions these featureless profiles extend out to large radial distances in the equatorial magnetosphere. Under exceptionally extended quiet conditions, there remains no evidence of a sharp 'knee' in the dayside density distribution up to $L = 8$. But this is not the usual situation since geomagnetic conditions are rarely quiet and unperturbed for one full day. Very often plasma density irregularities, enhancements and depressions are observed by satellites as well as by the whistler method.

Figure 4.14 illustrates six different categories of profile patterns identified by Horwitz *et al.* (1990a). These schematic patterns classify only large-scale plasmaspheric density structures, but not the density structures in the immediate vicinity of the LEIT and near the plasmapause. Some of these large-scale structures are probably due to partial detachment and to plasma elements corotating on 'pseudo-closed' streamlines.

Horwitz *et al.* (1986b, 1990) determined the relative frequency of occurrence of the different types of density profiles shown in Fig. 4.14. The most abundant ones are the multiple plateau profiles (category C) and featureless profiles (category A), both of which occur about 40% of the time.

Multiple plateau profiles had already been identified from whistler observations by Corcuff *et al.* (1972) (see Fig. 2.2). They are formed following a suc-

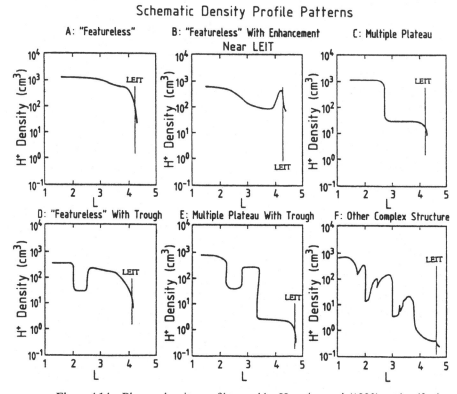

Figure 4.14 Plasma density profiles used by Horwitz *et al.* (1990) to classify the structures observed by DE-1. The low-energy ion transition (LEIT) shown at the end of each of these profiles represents the outer edge of detectable cold isotropic ions (after Horwitz *et al.*, 1990).

cession of recurrent substorm convection enhancements of decreasing amplitudes. The flat uniform density region between two steep vestigial plasmapauses are regions where plasma refilling is responsible for the gradually increasing equatorial density and total flux tube content. The profiles with significant troughs (D and E) appear about 10% of the time and are observed most commonly in the dusk–evening sector. From the statistical analysis of Horwitz *et al.* (1986b), it also appears that the trough widths are significantly smaller in the few cases observed around dawn and in the post-midnight sector than in other local time sectors (see Fig. 4.15).

Large quasi-periodic fluctuations in the electron number density are often encountered near and inside the plasmapause (LeDocq, Gurnett and Anderson, 1994). Their $-\frac{5}{3}$ power spectral slopes suggest the presence of well-developed two-dimensional magnetohydynamic turbulence in the frequency range from 2 mHz to 61 mHz.

Local time distribution of detached plasma elements

All across the dayside plasma trough region Chappell, Harris and Sharp (1971) observed 'detached' regions of high-density plasma. The distribution of these detached plasma elements along OGO-5 trajectories is shown in Fig. 4.17. The density in these detached plasma elements may be as high as two hundred ions/cm^3 in a normally low plasma trough background. There appears to be an

TROUGH REGIONS FOR TYPE D PROFILES
ALL Kp, LATITUDES, GEOCENTRIC DISTANCES
NO. OCCURRENCES = 37

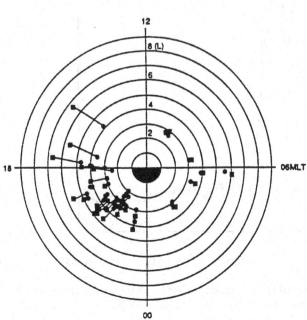

Figure 4.15 Distribution of widths (extents in *L*) of the trough regions for all density profiles of category D and E in Fig. 4.14 (after Horwitz *et al.*, 1986c).

accumulation of such plasma density enhancements in the afternoon–dusk sector. They are observed most frequently during moderate to disturbed magnetic activity conditions suggesting that they are generated near dusk by variations in the magnetospheric convection electric field as illustrated in Fig. 4.10b.

According to Chappell (1974) the plasma is initially 'detached' near dusk and drifts sunward. But the physical mechanism by which such a detachment can be achieved in the framework of the ideal MHD theory is not specified. The fact that the 'detached' regions are found more frequently close to the plasmapause near dusk and progressively further away as one goes to earlier local times (see Fig. 4.17) has been interpreted by Chappell (1974) as evidence that they are extracted out of the duskside bulge and drift sunward and perhaps noonward.

Using a less restrictive criterion to identify 'detached' plasma regions than that employed in Chappell's (1974) study, Kowalkowski and Lemaire (1979) have identified another class of outlying density regions in the post-midnight and dawn local time sectors. These elements are found just outside the plasmasphere after an enhancement of nightside convection activity. Figure 4.16 is one example among several similar cases identified in samples of OGO-5 data. These

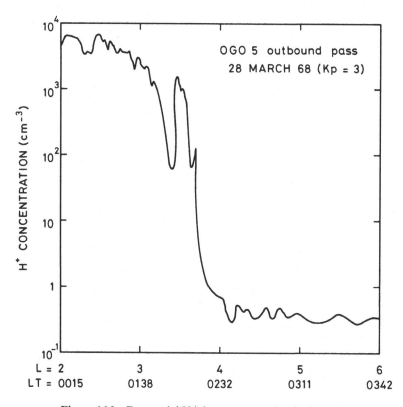

Figure 4.16   Equatorial H^+-ion concentration in the post-midnight sector observed by OGO-5 (after Kowalkowski and Lemaire, 1979).

nightside detached plasma elements have been interpreted by Lemaire (1985, 1987) as large pieces of the plasmasphere which are in the process of detachment due to the enhanced centrifugal effect associated with a sudden increase of the eastward convection velocity illustrated schematically in Fig. 4.10b. These large pieces of plasma, are assumed to become detached from the main body of the plasmasphere 'like icebergs from the icebank' (Lemaire, 1987, p. 116).

This interpretation is also supported by the results displayed in Fig. 4.15, indicating that the width of the troughs is significantly smaller at dawn and post-midnight than elsewhere. Indeed, it is where their detachment takes place that the separation between the elements and the plasmapause should be smallest. In Fig. 4.17 there are only a few cases of detached elements in the post-midnight and dawn to noon sectors, but this may be the consequence of the drastic selection criterium adopted in this study as well as to uneven sampling in local times.

Whether these plasma regions are attached 'tails' or detached 'blobs' is not easy to determine with single satellite measurements. But by comparing the radial and longitudinal extents of these detached regions, Chappell (1974)

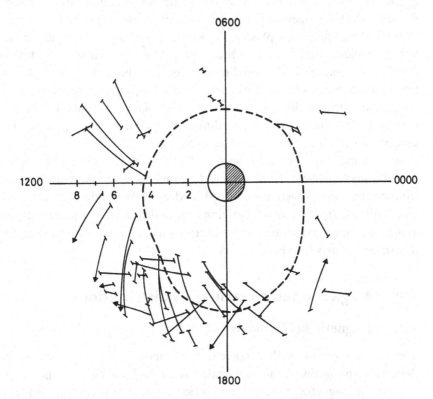

Figure 4.17 Portions of OGO-5 trajectories in (L, LT) coordinates along which detached regions of plasma with density greater than 10 ions/cm^3 were observed. The dashed line shows the average position of the plasmapause for comparison (after Chappell *et al.*, 1971a).

concluded from OGO-5 observations that there is no definite evidence of long tails that extend in longitude similar to those obtained in the ideal MHD models of Chen and Wolf (1972).

Carpenter *et al.* (1993) proposed that the outlying plasma elements could be detached or connected, depending on the cases considered. To identify which of them are detached or connected a 'fleet' of several satellites like those of the CLUSTER mission could be useful. Even more promising are new observational methods like the Global Magnetospheric Imaging proposed by Williams, Roelof and Mitchell (1992) and Mauk *et al.* (1995) as well as the Magnetospheric Radio Sounding Project described in Green *et al.* (1994) and proposed by Burch *et al.* (1995). Both projects aim to obtain 3-D pictures of the plasmasphere and of detached tails or blobs.

Other phenomena associated with detached elements

Since their discovery, detached plasma regions have been associated with many different magnetospheric phenomena. For instance strong extremely low frequency (ELF) hiss has been observed in detached regions by Kivelson and Russell (1973), long-period (300 s) oscillations in the magnetic field have been associated with detaching plasma; Pc1 and Pc2 oscillations in the magnetic field at synchronous orbit have also been interpreted by Barfield and McPherron (1972) as originating in detached plasma regions; the location of Pc4 and Pc5 micropulsations as observed by ground-based magnetometers appears to be connected with the distribution of detached plasma regions; spotty auroral precipitation equatorward of the diffuse belt in the evening sector may be associated with detached plasma elements.

All this indicates the importance of detached plasma regions drifting away from the plasmasphere. These detached plasma regions should be distinguished from plasma tails formed as described in ideal MHD models like that of Chen and Wolf (1972). These latter outlying tails of dense plasma become elongated and might be compared to plumes escaping from a pipe, stretching and being deformed by gusty convection flows.

4.10   Magnetic and electric field distributions

4.10.1   Magnetic field models

The convective flow in the magnetosphere and plasmaspheric dynamics are determined primarily by the large-scale electric field and magnetic field distributions in the magnetosphere. The magnetic field distribution is the vector superposition of multipole field components whose origin is inside the Earth, and external field components.

The internal magnetic field and its secular variation is well described by the

International Geomagnetic Reference Field (IGRF and DGRF) which is updated every five years by IAGA. Due to the dominance of the dipole term, the internal magnetic field component is usually approximated by a centered magnetic dipole whose magnetic moment is nearly anti-parallel to the rotation axis of the Earth. Since the dipole term is dominant and the influence of higher-order multipole terms vanishes more rapidly than the dipole term at large distances in the magnetosphere, this can be considered as a satisfactory first approximation in theoretical studies and in current numerical simulations (Fraser-Smith, 1987).

The external magnetic field is determined by the ring current, magnetopause currents, neutral sheet currents in the plasmasheet, field-aligned Birkeland currents, and diamagnetic currents circulating around and within plasma density irregularities. The motion of these diamagnetic plasma irregularities and the variability in time of all these current systems within the magnetosphere make the geomagnetic field highly time dependent and patchy.

There are several external magnetic field models for the magnetosphere, one of the latest being the model of Tsyganenko (1989). Other empirical models are those of McIlwain (1972), Mead and Fairfield (1975) and the Olson and Pfitzer's (1974, 1977) tilt-dependent model. Other time-dependent external magnetic field models have also been proposed and are reviewed by Pfitzer, Olson and Mogstud (1988) and Stern and Tsyganenko (1992). They are all based on large numbers of satellite measurements in the magnetosphere.

These empirical models are most useful to map accurately the fluxes of trapped radiation belt particles. To model the plasma density distribution in the plasmasphere a centered dipole is usually assumed as a satisfactory first approximation. In most numerical simulations of the dynamical flow of cold plasma in the magnetosphere the dipole approximation is used. However, already at geostationary orbit the noon–midnight asymmetry of the magnetic field is quite significant and needs to be taken into account. This local time aymmetry has been included in most external magnetic field models including the M2 model used by McIlwain (1972). The M2 model gives the magnetic field intensity in the equatorial plane. It is simple, analytical and based on measurements from geosynchronous satellites. Although it is generally not recognized in the modelling community, this model is rather appropriate to study the distribution of plasma and its convection in the equatorial plane of the plasmasphere and plasmatrough. Of course a 3-D model depending on solar-wind parameters would be preferable. But 'a truly accurate model of the Earth's magnetic field would be extremely complex and unwieldy to use. It would also produce distracting effects which would mask the basic properties of particle motion being sought' (McIlwain, 1986, p.187).

4.10.2 Electric models

Since 1961 a large number of electric field models have been published based on experimental data sets and on theoretical guesses or expectations. The most popular and first semi-empirical convection electric fields models are those of Nishida (1966), Brice (1967) and Volland (1973, 1975). Sojka, Rasmussen and Schunk (1986) have developed a semi-empirical model of the magnetospheric electric field which is not only dependent on the K_p-index but also on the interplanetary magnetic field (IMF) intensity and direction. This is an interesting extension of the Volland model, allowing one to model the two- and four-celled convection patterns observed over the polar cap respectively during southward and northward IMF.

Reviews of convection electric fields models exist in books like that by Lyons and Williams (1984) and in articles such as by Stern (1977), Blanc (1978), Blanc and Richmond (1980), Wolf (1983), Caudal and Blanc (1988) and del Pozo and Blanc (1994). We give here only a short overview, emphasizing those models which we consider currently to be the most popular, portable or useful in studies of plasmasphere dynamics and in studies of the formation of the plasmapause.

Different categories of models

It should first be pointed out that there are different categories of models available: the sometimes called 'first-principal', theoretical or physical models on the one hand, and the empirical models on the other. The first category is based either (i) on the solution of physical equations with assumed boundary conditions in the ionosphere and in the outer magnetosphere – the Rice University time-dependent electric field model belongs to this class – or (ii) on assumptions about the distribution of convection velocities in the magnetosphere; the uniform dawn–dusk electric field or tear-drop model as well as the 'shielded Volland–Stern' models belong to this latter class of empirical or semi-empirical models. The models of Sojka, Rasmussen and Schunk (1983), McIlwain (1972, 1974, 1986) are also semi-empirical ones.

Empirical E-field models are obtained by averaging experimental data collected over a wide range of L-values and local times. Empirical or semi-empirical models will be briefly discussed and used below.

Measuring electric fields in space

Double probe measurements have been used in the past to directly determine magnetospheric electric fields. But it must be pointed out that it is difficult to measure electric fields in space directly because of their relatively low value and because of technical reasons which have been discussed by Pedersen et al. (1985). A novel method for determining the electric field intensity in the magnetosphere was used on the GEOS-2 mission. It consisted of measuring the displacement due to the $\mathbf{E} \times \mathbf{B}/B^2$ drift velocity of an electron beam after one gyration in the ambient magnetic field. This technique was used by Beaumjohann, Haerendel and Melzner (1984, 1985) to study the electric field distribution at geosynchronous altitude.

Magnetospheric electric field intensities have also been inferred indirectly from measurements of the plasma bulk velocity $\mathbf{v} = \mathbf{E} \times \mathbf{B}/B^2$, when the magnetic field intensity B is known either from a model or from direct *in situ* measurements. Barium clouds releases, whistler observations, incoherent scatter radar measurements, and measurements of dynamical energy spectra of keV-electrons and protons are different alternative experimental techniques which have been used with some success to determine global electric field distributions in the magnetosphere (Carpenter and Stone, 1967; Carpenter *et al.*, 1972; Mozer and Serlin, 1969; Carpenter, Park and Miller, 1979; McIlwain, 1972, 1974, 1986; Maynard, Heppner and Egeland, 1982; Temerin *et al.*, 1982; Blanc, 1983a;b; Foster, 1983, 1984; Iversen *et al.*, 1984; Fontaine *et al.*, 1986; Tanskanen *et al.*, 1987; Block *et al.*, 1987; Fälthammar *et al.*, 1987; Boström, Koskinen and Holback, 1987; Boström *et al.*, 1988; Senior *et al.*, 1989; Peymirat and Fontaine, 1994).

Figure 4.18 shows the intensity and direction of the measured electric field component in the GEOS-1 spin plane for a pass through the plasmasphere (Pedersen *et al.*, 1985). The comparison of these measurements with two E-field models (dashed lines) clearly indicates that the model without corotation (curve 2) is a poor approximation. On the contrary, curve 1 corresponding to the corotation electric field provides a good fit to the GEOS-1 observations inside

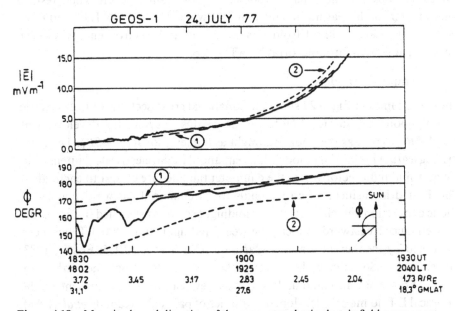

Figure 4.18 Magnitude and direction of the magnetospheric electric field component perpendicular to the spin axis of GEOS-1 measured during an inbound pass through the plasmasphere. Curve 1 is a model field representing perfect corotation. Curve 2 is a model without any corotation (after Pedersen *et al.*, 1985).

3.3 Earth radii. Although Maynard, Heppner and Egeland (1982) and Pedersen *et al.* (1985) have been able to confirm by such *in situ* double probe measurements that the inner plasmasphere is rotating with the angular velocity of the underlying ionosphere, considerable deviations from simple corotation are found further out. This is clearly illustrated in the lower panel of Fig. 4.18 indicating large angular variations of the E-field direction. Such large deviations from corotation are generally observed near the plasmapause during disturbed conditions (Maynard, Heppner and Egeland, 1982). Under disturbed conditions GEOS-2 measurements have shown that just inside the duskside plasmapause the electric field can become many times stronger than the corotational field, and be oppositely directed to it. These observations of strong subauroral electric fields in the duskside plasmasphere are in agreement with results inferred from earlier rocket-borne measurements (cf. Fahleson *et al.*, 1971) and with observations of Subauroral Ion Drifts (SAID) (Galperin, Ponomarov and Zosinova, 1973a; b, 1975; Smiddy *et al.*, 1977; Rich *et al.*, 1980 ; Maynard, Aggson and Heppner, 1980 ; Anderson, Heelis and Hanson, 1991).

During quiet geomagnetic conditions, the large-scale magnetospheric convection electric field is too weak to be measured with the double probes flown so far : $E < 0.3\,\mathrm{mV/m}$. Values as low as $0.1\,\mathrm{mV/m}$ have, however, been measured with the electron beam technique on GEOS-2. Using these results, Baumjohann *et al.* (1985) have determined the average distribution of magnetospheric convection electric fields in the equatorial plane at geostationary orbit. Their results confirm many of the features included in McIlwain's (1974, 1986) E3H and E5D models, which are deduced from dynamical ion and electron energy spectra measured at geosynchronous orbit by ATS-5&6.

Empirical models

The dashed lines in Fig. 5.27 show the equatorial cross-sections of E3H electric field equipotential surfaces. Note the dawn–dusk asymmetry developing at $R > 4$–$5R_\mathrm{E}$, as well as the noon–midnight asymmetry for $R > 4$–$5R_\mathrm{E}$. The most striking feature in this E3H model is the enhanced radial electric field intensity in the post-midnight sector where the equipotential lines are closest to each other. This E-field intensification corresponds to an enhanced eastward magnetospheric convection velocity in the post-midnight sector as sketched in Fig. 4.10b. This enhanced eastward velocity between midnight and 0400 LT has been confirmed by incoherent scatter radar data (Oliver *et al.*, 1983; Foster, 1983; Foster *et al.*, 1986a; 1986b; de la Beaujardière *et al.*, 1986; Fontaine *et al.*, 1986; Senior, Fontaine and Caudal, 1989). This key feature is also present in the empirical E-field models developed in a series of papers by Kamide *et al.* (1986), Marklund *et al.* (1987a; 1987b; 1988) and Blomberg and Marklund (1988a, 1988b).

The empirical ionospheric electric field models make use of (i) auroral UV-images from satellites, combined with (ii) *in situ* observations of electric fields, precipitating particles and magnetic fields, as well as (iii) relevant ground-based data. Figure 4.19 shows the equipotential distribution corresponding to the Marklund–Blomberg model projected into the equatorial plane. Note the similarity with the E3H model (Fig. 5.27) in the post-midnight sector. Although the enhanced eastward convection velocity is present in all empirical electric field models based on reliable observations (including McIlwain's most recent E5D model), it is, however, missing in simplified mathematical models like the 'tear-drop' model designed or used by most simulators.

The dotted line in Fig. 4.19 is the shape of the plasmapause as determined by Carpenter (1966) from early whistler observations. Note that the last closed equipotential (thick solid line) in the Marklund and Blomberg model has a bulge like Carpenter's (1966) plasmapause shape. The stagnation point at 1800 LT, between $L = 5$ and 6 of this empirical E-field coincidentally fits with the tip of the bulge of Carpenter's plasmapause shape!

A similiar point of singularity exists in the K_p-dependent E5D electric field model of McIlwain (1986). As for most earlier models this stagnation point of the E5D model is located in the dusk local time sector and its position moves closer to the Earth when K_p increases. This empirical model is probably more realistic, anyway more comprehensive (because it is K_p-dependent) and simpler from a mathematical point of view (the analytical expressions are simpler) than its 1974 version (E3H). It could be recommended as a reference for future simulations and data organization. The E5D model is displayed in Fig. 5.28. It has also been used by Lemaire (1996) for numerical simulations of the formation and deforma-

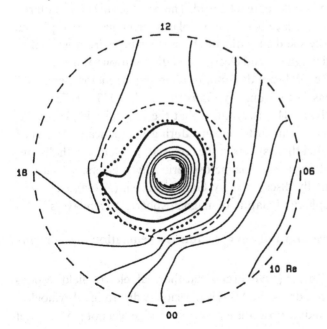

Figure 4.19 Equatorial cross-section of the electrostatic equipotential surfaces corresponding to the empirical model of Blomberg and Marklund (1988). The solid line corresponds to the last closed equipotential in this model. For comparison the equatorial plasmapause position deduced by Carpenter (1966) from whistler observations (after Marklund *et al.*, 1988).

tion of the plasmapause.

Unlike the time-dependent physical E-field model developed at Rice University, the E5D empirical model is easily portable and does not consume much CPU time; in addition to a stagnation point, it has many realistic features which are not present in the 'tear-drop model' or more generally the Volland–Stern family of models.

4.10.3   Limitations of current empirical models of the convection electric field

One common deficiency of all the electric field models described above is that they assume a smooth variation of the electric field intensity across the last closed equipotential which is also the Alfvén layer of zero energy electrons and ions. Block (1966), Karlson (1970, 1971) and Wolf (1970) have indicated that a charge separation electric field appears along Alfvén layers: i.e. at the inner edge of the plasmasheet or along substorm injection boundaries.

Grad-B drifts as a generating mechanism for Alfvén layers

In the presence of electrons and protons of non-zero energies, charge separation (polarization) electric fields necessarily appear within the magnetospheric plasma, in addition to the externally imposed electric field. Such a discontinous electric field distribution builds up along these surfaces due to the different grad-B, curvature and inertial drifts of positively and negatively charged particles in the non-uniform (dipole) magnetic field. The presence of such a polarization electric field at the interface between two plasma irregularities or plasma clouds changes drastically the drift paths of charged particles from what they would be in the external magnetic and electric field distributions alone.

Experimental evidence of abrupt discontinuities in the magnetospheric electric field distribution has been found by Galperin *et al.* (1973a, 1975). They discovered from COSMOS-184 observations that the ionospheric plasma exhibits rather abrupt changes in drift velocity during magnetically disturbed conditions. They associated these shears in the convection velocity with discontinuous electric distributions across magnetospheric Alfvèn layers. Subauroral ion drifts (SAID), originally discovered by Galperin *et al.* (1973a), were also assumed to be the ionospheric signature of magnetospheric Alfvén layers.

Difference in electron and ion gyroradii as a generation mechanism for Alfvén layers

Lemaire, Roth and De Keyser (1996) argue that the peak electric field responsible for the large westward drifts in SAIDs is principally due to the thermoelectric charge separation building up at the equatorward edge of a hot plasmasheet cloud intruding impulsively in the subauroral plasmatrough and plasmasphere. They propose a model for the equatorial voltage generator of a SAID based on the kinetic model of tangential discontinuities. According to their model the charge separation electric field is directed away from the Earth and confined in a narrow subauroral band of L-shells whose extent is a few times the gyroradius of 200–400 keV protons.

Unlike in the Alfvén layer model of Block (1966) and Karlson (1971) or Wolf (1970), the charge separation electric field in the Lemaire, Roth and De Keyser (1996) model is not only the consequence of different of grad-B drifts for the ions and electrons in the non-uniform (dipole) magnetic field, but is principally due to the tendency for ions to reach further out than the electrons as a consequence of their different gyroradii. The charge separation due to the difference of plasmasheet or ring current ions and electrons gyroradii is also neglected in the theoretical shielding model by Senior and Blanc (1984).

Both mechanisms producing non-smooth electric fields in the equatorial region of the magnetosphere do not exclude each other but are complementary. Small-scale charge separation electric fields like those occurring in tangential discontinuities modelled by Lemaire and Burlaga (1976), in Alfvén layers or in shocks are kinetic features which, like double layers, are not anticipated in current global models for magnetospheric electric fields.

4.11 Concluding remarks

A reading of Chapters 2, 3 and 4 points to the many gaps in our knowledge of the plasmasphere. We know little about the plasmapause as a boundary, or more correctly, as a boundary layer where hot and cold plasmas come into contact with each other. We have almost no observations (except those of Fig. 4.13) of the plasmapause in the process of formation, and are unable to describe the associated gradient of the radial electron density profiles. We are unable to predict the density levels just beyond the plasmapause, which are needed if the recovery dynamics are to be investigated on a realistic basis. We would like to know in detail how the plasmapause formation process is related to the physics of hot plasma penetration of the nightside magnetosphere during substorms, and thus to understand the extent to which the formation of a plasmapause profile at a new location is a local process.

As emphasized in this Chapter 4 and Chapter 2 (Section 2.5.1), most of the scientific community may be unaware that, during disturbed periods, as much or more plasma is lost from the outer plasmasphere (i.e. from the volume inside the plasmapause surface) as is lost by erosion from the region beyond the newly formed plasmapause. This may occur through a process of enhanced downward flow into the ionosphere; a good picture of this phenomenon, including its often quite sharp low-latitude limits, is not yet available to us. This density decrease effect may occur in restricted longitude regions, giving rise to longitudinal variations in the outer plasmaspheric density distribution. In general, the density decreases in the outer plasmasphere during disturbed periods, but the amount of the reduction may vary irregularly with longitude or local time in ways yet to be determined (Carpenter, 1995, personal communication).

The distribution of irregularities in the vicinity of the plasmapause has only begun to be explored, as well as the physics of their origin. Questions concerning the origin and distribution of field-aligned irregularities within and beyond the plasmasphere, including those that guide whistlers, remain largely unanswered.

It has become clear that dense cool plasmas are not efficiently evacuated from the afternoon–dusk outer magnetosphere. Is this effect related to the existence of large fluctuating electric fields on field lines threading the auroral oval? To what extent are the convection regions at ionospheric heights and at high latitudes decoupled from convection in the overlying regions of the magnetosphere?

There are rich, but only partially exploited, data sets that could be used to investigate many aspects of the problems mentioned above. Among these are the measurements of resonance frequencies in the local plasma from ISEE-1 and CRRES, both of which can be used to study the total density along highly eccentric orbits near the equator. Similar data, but from different orbits are available from GEOS-2, DE-1, AKEBONO. A new generation of low-energy particle analyzers, operating at multiple locations in synchronous orbit, offers a unique perspective on the energetics and dynamics of the outer plasmasphere and bulge region.

Among the many fruitful experimental approaches that should be considered in the future, one involving existing data and another imaging and radio soundings from satellites deserve special comment and attention.

The sheer size of the plasmasphere system, the elusiveness of plasmapause formation, and the dynamic and complex nature of the system's behavior suggest the need for imaging and radio sounding on a global basis. Imaging of the plasma through EUV methods should be done in conjunction with radio sounding in which plasma structures such as the plasmasphere and magnetopause are be probed remotely (Williams, Roelof and Mitchell, 1992; Green et al., 1994; Mauk et al., 1995; Burch et al., 1995).

Chapter 5

Theoretical aspects related to the plasmasphere

5.1   Introduction

Thus far we have reviewed observations of the plasmasphere and its properties (Chapters 1, 2 and 3). In Chapter 4 an overall phenomenological description of these properties was attempted. It remains to review in this Chapter the main steps followed since 1960 in our theoretical description and understanding of the plasmasphere, of its connection to the topside ionosphere, and of its outer boundary, the plasmapause.

We start with one of the first theoretical problems encountered by the whistler community: what is the distribution of plasma along the geomagnetic-field-aligned filamentary plasma ducts within which VLF waves propagate? This question along with hydrodynamical and kinetic models of the plasmaspheric refilling mechanism will be discussed in the first part of this Chapter. In the second part we review the mechanisms of convection and of plasma interchange motion in a magnetic field. Following historical order, we examine the theories that have been proposed for the formation of the plasmapause and for the dynamics of cold plasma in the inner magnetosphere.

There are subsidiary or complementary theoretical aspects that will not be addressed in this Chapter, examples being questions of propagation and ray tracing of VLF waves in the plasmasphere and the theory of wave–particle interactions. Such interactions were mentioned in Chapter 4 as a potential mechanism for particle pitch angle scattering and for heating the outer plasmasphere; these important physical processes themselves would deserve an entire monograph.

The basic theoretical concepts and ideas which flourished during the past thirty years have been discussed in a number of contributed articles and review papers; we are not sure that we have properly quoted them all, and any omission must be taken as unintentional.

5.2   Field-aligned and equatorial plasma density distributions

5.2.1   Introduction

Comprehensive time-dependent and three-dimensional models of the middle- and high-latitude ionosphere are now available (see Schunk, 1988a; Schunk and Sojka, 1996b). These models take into account photochemistry, recombination processes and production of various ions due to reactions with the neutral atmosphere; they are based on the transport equations for mass, momentum and energy for the various ions. Horizontal and vertical ion flows across and along magnetic field lines are determined by the convection electric field model adopted. The effects of counterstreaming of H^+ and He^+ along plasmaspheric tubes have been comprehensively studied by Bailey *et al.* (1977, 1990), Young *et al.* (1980), Bailey (1980), Richards, Schunk and Sojka (1983), Richards and Torr (1985), Förster and Jakowski (1988), Rasmussen and Schunk (1988), Singh (1988, 1990) and Singh and Torr (1990).

Self-consistent, coupled ionosphere–thermosphere models have been developed by several groups: Fuller-Rowell *et al.* (1987, 1996), Rees and Fuller-Rowell (1992), Roble *et al.* (1988), Roble (1996), Millward *et al.* (1996), Sojka and Schunk (1985), Schunk (1988a), Schunk and Sojka (1992, 1996a). These physical models have an advantage over the empirical models in that the response of the system to changing geophysical conditions can be modeled. But they have such other disadvanges as their need for large CPU time and the sensitivity of the solutions of many input parameters which are usually not well known.

Spherical harmonic models of the ionosphere–thermosphere based on fit functions to outputs from comprehensive numerical or physical models have been determined by Anderson, Forbes and Codrescu (1989), Daniell *et al.* (1990, 1996), Andersen, Decker and Valladares (1996). This type of model which carry names like FAIM (Fully Analytic Ionospheric Model) SLIM (Semi-Empirical Low-Latitude Ionospheric Model) or HLISM (High Latitude Ionospheric Specification Model) are computationally faster than the numerical models they are derived from.

A powerful tool developed by Richmond and Kamide (1988) to determine the state of high-latitude electrodynamic conditions in the ionosphere is the well-known AMIE model: (Assimilative Mapping of Ionospheric Electrodynamics). Richmond *et al.* (1990) applied this technique to combined magnetometer and incoherent scatter radar. The mathematical model of the middle- and high-latitude ionosphere developed at the Utah State University by Schunk (1988a) and co-workers is also a comprehensive tool to simulate the distributions of ionospheric parameters. While these modern electrodynamic and ionospheric models are important for understanding the coupling between the ionosphere and the plasmasphere, it is beyond the scope of this book to review them. The emphasis in the following paragraphs will mainly be on the high-altitude extension of the low and mid-latitude ionosphere; i.e. the plasmasphere.

Furthermore, ionospheric models driven by real-time magnetospheric inputs are also currently under development. These different complementary approaches have been recently reviewed and compared by Schunk and Sojka (1992, 1996b).

New and fundamental progress is currently underway to describe the changes of the particle velocity distribution of the ions in the transition region between the collision dominated ionosphere and the collisionless ion-exosphere. This effort is led by Barakat et al. (1990, 1994a, 1995), Barghouti et al. (1993, 1994) at the Utah State University, Logan, and by Wilson et al. (1992, 1993) at the University of Alabama, Huntsville. These new basic advances in the kinetic theory of the polar wind and post-exospheric models for plasmaspheric refilling are based on direct Monte-Carlo simulation method, hybrid fluid/particle-in-cell methods, and on numerical solutions of the fundamental Fokker–Planck equation (Lie-Svendsen and Rees, 1996; Pierrard, 1997).

The main constituents of the atmosphere at great heights are helium and hydrogen. Hydrogen ions are mainly formed by the charge exchange reaction (4.1). This reaction is accidentally resonant and proceeds at almost the same rate in both directions. Therefore, the main sink for H^+ ions is the reverse reaction. Throughout the F-region there are sufficient collisions to maintain H^+ in chemical equilibrium. However, in the topside ionosphere where the O^+ density falls below approximately $5 \times 10^4 \, cm^{-3}$, H^+ ions are able to diffuse along magnetic field lines. The direction of the H^+ diffusion velocity depends mainly on the relative densities of the species involved in the charge exchange reaction (4.1). When the plasmasphere is depleted, more H^+ ions are produced by the forward reaction than are removed by the reverse process. This leads to a net upward flow of H^+ into the plasmasphere. However, there is a limit to the rate at which the protonosphere can be replenished by the upward H^+ flux (Hanson and Ortenburger, 1961; Hanson and Patterson, 1963, 1964; Hanson et al., 1963). Geisler (1967) indicated that the most important factor limiting the magnitude of the upward H^+ flux is the neutral hydrogen density.

Hanson and Patterson (1963) and Hanson et al. (1963) provided a first estimate of the proton fluxes likely to be involved as a result of the diurnal variation of atomic hydrogen concentration in the exosphere. They considered the flow of hydrogen into and out of the protonosphere by the charge exchange process referred to above, the escape of hydrogen from the daytime exosphere associated with higher daytime temperatures, and also lateral flow of neutral hydrogen around the Earth due to the noon–midnight asymmetry in the distribution of hydrogen and of thermospheric temperatures. It is clear from this work that modifications to a simple equilibrium theory are required to describe completely the physics of the ion-exosphere in terms of diurnal changes, seasonal variations, plasmaspheric refilling and inter-hemispheric flow of ionization along magnetic field lines. Comprehensive model simulations including all these factors can be found in Bailey (1983), Förster and Jakowski (1988), Richards *et al.* (1983, 1988), Guiter *et al.* (1995a, b).

Before describing the more complicated non-equilibrium or dynamical theories, it is useful to consider the extent to which an equilibrium theory is adequate to describe the field-aligned distribution of the plasma in a rotating or non-rotating ion-exosphere, as discussed by Bauer (1962, 1963), Angerami and Thomas (1964), Eviatar, Lenchek and Singer (1964), Hartle (1969) and Lemaire (1976a). We first consider the case of diffusive equilibrium (DE), in which the plasma velocity distribution is Maxwellian and isotropic, and secondly the case of exospheric equilibrium (EE), in which the velocity distribution is anisotropic, without any net inter-hemispheric flow.

Due to their large mean free path compared to their gyroradius the charged particles tend to move longer distances along magnetic field lines than in the direction perpendicular to the magnetic field, B. This is why it is generally considered that the plasma density is distributed in a smooth manner along magnetic flux tubes while sharper density and temperature gradients can be maintained more easily in the direction transverse to magnetic field lines. When plasma is in hydrostatic equilibrium, its field aligned density distribution is determined by the gravitational potential. Besides the gravitational forces, which are proportional to the ion and electron masses, one must take into account the forces due to the ambipolar electric field, which has a non-zero component parallel to B. This polarization field is induced in all geophysical plasmas by the electric charge separation that develops because of the different gravitational forces acting on the electrons and on the heavier ions.

Inertial and thermo-electric effects also produce polarization electric fields in the ionosphere and magnetosphere. These ambipolar electrostatic fields play a key role in maintaining quasi-neutrality in the ionized gas. The parallel components of these polarization E-fields are neglected in the MHD approximation of plasma physics, where it is assumed that $\mathbf{E}.\mathbf{B} = 0$. Their intensity is generally small compared with the perpendicular electric field intensity, so that their effects can be ignored in calculations of drift paths of magnetospheric particles whose energies are 1–100 keV. However, as demonstrated in Section 5.2.2, neglecting the effect of these ambipolar E-fields for ionospheric particles with energies smaller than 1–2 eV necessarily implies the violation of charge neutrality of plasmas in the ionosphere and plasmasphere (Lemaire and Scherer, 1970,

1973). Hence the polarization E-field induced in geophysical plasmas by the gravitational force is important, and should not be ignored as it has been in early MHD models of the plasmasphere. The key role of this polarization electric field in accelerating the polar wind H^+ ions out of the ionosphere was pointed out by Dessler and Cloutier (1969) and by Lemaire and Scherer (1970), and was re-emphasized in a recent study by Lin et al. (1994).

5.2.2 The polarization electrostatic field

When diffusive equilibrium is reached in a neutral atmosphere, each atomic constituent is distributed with height in the gravitational field independently of the other species, according to its own atomic mass. In an ionized atmosphere this is not quite the case; the height distribution of a light ionic constituent is influenced by the distribution of the heavier charged particles. As already mentioned above and in Chapter 4, an electric field **E** builds up because the gravitational charge separation between the electrons and the heavier positive ions. This electrostatic field is directed upwards (i.e. opposite to the gravitational acceleration **g**) because of the tendency of positive ions to settle down at the base of the atmosphere with a density scale height inversely proportional to their mass, while the electrons tend to be distributed more uniformly with height due to their smaller gravitational potential energy. This polarization electric field, which was first introduced by Pannekoek (1922) and Rosseland (1924) in studies of ionized stellar atmospheres, always appears in any ionospheric and magnetospheric plasma. Mange (1960) gave an expression for the Pannekoek–Rosseland E-field for the ionosphere, when O^+ and H^+ are in hydrostatic equilibrium in the gravitational potential field.

Since the existence of this electric field is not generally discussed in text books of MHD plasma theory, it is useful to recall here how it can be determined. The vertical distribution of the partial pressure p_j for each charged particle species with atomic mass m_j and electric charge $Z_j e$ is obtained by solving the following set of coupled hydrostatic equations:

$$dp_j/dz = - m_j n_j g + Z_j n_j eE \qquad (5.1)$$

These hydrostatic equations can be integrated for each separate species, including the free electrons for which m_e can be neglected compared to the ionic masses. In the simplest case the temperatures of all ions and electrons are equal $(T_i = T_e = T)$ and independent of the altitude z. By using the perfect gas law, $p_j = kn_j T$, multiplying the equations (5.1) by Z_j and summing over j, it can then be verified that the charge separation electric field required to maintain quasi-neutrality $(\Sigma_i n_i Z_i = n_e)$ at all altitudes is given by:

$$e\mathbf{E} = - (\Sigma_j Z_j n_j m_j / \Sigma_j Z_j^2 n_j)\mathbf{g} \qquad (5.2)$$

This is the Pannekoek–Rosseland E-field. It can be shown that, to a good approximation, the intensity of this E-field is equal to $-(\text{grad } p_e)/en_e$, where n_e and p_e are the electron density and pressure, respectively. It corresponds to the ambipolar electric field when the vertical diffusion velocities of the ions with respect to the electrons are equal to zero, and when the plasma as a whole has no vertical bulk speed, i.e. when it is in hydrostatic equilibrium.

This electric field is directed upward; it accelerates the ions upwards and the electrons downwards. The electrostatic potential barrier for the electrons is much higher than their gravitational potential barrier. In the case of a purely hydrogen plasma, the total force (gravitational plus electrical) acting upon the electrons is then exactly equal to the total downward force acting upon the proton:

$$\mathbf{F}_{\text{tot,p}} = m_p\mathbf{g} + e\mathbf{E} = (m_p + m_e)\mathbf{g}/2 = m_e\mathbf{g} - e\mathbf{E} = \mathbf{F}_{\text{tot,e}}$$

As a matter of consequence, the density scale heights of both electrons and ions then become equal, by the requirement that the plasma must be quasi-neutral at all altitudes:

$$H_p \equiv -\,\mathrm{d}z/\mathrm{d}\ln n_p = kT_p/F_{\text{tot,p}} = kT_e/F_{\text{tot,e}} \equiv H_e$$

If the electrostatic polarization field should be smaller than the Pan-nekoek–Rosseland E-field then, the density scale height of the electrons would be larger than that of the positive charges, and consequently either the plasma would not be quasi-neutral at all altitudes (which is not acceptable), or it would not be in hydrostatic equilibrium as assumed here (according to equation 5.1.)

Ganguli and Palmadesso (1987) have derived a more general expression for this parallel ambipolar electric field which includes the effect of field-aligned flows of ions and electrons ($nv_{\|}$) as well as minor contributions due to $\delta v/\delta t$, the change of the bulk velocities ($v_{\|}$) due to collisions as modeled by Burgess (1969), and to anomalous transport effects associated with plasma turbulence as modeled by Mitchell and Palmadesso (1984).

$$\begin{aligned}
E_{\|} = {}& (m_s/en_eA)(\partial/\partial r)(n_pv_{p\|}^2 A - n_ev_{e\|}^2 A) \\
& -(k/e)[\partial T_{e\|}/\partial r + T_{e\|}\partial\ln n_e/\partial r + (T_{e\|} - T_{e\perp})\partial\ln A/\partial r] \\
& -n_{0+}m_eg_{\|}/en_e + (m_e/e)[\delta v_{e\|}/\delta t - (n_p/n_e)\delta v_{p\|}/\delta t]
\end{aligned} \tag{5.2'}$$

where $A(r)$ is the cross-section of the flux tube; n_e, n_p and n_{0+} are, respectively, the number densities of the electrons, H^+ and O^+ ions, $T_{e\|}$ and $T_{e\perp}$, respectively, the parallel and perpendicular electron temperatures. It can indeed be verified that equation (5.2') degenerates into Pannekoek–Rosseland's formula given by equation (5.2) when $v_{p\|} = v_{e\|} = 0$, $T_{e\|} = T_{e\perp} = T_{0+} = T_p$, $\delta v_{e\|}/\delta t = \delta v_p/\delta t = 0$.

It can also be verified from the gravitational and electric Poisson's equations that only a minute charge separation suffices to produce the electric field given by equation (5.2): $(n_p - n_e)/n_e \leq 4 \times 10^{-37}$ (see Appendix in Lemaire and Scherer, 1970). This demonstrates that quasi-neutrality in a space plasma is almost always a very good approximation.

Lemaire and Scherer showed in 1970 that to maintain local quasi-neutrality in an ionosphere where plasma is accelerated upwards, as in the polar wind, the electrostatic field must be larger than the Pannekoek–Rosseland field. Indeed, the inertial force $-nmdv/dt$ resulting from the upward acceleration of the ions (: first term in the right hand side of 5.2′) then has to be added to the gravitational force; both forces are proportional to m, the mass of the particles, while the electric force is proportional to their charge Ze. It is precisely this larger upward E-field which accelerates the light ions to supersonic speeds in the solar and polar wind flows as pointed out by Dessler and Cloutier (1969) and Lemaire and Scherer (1970, 1973, 1974). It is the same polarization E-field that accelerates ions upward in refilling plasmaspheric flux tubes.

Integration of equation (5.2) along a radial direction or along a magnetic field line gives the electrostatic potential difference between the bottom of the ionosphere and any other point. The difference of electric potential energy between the ionosphere and infinity is of the order of half the gravitational potential energy, i.e. 0.6 eV in the case of the Earth's ionosphere. Additional thermoelectric potential differences – as large as 5 kV – can be generated along auroral field lines due to charge separation between hot magnetospheric plasmas and the cold ionospheric plasma (Hultqvist, 1971; Knight, 1973; Fridman and Lemaire, 1980; Lemaire and Scherer, 1974, 1983; Pierrard, 1996).

The intensities of these parallel electric fields are usually small ($\sim \mu V/m$) compared to perpendicular magnetospheric convection or the corotation electric fields ($\sim mV/m$). Note, however, that by an appropriate Lorentz transformation of the frame of reference the convection and corotation electric fields can be reduced to an arbibrarily small value which is exactly equal to zero in the comoving frame of reference. This is not the case, however, for the parallel component of the polarization E-field. As a matter of consequence, $\mathbf{E}.\mathbf{B} \neq 0$ in any frame of reference. The relative largeness of the convection and corotation electric fields compared with the ambipolar or polarization parallel electric fields should not be considered a sufficient reason to disregard the effects of the latter, i.e. to 'justify' the pertinence of the ideal MHD approximation when mapping magnetospheric electric fields down into the ionosphere along 'equipotential magnetic field lines'. Indeed, without this small electric field ideal MHD plasma are not able to be quasi-neutral in a gravitational or any external field of forces. A similar conclusion has been reached recently by Newell and Meng (1993) in a different context.

5.2.3   The total field-aligned potential energy distribution

The total force acting on a neutral particle of mass m_j at radial distance r and latitude λ is the sum of the radial gravitational force, $f_g = -GMm_j/r^2$, and the centrifugal pseudo-force, $f_c = m_j\Omega^2 r$, which is perpendicular to the axis of rotation of the Earth. This total force can be derived from the total potential energy distribution

$$m_j\Phi = -m_jGM/r - m_j\Omega^2 r^2\cos^2\lambda/2 \qquad (5.3)$$

where M is the mass of the Earth, and G is the universal gravitational constant.

As indicated in the previous section, the gravitational forces induce a polarization electrostatic field in the plasma. The same is true for the centrifugal forces, of course. In the frame of reference rotating with the angular velocity Ω the electric field is given by $\mathbf{E} = -\nabla\Phi_E$, where the electrostatic potential energy, $e\Phi_E$, of a charge e is determined by integrating equation (5.2). It is given by

$$e\Phi_E = -(m_p - m_e)\Phi/2 \qquad (5.4)$$

when the plasma is in hydrostatic equilibrium and the proton and electron temperatures are equal and independent of altitude: $T_p = T_e = \text{constant}$.

When these temperatures are not equal or when they depend on altitude, there exists a similar but more complicated relationship between the electric potential and the gravitational potential (see Appendix in Lemaire and Scherer, 1970). In the more general case when the boundary conditions in both hemispheres are not symmetrical, interhemispheric flow will exist along magnetic flux tubes; the Pannekoek–Rosseland potential (5.4) and field (5.2) corresponding to hydrostatic equilibrium are then no longer applicable; in this case the electrostatic potential, $\Phi_E(z)$ must be determined as a function of altitude by solving iteratively the quasi-neutrality equation:

$$n_e(\Phi_E) - \Sigma_i Z_i n_i(\Phi_E) = 0 \qquad (5.5)$$

This algebraic equation can be solved for Φ_E by the Newton–Raphson or the secant methods for a set of altitude z. Such an iterative numerical procedure to solve the quasi-neutrality equation has been used by Lemaire and Scherer (1969; 1974) in their kinetic models of the polar wind. More recently, Lemaire *et al.* (1991) proposed a faster numerical method: i.e. integrate the derivative of the quasi-neutrality equation or the derivative of Poisson's equation, instead of solving iteratively the algebraic quasi-neutrality equation.

The distribution of the total potential energy of a charged particle along a centered dipole magnetic field line is obtained by replacing r by $LR_E\cos^2\lambda$ in equations (5.3) and (5.4), where LR_E is the equatorial radius of the magnetic field line. The dipole axis is assumed to be parallel to the axis of rotation of the Earth. This non-essential assumption can be relaxed, but at the expense of increased mathematical complexity in the formulation of the problem. Figure 5.1 illustrates the total potential energy of a proton along different magnetic field lines in a corotating protonosphere. The field aligned potential distribution has a maximum in the equatorial plane for field lines whose L-parameters are smaller than the critical value

$$L_c = (2GM/3\Omega^2 R_E^3)^{1/3} = (2/3)^{1/3} L_o \qquad (5.6)$$

where L_o is the equatorial distance in Earth radii of the geostationary orbit: $L_o = 6.6$ and $L_c = 5.78$ when Ω is equal to the angular velocity of the Earth; the value of L_c becomes smaller than 5.78 in the post-midnight local time sector when enhanced eastward magnetospheric convection velocities are present. This usually occurs during of magnetospheric substorms (see Fig. 4.10b). For instance, when the angular velocity Ω is enhanced by a factor of 3, the value of L_c is reduced from 5.78 to 2.78.

For $L > L_c$, the total potential energy has a minimum in the equatorial plane and two symmetrical maxima at the latitudes $\lambda_M = \arccos(L_c/L)^{3/8}$. Magnetic field lines for which $L > L_c$ traverse a surface at $\lambda = \pm \lambda_M$ that has been called the 'Roche Limit' of ionospheric plasma, due to the analogy with the Roche Limit in astrophysics (Lemaire, 1974, 1976a). Indeed, at this surface the total potential distribution has a maximum, like the total potential of a rotating binary star system that has a maximum at its Roche Limit surface (see Kopal,

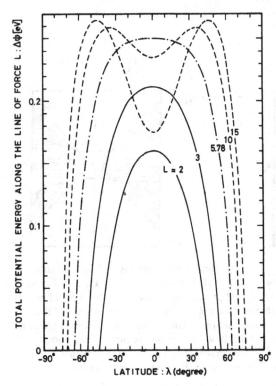

Figure 5.1 Total potential energy of a proton along different dipole magnetic field lines determined by their L-parameter, i.e. their equatorial distance. The potential energy is given in eV, as a function of the dipole latitude. The gravitational potential is predominant at low altitudes. For $L > 5.78$ the centrifugal effect become important at high altitude and produces the equatorial potential well shown by the broken lines. The electrostatic potential is assumed to be the Pannekoek–Rosseland potential (after Lemaire, 1974).

1989). However, since the similarity between the astrophysical and terrestrial exospheric situations is disputable, Lemaire (1985) proposed to call this surface the Zero Parallel Force (ZPF) surface. Indeed, it is the locus of all points where the projection of the total force parallel to the magnetic field direction changes sign and becomes equal to zero. This surface is illustrated in Fig. 5.2.

S. J. Bauer (1994, personal communication) pointed out to the authors that Alfvén (1967) had already given the expressions of the field-aligned component of the gravitational force and of the centrifugal force along dipole magnetic field lines. In Alfvén's article it is argued that when the plasma convection velocity $v_\varphi(r)$ is smaller than the local Keplerian orbital velocity $v_K(r)$ by a factor $(2/3)^{1/2}$, the parallel components of the gravitational and centrifugal forces balance each

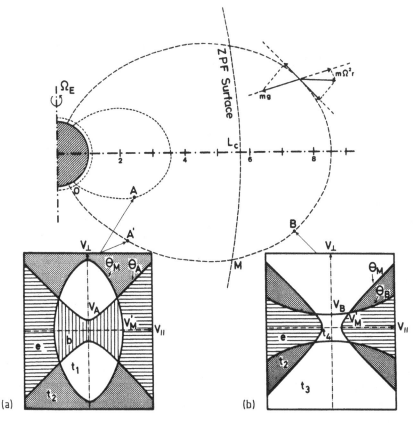

Figure 5.2   Two dipole magnetic field lines corresponding to $L = 3.5$ and 8.5 are illustrated by dashed lines. The meridional section of Zero Parallel Force (ZPF) surface where the field aligned projections of the gravitational and centrifugal forces balance each other is shown. L_c is the equatorial distance of this ZPF surface; $L_c = 5.78$ for an ion-exosphere corotating with the Earth's angular velocity. The two panels (a) and (b) show the different regions in the velocity space corresponding to all different classes of orbits of charged particles below (at A and A′) and above the ZPF surface (at B) (after Lemaire, 1976).

other. Therefore Alfvén considered that, in order to be in equilibrium, plasma trapped in a dipole magnetic field had to move around in azimuth (i.e. be convected in the direction perpendicular to the magnetic field lines) with a velocity smaller by a factor 1.22 than the local Keplerian velocity.

But, in a later paper Alfvén (1976) noted that, when 'the conductivity of plasma is infinite, all parts of the plasma must rotate with the same angular velocity as the central body'. In this case, at any point P with the coordinates (r,λ), where $r(\cos \lambda)^{2/3} < L_c$, a plasma element will fall down toward the central body. Otherwise, when $r(\cos \lambda)^{2/3} > L_c$, the centrifugal force dominates the gravitational force and the plasma element will 'fall down' toward the equatorial plane. The result is that when the convection electric field corresponds to the corotation E-field, the plasma element splits into two parts at the surface where $r = L_c(\cos \lambda)^{-2/3}$. One one side of this surface the plasma falls downwards, while on the other it moves upwards. This surface coincides with the Roche Limit or ZPF surface introduced two years earlier by Lemaire (1974, 1976a). The plasma along the inner portion of the flux tube is tight to the ionosphere while the plasma in the outer part tends to move away and to form a ring in the equatorial plane around the central body. Alfvén (1976) and Alfvén and Arhenius (1976) did not envisage the consequences of these centrifugal effects for the corotating terrestrial plasmasphere.

5.2.4   The diffusive equilibrium (DE) density distributions

When all ions and electrons are in hydrostatic, isothermal and diffusive equilibrium in the gravitational field and Pannekoek–Rosseland electric field, the field-aligned density of each particle is determined by

$$n_j(\lambda) = n_j(\lambda_0)\exp[-(m_j\Phi + Z_je\Phi_E)/kT_j] \qquad (5.7)$$

$n_j(\lambda_0)$ is the density of the particle (Z_j, m_j) at the reference latitude λ_0 where the magnetic field line L traverses the reference level $r = r_0$. When the electric potential is given by equation (5.4), the number density of the electrons is everywhere equal to the total number density of positive charges i.e. the plasma is then quasi-neutral. A density distribution decreasing exponentially as a function of the potential energy of the particles is said to be in barometric equilibrium or in diffusive equilibrium when several different ion species are present.

Angerami and Thomas (1964) analyzed in detail the properties of these equilibrium solutions for different relative concentrations of O^+, He^+ and H^+ specified at a reference altitude of 500 km. They extended their study to include more general cases in which ion and electron temperatures vary with altitude. To describe these more general models they found it useful to define a geopotential height z which is then used as the independent variable to integrate the coupled set of hydrostratic equations (5.1) for the electrons and ions. Their 'temperature-modified geopotential height' is defined as

$$z = \int_0^{s'} f(s)t(s)ds/g_0 \qquad (5.8)$$

where $f(s)$ is the field-aligned component of the sum of the gravitational and centrifugal forces, $t(s) = T_{jo}/T_j(s)$ depends on the distance s along the magnetic field line and is inversely proportional to the particle temperature, and g_0 is the gravitational acceleration at the reference level r_0. This new variable z is a generalization of the geopotential height introduced earlier by Bauer (1962), but, the definition (5.8) is more general since it incorporates the variation of the temperature along the magnetic field line and the contribution of the centrifugal force. This geopotential height is a useful coordinate for study of field-aligned plasma flow and interhemispheric ionization transport. It should be reintroduced in modern theoretical studies of the plasmasphere refilling process.

Figure 5.3 illustrates the field-aligned distribution of the O^+, He^+, H^+ ions and electrons densities as a function of geopotential height z (left scale), and as a function of altitude h (right scale). In this case $t(s)$ was assumed independent of s, the temperatures of all particles were assumed to be equal to 1500 K, and the ionic composition at the base was assumed to be $[He^+]/[O^+] = 6.3 \times 10^{-3}$, $[H^+]/[O^+] = 1.6 \times 10^{-4}$. Note that the concentration of the heaviest ions (O^+) is everywhere a decreasing function of height, and that, at $z < 400$ km, its density scale height H_O is twice as large as the density scale height of neutral

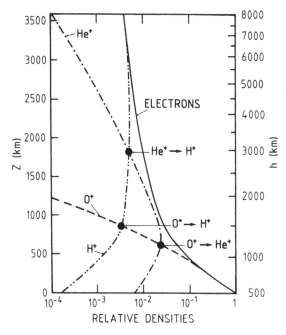

Figure 5.3 Relative electron and ion densities along a magnetic field line as a function of the temperature-modified geopotential height z for a constant temperature (1500 K). The altitude h is given by the right-hand scale. The electrons and ions are assumed to be at the same temperature, and the ionic composition at the base (500 km altitude) is $[He^+]/[O^+] = 6.3 \times 10^{-3}$, $[H^+]/[O^+] = 1.6 \times 10^{-4}$ (after Angerami and Thomas, 1964).

oxygen atoms having the same temperature. This is a consequence of the additional electric force given by equation (5.2) and opposing the gravitational force. At low altitude He^+ and H^+ ions are minor constituents; their concentrations increase with height for $h < 1200$ km. In Fig. 5.3 a dot indicates the altitude where He^+ becomes more important than O^+ (at $h = 1200$ km); H^+ predominates over He^+ at altitudes above 3000 km.

The electron concentration (solid line in Fig. 5.3) is equal to the sum of all the concentrations of the singly ionized ions O^+, H^+ and He^+. Similar electron density distributions corresponding to isothermal diffusive equilibrium have been used to interpret whistler observations through calculations of the time of propagation of VLF waves in field-aligned whistler ducts.

Figure 5.4a shows how the distribution of electrons in a magnetic flux tube changes when the temperatures and/or the relative concentrations of the ions at 500 km are varied as indicated in Table 5.1. Curve d corresponds to a smaller percentage of He^+ and H^+ at the base level, so that O^+ is predominant over a greater height range. Thus the electron density maintains a high rate of decrease over a greater height range, and is smaller for the same z than in curve c. This figure indicates that a change in the relative ion composition at the base level is far more important than a temperature gradient in determining the density distribution of the electrons at high altitude in magnetospheric flux tubes. We wish to emphasize this conclusion obtained more than thirty years ago by Angerami and Thomas (1964), since it is often overlooked in discussions of modern models of field-aligned plasma density distributions.

The observed ion density distribution measured on the ascent of a Blue Scout rocket by Sagalyn and Smiddy (1964) by means of a spherical electrostatic analyser is shown in Fig. 5.4b. Note the marked 'kink' in the ion distribution at 650 km. At this altitude, where the slope of ion density is changing by a factor of $m_{O^+}/m_{H^+} = 16$, the dominant ion species changes from O^+ at low altitude to H^+ above 650 km. The existence of such a sharp 'kink' is indicative that the He^+ ions remain everywhere a minor ionic constituant, unlike in Figs. 5.3 and 5.4a. Indeed, in these latter figures there is a range of altitudes where the He^+ ion density is larger than the O^+ and H^+ ion densities.

At low altitudes where O^+ is predominant for all curves, the slopes of $n_e(z)$ are proportional to the temperature. At higher altitude the effect of ion composition becomes relatively more important, and the slopes (or electron density scale heights) are no longer proportional to temperature. This is also illustrated in Fig. 5.4a where the reduced slope of curve e at low altitude is due to the larger temperature (2000 K instead of 1500 K). But, due to the reduced H^+ and He^+ relative concentrations, the He^+ ions become predominant between $h = 2000$ km and 8000 km. This implies that it could be misleading to infer the electron temperature from measured electron density scale height. Furthermore, inferring heating processes at high altitude from large values of the electron

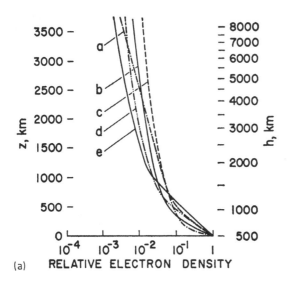

(a)

Figure 5.4a Relative electron density along a line force as a function of the temperature-modified geopotential height z for isothermal ion-exospheres in diffusive equilibrium. The corresponding altitude (h) is given on the right-hand scale. The compositions at the base level and the temperatures corresponding to the curves a–e are shown in Table 5.1 (after Angerami and Thomas, 1964).

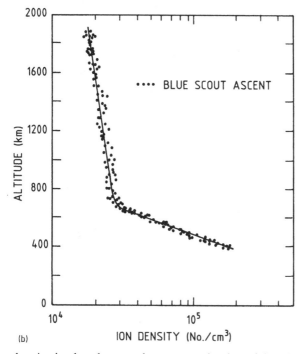

(b)

Figure 5.4b Altitude versus ion density measurements during the ascent flight of a Blue Scout are shown by dots. A sharp change in the slope of the ion density distribution is observed near 700 km (after Sagalyn and Smiddy, 1964).

density in the plasmasphere, may also be misleading when the relative concentrations of O^+, He^+ and H^+ ions at the low-altitude reference level are not well known.

Similar multi-ionic hydrostatic models were calculated earlier by Bauer (1962; 1963) without the centrifugal effect which tends to increase slightly the electron density at high levels. For field lines whose invariant latitudes are smaller than 55° (i.e. $L < 3$) the increase of equatorial density n_{eq} due to

Table 5.1 *Boundary conditions at 500 km corresponding to curves of Fig. 5.4.*

Curve	T, K	$[He^+]/[O^+]$	$[H^+]/[O^+]$
a	500	6.3×10^{-2}	1.6×10^{-2}
b	1000	2.0×10^{-2}	1.6×10^{-3}
c	1500	2.0×10^{-2}	1.6×10^{-3}
d	1500	6.3×10^{-3}	1.6×10^{-4}
e	2000	2.0×10^{-3}	1.6×10^{-5}

centrifugal effects is less than 20%. But for $L > L_c = 5.78$, the increase of the diffusive equilibrium value of n_{eq} and of total electron content of a magnetic flux tube N_T become more significant when the angular rotation speed of the plasmasphere is taken into account. A larger amount of cold ionospheric plasma can then accumulate in the equatorial potential well. This centrifugal effect is most significant in the post-midnight local time sector where eastward magnetospheric convection is suddenly enhanced for $L > 4$–5 at the onset of substorms.

When a field-aligned plasma element is convected from the dayside into the post-midnight LT sector at $L > 4$–5 in the equatorial region, it falls within an equatorial potential well that is deepest after local midnight; more particles can therefore be scattered and accumulate on trapped orbits confined near the equatorial plane. It has been suggested by Lemaire (1985) that the deepening of the equatorial potential well when a flux tube is convected past the post-midnight sector can produce the upwelling ionization flow out of the mid-latitude ionosphere into the plasmatrough. This mechanism should contribute to the formation of the light-ion trough in the nighttime upper ionosphere as discussed in Section 5.5.4.

Although various other field-aligned electron density distributions (see Roth, 1975) have been proposed in the early analysis of whistler f–t spectrograms, the Diffusive Equilibrium (DE) model is usually adopted for field lines located inside the plasmasphere. The appropriateness of the DE model was first pointed out by Angerami and Carpenter (1966), and by Smith and Angerami (1968). It was further supported by Park (1972), and has been demonstrated in comparisons with *in situ* electron density measurements by Carpenter *et al.* (1981).

Olsen, Scott and Boardsen (1994) examined the distribution functions of trapped H^+ ions along the $L = 4.6$ drift shell from DE-1 data. They found that the changing bi-Maxwellian character of the velocity distribution functions along magnetic field line showed reasonable agreement with a simple mapping procedure based on Liouville's theorem.

5.2.5   The exospheric equilibrium (EE) density distributions

Outside the plasmasphere the density is more than one order of magnitude lower than inside the plasmapause, usually less than 50 electrons and ions per cm^3 at $L \sim 4$ in the equatorial plane. The Coulomb collision time of a thermal proton (0.25 eV) is longer than 2 hours when $n_e < 10\,\mathrm{cm}^{-3}$; two hours is also the time (free-flight-time) needed for a proton of this energy to spiral along the field line $L = 4$ from one hemisphere to the other, assuming no deflections by Coulomb collisions or wave–particle interactions. The free-flight-times and collision times of thermal electrons are about $(m_+/m_e)^{1/2}$ or 42 times smaller, but the expected number of collisions during travel from one hemisphere to the other is almost the same for both the thermal electrons and H$^+$ ions in the plasmatrough (see Tables F1 and F2 in Lemaire, 1985). Therefore, plasma of ionospheric origin is almost collisionless in the plasmatrough outside the plasmasphere. Of course, particles with energies larger than 0.25 eV are even more collisionless, since the Coulomb collision cross-section is a rapidly decreasing function of kinetic energy (Spitzer, 1956). Nevertheless, the importance of Coulomb collisions cannot be ignored in the plasmatrough, and certainly not in the plasmasphere. The significance of this process, as compared to presumed wave–particle interactions even for the more energetic Ring Current particles, has been re-emphasized recently by Fok et al. (1993).

When the background plasma density is low, as in the plasmatrough, particles can move long distances along magnetic field lines without being deflected significantly. Their orbits can then be organized into the different classes that are illustrated in the two panels of Fig. 5.2. These different classes are (e) 'escaping' particles, which have enough kinetic energy to go over the maximum of the total potential barrier; (b) 'ballistic' particles, which do not have enough energy to do so; they fall back into the ionosphere. There are also various categories of trapped particles (t_1, t_2, t_3, t_4) which have mirror points either in the same hemisphere or in both hemispheres, or which are trapped in the equatorial potential well for field lines with $L > L_c$. The t_3, t_4 categories of particles do not exist in a non-rotating ion-exosphere. A detailed description of all these classes of particle orbits is given in Lemaire (1976a, 1985).

When all these orbits are populated by ions and electrons whose velocity distributions are Maxwellian and isotropic, the field-aligned density distribution is the same as the DE model discussed above. Such a state of equilibrium, corresponding to detailed balance between the particles in all classes of orbits, can be maintained by Coulomb collisions inside the plasmasphere; the flux tubes are then considered to be saturated. However, Lemaire (1985, 1989) emphasized that when one of these categories of orbits is not populated (e.g. when for instance the t_3, t_4 trapped particles are missing or are under-represented) Coulomb pitch angle deflections will scatter escaping (e) and trapped t_2 particles into those orbits. The pitch angle diffusion process by Coulomb collisions can be enhanced by the effect of wave–particle interactions, provided, of course, that large enough amplitude waves are continuously available to provide this additional scattering mechanism.

Pitch angle scattering mechanisms, which contribute to the population of the t_3, t_4 orbits, eventually cause more and more ions and electrons to flow up and to become stored into the protonosphere. Conversely, an excess of trapped particles in the equatorial region of the nighttime ion-exosphere causes ionisation to flow downward along magnetic field lines, as a result of pitch angle scattering.

The upper solid line in Fig. 5.5 corresponds to DE equilibrium similar to that illustrated in Figs 5.3 and 5.4a, and applies along the field line $L = 6$ to a rotating ion-exosphere with a constant temperature of 3000 K; the H^+ and electron densities at the exosbase altitude (1000 km) are assumed to be equal to $10^4 \, cm^{-3}$. The different shadings in Fig. 5.5 illustrate the contributions of the

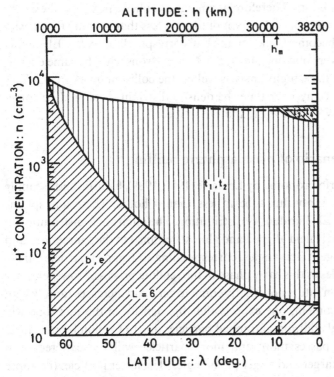

Figure 5.5   Electron and H^+ ion density distribution along the field line $L = 6$ versus dipole latitude (λ). The upper solid line corresponds to a corotating protonosphere in diffusive equilibrium (DE); the lower solid line corresponds to the exospheric equilibrium (EE) when ballistic (b) and escaping (e) particle orbits are populated, but no trapped orbits (t_1, t_2, t_3, t_4). The broken lines corespond to models in which the effect of rotation is ignored (after Lemaire, 1976).

different categories of particles according to their classes of orbits. The broken lines show the corresponding distributions when the centrifugal effects are neglected, as in the work of Eviatar *et al.* (1964). The lower solid line represents the field-aligned density corresponding to Exospheric Equilibrium (EE). In this extreme case the protonospheric density is formed only of ballistic (b) and escaping (e) particles evaporated directly from below the exobase level; in this case no trapped particles are assumed to be present in the ion-exosphere. It has been shown by Lemaire (1976a) that this EE density distribution decreases with radial distance approximately as R^{-4}, where R is the radial distance. The R^{-4}, field-aligned density model was used in early interpretations of spectrograms of whistlers that propagated in the plasmatrough beyond the plasmapause (Angerami, 1966).

   The EE distribution should be considered as a kind of ideal minimum density profile, below which the ion-exosphere cannot be maintained much longer than one free-flight-time for an ionospheric ion moving from one hemisphere to the other. Along field lines $L = 4$ this flight time is 2 hours for a proton of $0.25\,\text{eV}$ while at $L = 6$ it is 4 hours. Therefore, if a magnetospheric flux tube should be completely empty at some initial instant of time, in less than 4 hours the density distribution would become higher than that corresponding to the EE model illustrated by the lower solid line of Fig. 5.5. For electrons with the same energy, it can be shown that these flight times as well as the collision times are 42 times smaller. Therefore, at any later time the density distribution should be somewhere between the two extreme EE and DE models.

5.2.6   Kinetic scenario of plasmaspheric refilling

In the refilling scenario proposed by Lemaire (1985, 1989), ballistic and escaping particles invade an empty flux tube in less than 4 hours. These ionospheric particles build up an EE-like density distribution. The pitch angle distribution of the ions is then strongly field aligned and almost confined to the source cone and the loss cone. At larger pitch angles, corresponding to trapped particles, there are almost no particles in this early phase of the refilling process. However, since the Coulomb collision rate is not strictly equal to zero and increases rapidly with the total flux tube content, one expects that the density in the flux tube will increase by continual addition of trapped particles due to Coulomb pitch angle scattering. As time passes more and more particles will be scattered onto trapped orbits with larger and larger pitch angles. The faster particles (i.e. those forming the superthermal tail of the velocity distribution function) will be scattered more slowly than the thermal particles of lower energy. Indeed, the Coulomb collision cross-section is a rapidly decreasing function of the relative speed between colliding charged particles.

   It is therefore expected that the pitch angle distribution of particles with the lowest energies will become isotropic first. Later on, the pitch angles of super-thermal particles will also tend to become more isotropic (Lemaire, 1989). The switch over from a field-aligned to an isotropic pitch angle distribution should occur for 1–2 eV particles when the cold background density has increased to a value of 10–50 cm^{-3}. At higher energies it takes places for higher density levels. These views are supported by DE-1 observations published by Sojka et al. (1983) and Nagai et al. (1985). Indeed, these observations indicate that field aligned superthermal ion beams survive in plasmatrough flux tubes as long as the equatorial density does not exceed ~ 20 cm^{-3} at geosynchronous orbit. Similar observations are reported by Carpenter et al. (1993).

Monte Carlo simulations of the effect of Coulomb interactions on the velocity distribution function of ionospheric ions streaming in an empty ion-exosphere have recently been reported by Barakat et al. (1990) and by Barghouthi et al. (1990, 1993). These simulations have shown that, at the exobase, where the mean free path of ions becomes equal to the plasma density scale height, the H$^+$ velocity distribution function changes drastically from a nearly isotropic Max-wellian (in the collision-dominated region) to one that is field aligned and resembles a 'kidney bean embedded in a Maxwellian'. This is illustrated in Fig. 5.6 where z is a collisional depth defined by

$$z = \int_r^x (v/v_T)\mathrm{d}h$$

v_T is the thermal velocity and v the Coulomb collision frequency of the test particles with the field particles. z is analogous to the optical depth in the theory of stellar atmospheres. These quantitative results were confirmed and expanded by Wilson (1992) and Wilson et al. (1992) using a semi-kinetic particle-in-cell model to describe the gradual plasmaspheric refilling mechanism. The plasma accumulation at high altitudes occurs through collisional thermalization and pitch angle scattering controlled by the rate of velocity dependent Coulomb collisions, as suggested by Lemaire (1985, 1989).

Fig. 5.7a, b, c, d, from Wilson et al. (1993), show field-aligned distributions of H$^+$ density, bulk speed, parallel temperature, and perpendicular temperature, respectively, as functions of latitude for different elapsed times ranging from 1 to 48 hours. The curves illustrate the refilling of a fully depleted ($L = 4.5$) flux tube with non-symmetric boundary conditions in opposite hemispheres. In the first hour, two field-aligned streams emerge from the northern and southern hemi-spheres, interpenetrate, and then flow into the conjugate ionosphere. During the first hour the flow along the field line is highly asymmetric, with a net transfer of plasma from north to south. This is a consequence of the greater amount of O$^+$ in the northern topside ionosphere assumed in this case study.

Like Barghouti et al. (1990, 1993), Wilson (1992) and Wilson et al. (1993) determined the velocity distribution, $f(v_\parallel, v_\perp, t)$ for the ions beams (see Fig. 5.8).

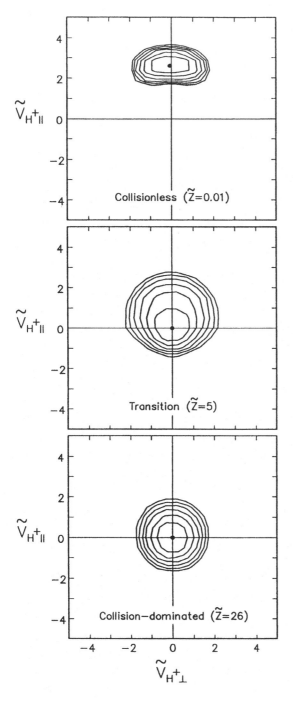

Figure 5.6 The velocity distribution function (VDF) of H^+ ions diffusing along magnetic field lines through a background of O^+ ions in hydrostatic equilibrium. These results are obtained by a Monte Carlo method simulation. The panels represent equidensity contours in velocity space $(v_\perp, v_\parallel)$ at three different altitudes in the polar wind. The lowest panel shows the nearly Maxwellian VDF in the collision dominated region ($z = 26$); the top panel shows the VDF of the outflowing H^+ ions in the collisionless region ($z = 0.01$); the mid-panel corresponds to an altitude in the middle of the transition region between the collision-dominated and collisionless regions (after Barghouti et al., 1993).

Although different numerical methods were used in both independent studies, it is gratifying to note that they reach basically the same physical conclusions. Within a few hours after the refilling starts, a significant number of ions from the northern hemisphere return after reflection in the southern hemisphere, and the phase space distributions become nearly symmetric around the equator. The gap in phase space between the two counterstreaming beams (see Fig. 5.8) gradually fills from the ionosphere toward the equator as particles are collisionally scattered onto trapped trajectories, with mirror points further removed from the ionosphere. Most of these collisions occur at low altitudes. When the gap is completely filled, the pitch angle distribution has a maximum at 90°; this occurs after about 34 hours for $L = 4.5$. By the time this gap is filled, the

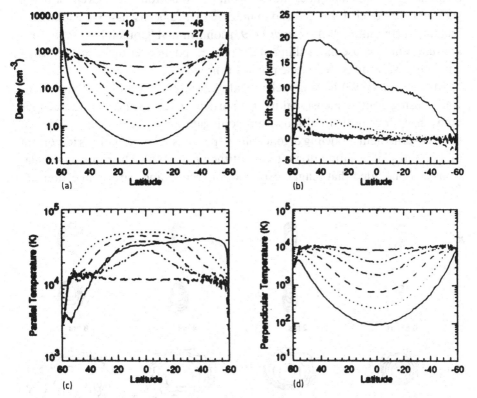

Figure 5.7. Plots of (a) H^+ density, (b) drift speed, (c) parallel temperature, and (d) perpendicular temperature as functions of latitude along a flux tube at $L = 4.5$. These snapshot distributions correspond to times of 1, 4, 10, 18, 27 and 48 hours after the initiation of refilling. These results are obtained with a hybrid particle-in-cell numerical code. It shows how an empty flux tube of the plasmasphere refills as a result ionospheric evaporation as well as Coulomb pitch angle scattering and thermalization. The conditions at the lower boundary (500 km) in both hemispheres were assumed to be non-symmetrical and adopted from the MSIS-86 neutral atmosphere model and IRI ionospheric model (after Wilson *et al.*, 1993).

ions are truly subsonic everywhere, with parallel and perpendicular temperatures nearly equal along much of the flux tube.

The evolution of the drift velocity of H^+ ions is shown in Fig. 5.7b at the times indicated. Although the drift velocity (in panel b), density (a), and parallel (c) and perpendicular (d) temperatures look symmetric about the equator, there are differences at the ionospheric ends of the flux tube due to the different boundary conditions in the two hemispheres.

During the refilling process the density decreases smoothly toward the equator where the plasma density is minimum. In the curves of Figs 5.7a and b there is no sign of propagating shock waves. Therefore, according to the kinetic and semi-kinetic scenarios discussed in this section, plasmaspheric refilling occurs from the base of the flux tube, and not from top to bottom as believed for many years and as modeled with hydrodynamical codes to be discussed in a following section. Furthermore, Wilson et al. (1993) confirm that 'Coulomb self-collisions are sufficient to cause the refilling of the plasmasphere over time scales consistent with observations. No exotic process is needed'.

Miller et al. (1993) have modeled field-aligned plasma flows in the plasmasphere using a one-dimensional hybrid particle code to study the interactions between upflowing thermal ions from conjugate ionospheres. They point out that self-consistent modeling of ionosphere–plasmasphere coupling is important and requires information down to an altitude of 200 km. Their kinetic simulations demonstrated that magnetospheric convection and particle injection can

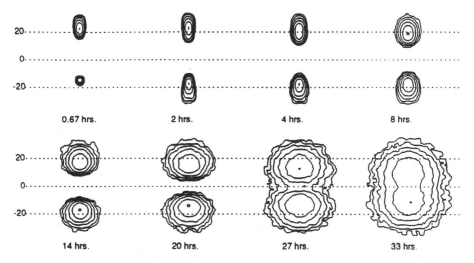

Figure 5.8 H^+ ion velocity distribution ($v_\parallel$ versus $v_\perp$ in km/s) at the equator along a magnetic field line at $L = 4.5$, and, at the indicated times (from 0.67 h to 33 h after the initiation of refilling). These plots are made by binning simulation ions according to their parallel and perpendicular components of velocity. The dashed horizontal lines indicate different values of $v_\parallel$ (after Wilson et al., 1993).

change the initial conditions for plasmaspheric refilling on time scales shorter than one hour. But over long time scales (days) this short time scale information is lost.

The effects of magnetic field line divergence, and of the external body forces were simulated separately in the study by Barghouti *et al.* (1993). To illustrate the effect of the collision model on the shape of the velocity distribution function these authors used alternatively a 'Maxwell molecule' collision model and the Coulomb interaction model.

In subsequent Monte Carlo simulations Barakat and Barghouti (1994a;b) examined the effect of wave–particle interactions on the velocity distribution of polar wind H^+ and O^+ ions flowing out of the ionosphere along magnetic field lines (see also Barghouti *et al.*, 1994; Barakat *et al.* 1995a and the work by Chang *et al.*, 1986; Retterer *et al.*, 1987).

Monte Carlo simulations have also been employed by Yasseen *et al.* (1989) and Tam, Yasseen and Chang (1995) to determine the velocity distribution of photoelectrons (treated as test particles) moving through a background of H^+, O^+, and the bulk of thermal electrons which all participate in a current-free polar wind type ionization flow. In their latest hybrid simulation code the evolution of the O^+ and H^+ velocity distributions are also determined by the Monte Carlo method, while the bulk of thermal electrons is treated as a fluid. Their results agree with observations in various aspects. They demonstrate for instance that a temperature anisotropy develops between upwardly and downwardly moving electrons. They find that $T_{e,up} > T_{e,down}$ due to the upward moving photoelectrons. This produces an upward heat flux for the total electron population.

Although, these Monte Carlo simulations are demanding large amounts of CPU time, they are illuminating and useful to test numerical models which are now being developed by Lie-Svendsen and Rees (1996) and Pierrard (1997). These authors calculated numerical solutions of the Fokker–Planck equation for the velocity distribution of H^+ ions moving along diverging magnetic field through a background of O^+ ions. Both groups of investigators obtain solutions which are similar to those of the Monte Carlo simulations, although they employ different boundary conditions at the bottom of the transition layer, as well as different mathematical methods to determine the solution of the Fokker–Planck equation. The former use the finite difference method, while the latter employ a generalized spectral method introduced by Shizgal and Blackmore (1984) and Shizgal, Weinert and Lemaire (1986). These powerful new numerical tools enable determination of the long-awaited post-exospheric approximations for ion velocity distributions in the transition layers between the collision-dominated ionosphere and collisionless magnetosphere (see Fahr and Shizgal, 1983). All these novel tools and methods, including the Monte Carlo simulation and particle-in-cell method, have now opened the door to modern plasma kinetic theory which might overrule fluid, moments, or other MHD approximations of the transport equations used for more than two decades to model space plasma distributions.

5.2.7   Plasmaspheric temperatures

During the early phase of the refilling process the parallel temperature at high altitude is rather large $2\text{--}4 \times 10^4$ K, when compared with 3000 K at the altitude of the reference level of 500 km (see Fig. 5.7c). The steep temperature gradients at low altitudes are due to the acceleration of the H^+ ions by the parallel component of the ambipolar electric field (5.2) when the ions leave the ionosphere. But in addition to this electric acceleration effect, noted by Wilson *et al.* (1993), it should be pointed out here, that steep temperature gradients are also expected in the exobase region as a consequence of a 'velocity filtration effect' (Pierrard and Lemaire, 1996). This effect is due to the decrease of the Coulomb collision cross-section when the relative velocity of the colliding ions increases, and to the non-Maxwellian character of the velocity distribution function. Scudder (1992a,b) has emphasized that the 'velocity filtration effect' can explain the 'heating' of the solar corona, where high temperatures are observed above the much cooler photosphere–chromosphere. Based on these results and on the results illustrated in Fig. 5.7c, Pierrard and Lemaire (1996) infer that the high ion temperatures observed in refilling flux tubes result from the tendency of the ions with larger velocities to experience less Coulomb drag and thus to escape more easily out of the collision-dominated ionosphere into the outer plasmasphere, where they form a population of superthermal ions.

From Fig. 5.7c it can be seen that the highest parallel temperatures are observed in the early stages of the refilling process, i.e. when the flux tube is almost empty. As time goes by and the density builds up, the temperature decreases and tends to become equal to that of the ionospheric heat bath. This is clearly illustrated in Figs. 5.7c and d. Fig. 5.7d shows the field-aligned distribution of the perpendicular temperature, $T_\perp$. The perpendicular temperature at early times shows the B^{-1} drop-off associated with conservation of the particle magnetic moment (adiabatic cooling). As time progresses, this temperature increases irreversibly and eventually tends to equal the exobase temperature. As the plasma density builds up the large initial temperature anisotropy, $T_\parallel/T_\perp$ gradually decreases and tends to $T_\parallel/T_\perp = 1$.

In Section 4.3.2 other heating mechanisms have been proposed to explain the large ion temperatures ($> 10\,000$ K) observed near the plasmapause at high altitudes. For a most recent review of this question see Comfort (1996).

Of course, part of the observed isotropization and 'heating' of the ion velocity distribution may also be a consequence of wave–particle interactions, as is often presumed. The effect of wave–particle interactions on the velocity distributions of H^+ and O^+ ions have been simulated by Chang et al. (1986), Retterer et al. (1987), Barakat and Barghouti (1995), Barakat et al. (1995) with their Monte Carlo simulation codes. But how much is contributed by wave–particle interaction mechanisms in addition to the unavoidable effects of Coulomb collisions remains to be substantiated. The importance of these additional scattering mechanisms depends on the intensity and spectrum of the electrostatic noise sporadically or permanently present in the medium.

Note also from Figs. 5.7c and d that the pitch angle distribution is most anisotropic at highest altitudes. Furthermore, it is there that the temperature anisotropy decreases most slowly due to the low rate of Coulomb collisions. Opposite effects are expected when the scattering and heating of plasmaspheric ions by wave–particle interactions are confined near the equatorial plane. Lin et al. (1992) also examined the effects of wave–particle interactions on the ion velocity and pitch angle distributions near the equator. They postulate an ad hoc 'but reasonable' wave spectrum of sufficiently high intensity to scatter a large number of ions evaporating out of the ionosphere. For high enough wave field intensities in the appropriate range of wave frequencies, Lin et al. (1992) showed that the additional scattering and heating mechanisms by waves will dominate the effects of Coulomb collisions. It remains to be confirmed experimentally that the postulated wave power is indeed 'reasonable' and that it is a permanent feature in the equatorial region.

Besides the kinetic and semi-kinetic models or Monte Carlo simulations or particle-in-cell simulations discussed in the two previous sections, a variety of hydrodynamical models have been proposed in the past to describe plasmaspheric refilling and the polar wind flow. For completeness we will now briefly review the main results based on these hydrodynamical models.

5.2.8   Time-dependent hydrodynamical models for field-aligned plasma density distribution

Numerous theoretical models describing the outflow of ionospheric plasma at polar and mid-latitudes have been based on numerical integration of hydrodynamical transport equations or moment equations. Unlike the case of kinetic models, where plasma is described by the velocity distribution functions of its different particle species, in hydrodynamical models the plasma is described in terms of the total number density or of the partial density of the electrons and different ion species, of their bulk speeds, parallel and perpendicular temperatures, parallel and perpendicular heat flow tensor components, etc. The spatial distributions of these macroscopic quantities $(n_i, v_i, T_i \ldots)$ are obtained by solving the moment equations for each particle species. These moment equations are obtained by multiplying the Boltzmann equation by various velocity moments and by integrating over velocity space. The result is a hierarchy of coupled differential equations – the transport or fluid equations – which describe the spatio-temporal variation of the moments of the velocity distribution functions of the electrons and ions species.

Hydrodynamical transport models of varying degrees of sophistication have been used for decades to describe ionospheric and interplanetary space plasmas. By combining the moment equations for the electrons with those for the ions (generally both species are assumed to have a common temperature and a common bulk velocity $\mathbf{u} = \mathbf{E} \times \mathbf{B}/B^2$), one obtains the magnetohydrodynamic (MHD) approximation of the transport equations. Furthermore, in the case of 'ideal MHD' it is postulated that $\mathbf{E.B} = 0$, i.e. that the electric field $\mathbf{E}$ has no component parallel to the magnetic field $\mathbf{B}$. The reason traditionally given to validate this approximation is that the electrical conductivity of space plasmas, σ, is 'infinitely large', so that, according to Ohm's law ($j_{\parallel} = \sigma E_{\parallel}$), the smallest parallel electric field would imply 'infinitely large values' for the field-aligned current $j_{\parallel}$. Of course this is based on the postulate that Ohm's law remains applicable in collisionless plasmas! This is not quite the case in the magnetospheric plasma, due to the low, but finite, Coulomb collision frequency. Wave–particle interactions, generally speaking, produce a linear, ohmic type of anomalous resistivity, but confirming experimental evidence is needed to assess their importance in scattering and heating the bulk of cold plasma in the ionosphere and plasmasphere.

In the absence of Coulomb collisions or wave–particle interactions, it has been shown by Knight (1973), Lemaire and Scherer (1974b; 1983), Fridman and Lemaire (1980) and Pierrard (1996) that the field-aligned current density, $j_{\parallel}$, is not determined by the parallel electric field, as inferred by Ohm's law, but by the total field-aligned potential drop, ΔV.

Ideal MHD transport equations have been applied with mixed success in the study of a wide variety of space plasma physics problems. They should not be equated with the more comprehensive multi-fluid moment equations and hydrodynamic transport equations consisting of coupled sets of continuity, momentum and energy equations for each particle species separately. Reviews of different approximations for the multi-fluid transport equations used over the last three decades to study the solar wind, the polar wind and the refilling of empty plasmaspheric flux tubes can be found in the papers by Raitt and Schunk (1983), Schunk (1988b), Gombosi and Nagy (1988), Gombosi and Rasmussen (1991), Singh and Horwitz (1992), Guiter, Gombosi and Rasmussen (1995) and Ganguli (1996).

Comparisons of transport models and kinetic or semi-kinetic models can be found in Holzer, Fedder and Banks (1971), Lemaire (1978), Schunk and Watkins (1979, 1981) and Demars and Schunk (1986, 1987, 1991). We will not review here all this material, and will limit the rest of this section to a brief overview of how these models have been applied in the past to describe the refilling of empty plasmaspheric flux tubes (Horwitz and Singh, 1991; Singh and Horwitz, 1992).

The first time-dependent plasmaspheric flux tube refilling model was proposed by Banks, Nagy and Axford (1971). It is known as the 'two-stage refilling scenario' and was expanded until the late 1980s within the framework of various hydrodynamical approximations (Singh, 1988, 1990). According to this scenario a polar-wind-like supersonic flow is driven out of the ionosphere by a large pressure gradient parallel to the magnetic field lines. These flows from the conjugate hemispheres collide at the equator. A pair of shocks are formed as a consequence of the collision in the equatoral region. The shocks propagate downward, one in each hemisphere. Between the shocks the plasma is dense and warm, while below the shocks the upward flows are supersonic. This corresponds to the well-known scenario of refilling from 'top to bottom'. It was further proposed that when the shocks reach the ionosphere, a second stage of refilling might follow, with upward subsonic flows lasting several days. Figure 5.9a shows the electron density distribution as a function of dipole latitude along a magnetic field line obtained with such a single-stream hydrodynamic model. The different curves show the evolution with time as the flux tube fills from top to bottom.

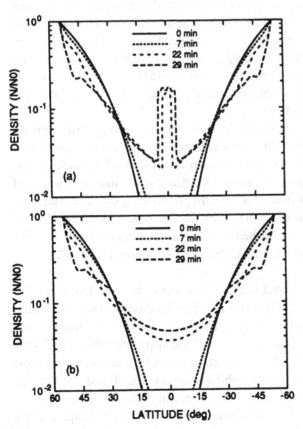

Figure 5.9 Field-aligned electron density distributions as a function of dipole latitude for (a) a single-stream hydrodynamic model, and (b) a two-stream hydrodynamic plasmasphere refilling model. The different curves show the evolution with time as the flux tube fills up. The solid line shows the electron density distribution when refilling started at $t = 0$ (after Rasmussen and Schunk, 1988).

As early as 1972 Schulz and Koons (1972) suggested that the ion beams colliding at the equator may not form shocks, but might excite waves that scatter the pitch angles of the ions flowing along the magnetic field lines. If the pitch angle enlarges sufficiently, the ions will be trapped in the equatorial region, possibly leading to the creation of a relatively dense plasma in the equatorial region as in the model of Banks et al.. If the density becomes sufficiently great, Coulomb collisions become an important factor in further thermalizing the plasma flow (Grebowsky, 1972; Khazanov et al., 1984; Singh et al., 1986; Guiter and Gombosi, 1990).

The early hydrodynamical refilling models were single-stream fluid models, in which the flows originating from conjugate ionospheres are treated as one fluid. Pairs of shocks are assumed to form because the average flow velocity upon merging of the symmetric flows at the equator is equal to zero. This is a rather artificial stoppage of the plasma flow. In subsequent hydrodynamic models the flows from conjugate ionospheres were treated as two separate fluids (Singh, 1988; Rasmussen and Schunk, 1988). In these improved two-stream models the oppositely directed plasma flows do not interact at the equator, but penetrate into the conjugate hemisphere where the streams encounter a cold plasma with increasing density. In this case, illustrated in Fig. 5.9b, plasma does not pile up in the equatorial region as in the single-stream model displayed in Fig. 5.9a. The interstream Coulomb drag and the electric and pressure forces slow down the fast ion beams, leading to the formation of shocks in each hemisphere. Such shocks first propagate upward toward the equator and then onward to the opposite ionosphere, where the flow supporting the hydrodynamic shocks originated.

Although this alternative model and the earlier 'top to bottom' plasmaspheric refilling scenario have been extensively developed and debated, Horwitz and Singh (1991) consider that 'it has not been satisfactorily resolved which is the most likely to occur in the magnetosphere. No clear picture has yet emerged of how, when and where the downward or upward propagating shocks would form if, indeed, they do so at all'. It is useful to recall here that in the kinetic scenarios and Monte Carlo simulations described above, no such shocks are formed in the early stages of plasmaspheric refilling (Lemaire, 1989; Wilson et al., 1992; Lin et al., 1992).

Furthermore, it must be noted that all these complex time-dependent hydrodynamic refilling scenarios depend on the plasma distribution at the initial time, $t = 0$. Usually it is assumed that a flux tube is completely empty above some arbitrarily chosen reference level in both hemispheres (500 or 1000 km). Below these reference altitudes the ionosphere is assumed to be an inexhaustible reservoir of particles and energy out of which the supporting flow is constant; i.e. independent of time. This may well be an unrealistic set of assumptions and rather artificial initial conditions. Unfortunately, there are no comprehensive (snapshot) observations of the initial plasma distribution in a flux tube at the time it starts to be refilled. Nor are there global and high-time-resolution observations during the refilling process itself.

Furthermore, we have no comprehensive observations nor theoretical models describing how a flux tube of the outer plasmasphere is emptied during substorm events. There may be several alternative mechanisms, each of which could lead to a different refilling scenario. To elucidate these critical issues single satellite observations are inadequate. Global imaging of the magnetosphere, as proposed at NASA by Williams, Roelof and Mitchell (1992) and Burch *et al.* (1995) are definitely needed to study the time evolution of the plasma distribution in a flux tube when it is emptied and when it is refilled.

Until new sounding methods have provided global observations of the magnetosphere, it will not be possible to test the many alternative hydrodynamic or kinetic models proposed so far to describe the field-aligned distribution of plasma in the magnetosphere. Fortunately for the interpretation of whistler measurements, the uncertainty in the field-aligned model of the distribution of ionization has relatively little effect on the calculation of the total tube content: it is less than a factor 1.5 (Angerami and Carpenter, 1966). It has little effect on the determination of the slope of the equatorial density profiles from whistler observations. It introduces, however, an uncertainty by a factor of 2 in the absolute value of the equatorial density (Carpenter and Smith, 1964). The lack of accurate knowledge of the field-aligned distribution of electron densities introduces also an uncertainty of the order of $0.15R_E$ in the determination of the equatorial radius (Angerami and Carpenter, 1966).

5.3   Equatorial plasma distribution

In the previous pages we have described various static and dynamic plasma density distributions along magnetic field lines, assuming corotation but no radial cross-L convection. For the interpretation of whistler spectrograms only relative density profiles along magnetic field lines are needed to determine values of the equatorial density. Once such a profile is adopted the equatorial density may be determined from the measurements. In this section we discuss theoretical models of the equatorial density profile as a function of L and compare them with whistler and *in situ* satellite measurements. We start with the diffusive equilibrium model and then discuss results corresponding to the exospheric equilibrium model. Finally we examine what could be expected from a theoretical point of view in the case of cross-L radial drift motions.

5.3.1   The diffusive equilibrium equatorial density profiles

Figure 5.10 shows the equatorial density as a function of L for three values of the angular rotation velocity Ω. The H^+ and electron densities at the reference altitude (1000 km) are arbitrarily assumed to be 10^3 cm^{-3}. Currently observed values could be a factor of 10 larger. This does not change however the qualitative features outlined below. The broken lines correspond to a non-rotating ion-exosphere, the solid lines to a corotating protonosphere, and the dotted lines to an angular rotation velocity three times larger. Note that when the angular velocity is enhanced by a factor of 3 the ZPF (zero parallel force) surface approaches closer to the Earth ($L_c = 2.78$), the field-aligned density distribution at $L = 8$ has then a minimum at $\lambda_M = 48°$ and the diffusive equilibrium (DE) equatorial density profile has a minimum at $L = 3.17$. Beyond this minimum value the plasma density for DE models increases indefinitely with L. The kinetic pressure at large distances becomes arbitrarily large and eventually exceeds the plasma pressure observed in the interplanetary medium whenever $\Omega \neq 0$. Therefore, unless a wall is placed at large distances to withstand this excess kinetic pressure, the DE density distribution is convectively unstable. Either Ω decreases with L or a radial expansion of the thermal plasma will result due to the unbalanced mechanical pressure at large radial distances. A similar argument was employed by Parker (1958) to infer the radial expansion of the solar coronal plasma, previously assumed to be in hydrostatic equilibrium.

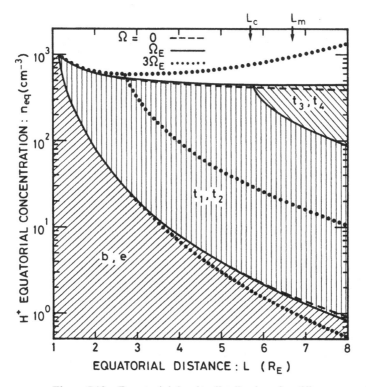

Figure 5.10 Equatorial density distributions for different exospheric models. The broken lines correspond to the non-rotating case; the solid line is for corotation; the dotted lines correspond to an angular velocity three times larger than the Earth's rotation velocity (after Lemaire, 1985).

In the models illustrated in Fig. 5.10, the H^+ density and temperature at the reference altitude (1000 km) are assumed to be independent of L, i.e. independent of the invariant latitude at the exobase level. This is a rather poor first approximation; observations recalled in Sections 2.5.1 and 4.8.3 indicate that the light ion densities have deep troughs at mid-latitude and that in general the ion composition changes significantly with invariant latitude such that, for example, the H^+ and He^+ concentrations at 1000–2000 km altitude decrease when the latitude increases (Bauer, 1963; Thomas and Sader, 1964; Rycroft and Alexander, 1969; Taylor $et\ al.$, 1972). Furthermore, the exobase temperature changes with latitude as well as with local time and season (Harris and Priester, 1962; Rycroft and Alexander, 1969; Brace and Theis, 1974; Titheridge, 1976a; Denby $et\ al.$, 1980).

Using realistic ion composition and temperature distributions at the exobase level versus invariant latitudes, Angerami and Thomas (1964) determined theoretical equatorial density distributions as functions of L for winter as well as summer days and nights. Figure 5.11 shows typical diurnal and seasonal changes for a given temperature and base composition. All these distributions have a steep gradient below $L = 2$, but most of them become rather flat beyond $L = 4$. Note, however, the rather gradual fall-off of the electron density with radial distance illustrated by curve d, for summer night conditions; this gradual falloff is attributable to the rapid decrease of the exobase H^+ concentration with invariant latitude over the range 50–65°.

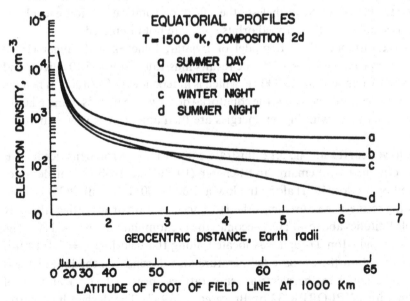

Figure 5.11 Calculated equatorial electron profiles based on the electron densities measured by Alouette at an altitude of 1000 km under different conditions. The seasonal and diurnal variations are shown by the curves a, b, c and d. (after Angerami and Thomas, 1964).

Angerami and Thomas (1964) compared their theoretical results with equatorial density profiles obtained from whistler observations and concluded that while their theoretical predictions were supported by these experimental data for summer night conditions, for most other conditions the slope of the density gradient predicted by DE models was generally small compared to the slopes observed by whistlers. In other words the DE models predicted equatorial density gradients that were smaller than those previously (and since) observed by satellites and whistlers in the plasmasphere beyond $L = 4$ (see also Carpenter and Anderson, 1992).

Rycroft and Alexander (1969) calculated similar DE equatorial electron density profiles for winter nights and summer days based on realistic temperature and ion composition distributions observed at 900 km altitudes. Their theoretical DE models confirm the general features outlined above.

5.3.2 The exospheric equilibrium equatorial density profiles

The lower curves in Fig. 5.10 correspond to equatorial distributions in a protonosphere which would be populated only by ballistic (b) and escaping (e) protons or electrons. The broken, solid and dotted lines correspond to models with different angular rotation velocities. The broken curve corresponds to the non-rotating model of Eviatar, Lenchek and Singer (1964). It can be seen that the equatorial density decreases at large radial distances when the angular velocity Ω increases. Nevertheless, the effect of rotation on ion-exospheric models is small when the trapped orbits t_1–t_4, are not populated.

At large distances the EE model of Eviatar, Lenchek and Singer (1964) decreases approximately as L^{-4}. The results illustrated in Fig. 5.10 correspond to an exobase temperature of 3000 K which is independent of latitude. When the exobase temperature increases, the equatorial density also increases and a larger number of particles with higher energies are then able to reach the equatorial region.

The lower curves in Fig. 5.12 illustrate equatorial electron densities for EE models obtained by Lemaire and Scherer (1974a) for exobase ion densities observed by Taylor (1971) along the low altitude (700–1100 km) OGO 6 orbit. According to these measurements the H^+ concentration drops rather abruptly at dipole latitudes above 40°; O^+ becomes the predominant ion beyond 50°. The associated Light Ion Trough was located along the field line $L = 3.4$. In this model calculation the exobase temperature also depends on the latitude: it is taken to be equal to the electron temperatures measured at an altitude of 1000 km with EXPLORER 32 by Brace et al. (1967). The dashed lines correspond to an (H^+, e^-) ion-exosphere, while the solid line corresponds to a multi-ion exosphere (O^+, H^+, e^-). Despite the negligible density of O^+ ions at very high altitudes, their relative concentration at 1000 km strongly controls the value of the electron and proton densities in the equatorial plane (see Section 5.2.4).

The solid dots in Fig.5.12 represent equatorial H^+ ion density measured with OGO 5 in the nightside region (Chappell, Harris and Sharp, 1970a). Although the OGO-5 ion densities, shown in Fig. 5.12, are higher outside the plasmapause and lower inside than presented in papers by Chappell *et al.* (1970a, b; 1971a,b) – presumably due to recalibration of the data – these data are expected to be good

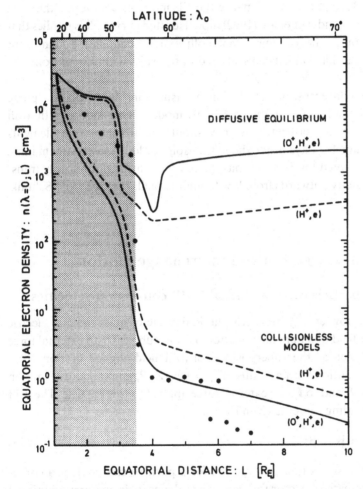

Figure 5.12   Equatorial density distributions corresponding to diffusive equilibrium (upper curves) and exospheric equilibrium (lower curves) for a (H^+,e) ion-exosphere (broken lines) and for a multi-ion-exosphere (O^+,H^+,e). In all cases the concentrations at the exobase have been determined from OGO 6 observations by Taylor (1971); the exobase temperatures are also latitude dependent and taken from EXPLORER 32 data by Brace *et al.* (1967). The dots correspond to OGO 5 observations provided by Chappell (1978, personal communication) through NSSDC (after Lemaire and Scherer, 1974).

enough to identify the location of the density gradient at $L = 3.5$ which corresponds to the plasmapause.

These simple models can easily be compared with the equatorial density profiles deduced by Carpenter and Anderson (1992) from ISEE and whistler observations (see figs 2.3, 2.15, 2.18, 2.22, 2.25, 4.9, or eqs. 4.3b, 4.4, 4.5a and b). Inside the shaded plasmasphere area, the observed densities are significantly larger than the EE model predictions. This excess density is due to a large number of particles that have been scattered by Coulomb collisions onto trapped orbits. Beyond the plasmapause the EE model is also lower than the values of Carpenter and Anderson (1992) model (dotted curve). This implies that in the plasmatrough the plasma is not collisionless, and that a significant fraction of particles have also been scattered onto orbits with mirror points at high altitudes.

Note that the concentrations of H^+ ions measured inside the plasmasphere are much lower than those predicted by DE models. This indicates that all trapped orbits are not saturated as they should be when plasma elements corotate more than 4–5 days inside the plasmasphere along closed stream lines. This may be accounted for by a plasmaspheric wind which continuously transports plasma radially outwards from low to high L-shells (Lemaire and Schunk, 1992, 1994).

5.4   Plasma convection and interchange motion

5.4.1   Single particle drifts and ideal MHD convection velocities

The motion of a low energy magnetospheric particle of mass (m) and electric charge (Ze), in a field of magnetic induction $(\mathbf{B})$ and subject to an external force $(\mathbf{F})$ can be regarded as the superposition of (i) a field-aligned motion, (ii) a circular Larmor motion with a radius of $r_L = mv_\perp/ZeB$, and (iii) a drift motion perpendicular to $\mathbf{B}$ and to $\mathbf{F}$, the external force applied to this particle. The drift velocity of the guiding center is given by

$$\mathbf{V}_F = \mathbf{F} \times \mathbf{B}/ZeB^2 \qquad (5.9)$$

Figure 5.13 illustrates this drift motion for a positive charge. It is a consequence of the increased Larmor radius r_L when the particle acquires additional momentum $(mv_\perp)$ by moving downward in the direction of $\mathbf{F}$. On the contrary, when the particle moves upward it loses energy, its momentum decreases and as a consequence the radius of curvature of the trajectory r_L becomes shorter. When averaged over a Larmor cycle the position of the guiding centre is shifted to the left with the velocity $\mathbf{V}_F$ given by equation (5.9). For a negative charge the drift is in the opposite direction.

When **F** is the gravitational force (mg) and **B** the geomagnetic field in the equatorial plane, the drift velocity is directed to the east for positive ions ($Z > 0$) and toward the west for the electrons ($Z = -1$). The gravitational drift velocity V_g for a proton at $4R_E$ in the equatorial plane where $B = 480$ nT and $g = 0.6$ m/s^2 is equal to 0.013 m/s; it is 1837 times smaller for the electrons. These gravitational drift velocities are orders of magnitude smaller than the drift velocity resulting from magnetospheric electric fields (**F** = Ze**E**), and therefore they are generally neglected. However, it will be shown below that despite their smallness, these drifts are of key importance because they induce polarization charges in plasma density irregularities and change the local electric field distribution in the vicinity of such irregularities.

The electric drift velocity (also called 'MHD convection velocity' or 'velocity of magnetic field lines') can be deduced from equation (5.9) by replacing **F** in this equation by Ze**E**:

$$V_E = E \times B/B^2 \tag{5.10}$$

This electric drift velocity has the same direction for positive and negative charges; its value is independent of the charge and mass of the particle. $V_E = 1850$ m/s when $E = 0.89$ mV/m; this is equal to the corotation electric field intensity at $4R_E$.

The Pannekoek–Rosseland electric field (5.2) is antiparallel to the gravitational force. It produces in the protonosphere an eastward drift velocity equal to half of the gravitational drift velocity, V_g. This small electric polarization drift velocity is in the same direction and equal for the protons and electrons.

Due to their Larmor motion particles have a magnetic moment ($\mu = mv_\perp^2/2B$) and they experience additional drifts in the inhomogeneous geomagnetic field.

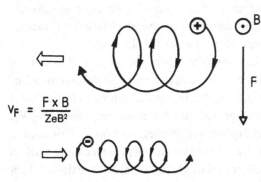

$$V_F = \frac{F \times B}{ZeB^2}$$

Figure 5.13 The upper curve represents the trajectory of a positive charge in a uniform magnetic field **B**. The force **F**, accelerates the particle downward; it increases its velocity and its Larmor gyroradius. When the particle moves back in the direction opposite to **F**, it is decelerated and its gyroradius decreases. The net displacement of the guiding centre after one Larmor rotation is then to the left. The drift velocity is equal to $V_F = F \times B/ZeB^2$. Negative charges drift in the opposite direction when **F** is directed downward. This is illustrated by the lower trajectory.

These drifts can also be deduced from equation (5.9) by replacing $\mathbf{F}$ by $\mu \nabla B$ and by $mv_{\parallel}^2/R_c$, where R_c is the radius of curvature of the magnetic field line and ∇B is the gradient of the magnetic induction. Unlike the electric drift these magnetic drifts, $\mathbf{V}_B$ (gradient-B drift) and $\mathbf{V}_c$ (curvature drift), depend on the electric charge of the particle as well as on its perpendicular energy and parallel energy, respectively. For a proton in the geomagnetic field both drifts are in the westward direction: $\mathbf{V}_B + \mathbf{V}_c = 0.05 \, \mathrm{m/s}$ for protons of $0.25 \, \mathrm{eV}$ at $4 R_E$ in the equatorial plane. For electrons of the same energy the sum of the gradient-B drift and curvature drift is equal to $0.05 \, \mathrm{m/s}$ but is directed toward the east.

When the thermal energy of the plasma particles is small (i.e. in the cold plasma approximation) these gradient-B and curvature drifts are neglected. Moreover, the gravitational drift, always present, is also generally neglected in studies of plasmaspheric and magnetospheric convection, 'because it is five orders of magnitude smaller than the electric drift velocity'. All charges then drift together in the same direction with the same velocity, $\mathbf{V}_E$. This assumption of the ideal MHD approximation of plasma physics implies that all ion species have the same convection velocity as the electrons!

In the ideal MHD approximation of plasma physics, one is used to saying that magnetic field lines are 'moving' with the electric convection velocity $\mathbf{V}_E$; it is then considered that the plasma ions and electrons are 'glued to the moving magnetic field lines distribution'. This concept was originally introduced by Alfvén in the early days of plasma physics (see Alfvén and Fälthammar, 1963), but later on, noting the usage made of the frozen-in-field concept, he warned the space plasma community of the potential danger of its misuse (Alfvén, 1976).

In the ideal MHD approximation, cold plasma streams parallel to the equipotential surfaces of the background magnetospheric electric field. Fig. 5.14 shows the equatorial equipotential lines of Brice's (1967) 'best estimate' electric field model. The arrows indicate the direction of the electric drift velocities which, in this approximation, are equal for all ions and electrons!

The limitations of the frozen-in-field concept can be recognized when a finite plasma element (a plasmoid) is considered instead of a uniform plasma distribution along electric equipotential lines. When a plasma element has a finite extent – i.e. a non-uniform density – the small gravitational and magnetic drifts ($\mathbf{V}_g$, $\mathbf{V}_B$, $\mathbf{V}_c$) slightly separate the positive charges from the negative ones. This charge separation produces a polarization electric field inside the plasma element which has to be added to the external (background or global scale) electric field. Like a dielectric placed in an external E-field, plasma irregularities in an external gravitational field and in a non-uniform magnetic field become polarized; this natural charge separation generates an additional electrostatic field inside the plasmoid as well as in its immediate surroundings.

Such polarization electric fields are ignored in ideal MHD theories of magnetospheric convection. In such approximations all plasma elements (e.g. whistler ducts) are assumed to be convected with the same convection velocity along the equipotential lines of the external electric field **E**. The effects of polarization electric fields produced locally by plasma density enhancements or depressions are then ignored. The reason for ignoring these polarization electric fields generally voiced in the 1970s was that the gravitational drift is five orders of magnitude smaller than the electric convection velocity! Unfortunately, such an argument overlooks the fact that a minute charge separation in a plasma can produce polarization electric fields which are comparable in strength with the external E-field intensity induced in the magnetosphere by the solar wind dynamo or by the corotation of the ionosphere. This is a consequence of the large dielectric constant of a magnetized plasmas.

There is sometimes a danger in oversimplifying the mathematical description of physical phenomenae with first approximation theories that ignore small effects which nevertheless play key roles. A well known example is in the field of plate tectonics, where it took quite some time for the geophysical community to recognize that continental drift, although extremely small, has many geophysical consequences and can no longer be ignored in modern geophysical theories.

In the next section we indicate why the 'small' gravitational drift plays a key role in plasmaspheric dynamics and specifically in the behavior of plasma

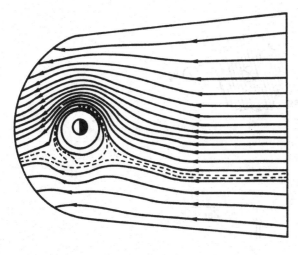

Figure 5.14 Streamlines in the equatorial plane obtained by combining a presumed solar wind driven convection pattern and the corotational flow pattern of ionospheric origin. The streamlines are parallel to equipotentials of a 'best estimate' electric field model designed by Brice (1967). The potential differences between solid lines are 3 kV. The dashed lines give intermediate equipotentials at 1 kV intervals. The arrows indicate the direction of the electric drift velocity. Equipotentials closed within the magnetosphere are expected to be inside the whistler knee or plasmapause surface. According to early theories for the formation of the plasmapause the last closed equipotential (the closed dashed line) corresponds to the location of the whistler 'knee' (after Brice, 1967).

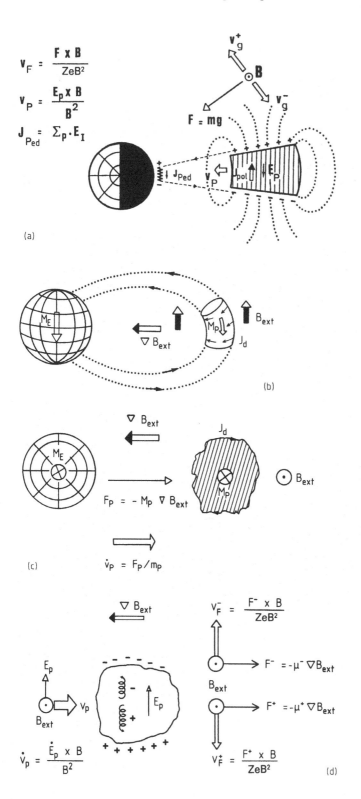

(a)

(b)

(c)

(d)

density irregularities in the magnetosphere. The same demonstration holds for the centrifugal force as well as for the magnetic forces producing gradient-B and curvature drifts.

5.4.2 Plasma interchange velocity

Gravitational drift and polarization electric field

Let us now consider a plasma density enhancement corotating with the Earth's angular velocity Ω_E, and examine the effects induced by the gravitational force. Figure 5.15a illustrates the equatorial section of an isolated plasma element moving across a dipole magnetic field, **B**, at an equatorial distance smaller than $6.6R_E$, where the gravitational force is larger then the centrifugal pseudo-force. In the corotating frame of reference the protons inside the density enhancement drift eastward with the velocity

$$\mathbf{V}_g = m_p \mathbf{g} \times \mathbf{B}/eB^2 \tag{5.11a}$$

due to the gravitational force.

The electrons drift in the opposite directions with a velocity m_p/m_e times smaller. These opposite drifts generate a polarization current ($\mathbf{J}_{pol}$) inside the plasma element and contribute to the build up of a positive electric charge density on the eastward boundary of the plasma element and an equal density of negative charges on its westward edge. Due to these drifts positive and negative charges are separated by a distance Δx which increases with time in the

Figure 5.15 (a) Illustration of a cold plasma element falling across geomagnetic field lines in the gravitational field. The gravitational drifts of the ions and electrons are in opposite directions and give rise to a polarization electric current ($\mathbf{J}_{pol}$) and a polarization electric field $\mathbf{E}_p$. The resulting electric drift $\mathbf{V}_p = \mathbf{E}_p \times \mathbf{B}/B^2$ is parallel to the gravitational force. The coupling of the plasma element to the resistive ionosphere limits the value of the electrostatic potential difference across the element of plasma. The intensity of the depolarization currents ($\mathbf{J}_{Ped}$) is proportional to the value of the integrated Pedersen conductivity (Σ_p). Therefore the value of Σ_p limits the free fall velocity of the plasma element in the gravitational field. The dotted lines are the electric field lines corresponding to the polarization electric field. (b) Illustration of a 3-D plasma element in the geomagnetic field; $\mathbf{M}_E$ is the magnetic dipole moment of the Earth; $\mathbf{M}_p$ is the magnetic dipole moment of the diamagnetic current system circulating inside and at the surface of the plasmoid. $\mathbf{M}_p$ is also equal to the sum of the magnetic moments of all ions and electrons contained in the plasmoid. (c) and (d) Dipole–dipole interaction between the Earth magnetic dipole $\mathbf{M}_E$ and $\mathbf{M}_p$. When both dipoles are parallel to each other the grad-B force, $\mathbf{F}_p$, accelerates the plasmoid away from the Earth. When the integrated Pedersen conductivity is small the acceleration of the plasma element is proportional to $\mathbf{F}_p$, and the maximum interchange velocity can then become large (after Lemaire, 1975, 1985).

east–west direction. As indicated in Section 5.4.1. the perpendicular component of the Pannekoek–Rosseland or ambipolar electric field given by equation (5.2) contributes to additional drifts of protons and electrons. But since these electric drift velocities have the same value and are in the same direction, they do not contribute a net polarization current: i.e. they do not contribute to separate positive and negative charges like the gravitational drifts. The rate of change of Δx is determined by the drift velocities of the protons (p) and electrons (e).

$$d(\Delta x)/dt = \mathbf{V}_{g,p} + \mathbf{V}_{g,e} \simeq (m_p + m_e)\mathbf{g} \times \mathbf{B}/eB^2 \tag{5.12}$$

As a result of this charge separation the plasma element becomes polarized. The polarization (dipole moment per unit volume) is

$$\mathbf{P} = ne\Delta\mathbf{x} \tag{5.13a}$$

where n is the electron or ion density. The polarization increases with time at a rate proportional to $d(\Delta x)/dt$ determined by equation (5.12). The space charge density produced by this charge separation mechanism is related to the divergence of the polarization electric field by

$$\rho = -\nabla.\mathbf{P} \tag{5.13b}$$

As a consequence of the dielectric property of magnetized plasmas, the polarization electric field itself produces a counterpolarization, which tends to reduce the actual electric field intensity, $\mathbf{E}$, inside the plasma element. Using Maxwell's equations and following the demonstration of Longmire (1963, p.75) and Chandrasekhar (1960, p. 71) it can be shown that

$$\mathbf{E} = -\mathbf{P}/(\varepsilon_\perp/\varepsilon_o - 1)\varepsilon_o \tag{5.13c}$$

where ε_o and $\varepsilon_\perp$ are respectively the permittivity of empty space and of the plasma in the direction perpendicular to $\mathbf{B}$. The DC perpendicular dielectric constant of the plasma is given by

$$K_\perp = \varepsilon_\perp/\varepsilon_o = 1 + [n(m_p + m_e)c^2/(B^2/\mu_o)] \tag{5.13d}$$

where μ_o is the permeability of empty space: $\mu_o = 1/\varepsilon_o c^2$. For $n = 300\,\mathrm{cm}^{-3}$, $B = 500\,\mathrm{nT}$ it can be verified that $K_\perp = 2 \times 10^5$. Thus the dielectric constant of a magnetospheric plasma is much larger than unity, since $nm_p c^2 \gg B^2/\mu_o$ (Longmire, 1963, p. 31).

Using equations (5.13a, c and d) one obtains:

$$\mathbf{E} = -eB^2\Delta\mathbf{x}/(m_p + m_e) \tag{5.14}$$

Electric drift velocity

The electric drift associated with $\mathbf{E}$ is equal to

$$\mathbf{V}_p = \mathbf{E}_p \times \mathbf{B}/B^2 = -e\Delta\mathbf{x} \times \mathbf{B}/(m_p + m_e) \tag{5.15}$$

Since the displacement Δx increases with time at a rate given by equation (5.12), the rate of change of the drift velocity (5.15) is given by

$$d(V_P)/dt = -(g \times B) \times B/B^2 = g - g_{\parallel} = g_{\perp} \tag{5.16}$$

where $g_{\parallel}$ and $g_{\perp}$ are, respectively, the parallel and perpendicular components of the gravitational acceleration with respect to the direction of the magnetic field. In the equatorial plane $g_{\perp} = g$. Thus the plasma element falls in the gravitational field across magnetic field lines. Its acceleration is equal to g, as in the case of a body formed of neutral atoms; this means that a plasma element (or a plasmoid) does not behave as a collection of non-interacting single charged particles. A collisionless plasmoid gets accelerated in a gravitational field just as does any other piece of material that is globally neutral, whether it moves across a magnetic field or not. In the ideal MHD approximation this same plasmoid would have been 'frozen-in' along the 'moving' magnetic field lines and be unable to drift otherwise than parallel to the equipotential surfaces of the external electric field distribution.

A similar demonstration can be given for the case of density enhancements embedded in a lower-density plasma background. In this case one must consider the polarization electric field produced by the plasma outside the surface of the plasmoid. The polarization electric field has a dipole component illustrated in Fig. 5.15a by the dotted electric field lines. As a consequence of this polarization electric field the plasma outside is convected around the moving density irregularity, like air flowing around a rain droplet falling in the gravitational field. Since in this transport process plasma from the background is interchanged with the plasma forming the plasmoid, the velocity of the plasma element is called the 'plasma interchange velocity'.

It must be added that a 3-D plasma element slips across magnetic field lines and is accelerated as described above for a cylindrical plasmoid, provided that charges can accumulate also at its surfaces which are normal to the magnetic field lines traversing the volume of the plasmoid. At the interface with the moving 3-D plasma element and the external background medium weak-double-layers must be developing, as described in Section 5.4.5. This is an additional ingredient and plasma kinetic process that is not take into account in the ideal MHD theory of magnetized plasmas.

Maximum interchange velocity

The plasma interchange velocity considered here should not be compared with the velocity of solitons, which are a different electrodynamic phenomenon. In the case of solitons the background electromagnetic field experiences a particular solitary fluctuation that propagates with a characteristic velocity, but plasma is not transported with the solitary wave from one place to another. On the contrary, in the case of interchange motion, a plasma element moves as a whole through the external magnetic and electric field distributions that are themselves temporarily and locally perturbed due to the presence of the moving diamagnetic and dielectric plasmoid.

As in the case of the rain droplet falling through the air, the value of the plasma interchange velocity cannot increase indefinitely. The drop of water will achieve a maximum velocity determined by the viscosity of the external fluid. The viscosity of the magnetospheric plasma is small and can be neglected; in the case of a plasmoid falling in the gravitational field, the maximum value of the free fall velocity is determined by the electrical resistivity of the ionospheric plasma or of the 'walls' within which the magnetic field lines are rooted. This is illustrated in Fig. 5.15a where field-aligned currents connecting the edges of the plasma element and the ionosphere are closed in the E-region by a horizontal Pedersen current $\mathbf{J}_{\mathrm{Ped}}$. The intensity of these transverse currents is proportional to the value of the integrated Pedersen conductivity (Σ_p). When the electrical conductivity in the lower ionosphere is small, the Pedersen currents are reduced and the intensity of the field-aligned currents is correspondingly reduced. These Birkeland currents are then too small to carry away the polarization charges accumulating at the surface of the plasma element and thus to short circuit the electric potential difference building up steadily between its eastward and westward edges. On the contrary, when Σ_p is large, the depolarization currents circulating through the ionospheric 'walls' are equal to the polarization currents flowing inside the plasma element. In this case the polarization charges are removed as fast as they accumulate at the surface.

The polarization currents, $\mathbf{J}_{\mathrm{pol}}$, driven by the gravitational drifts, flow opposite to E. Therefore, $\mathbf{J}_{\mathrm{pol}}.\mathbf{E} < 0$, as is the case in electric potential generators. The plasma element in the gravitational field behaves like an EMF generator (battery), while the resistive ionosphere represents the ohmic load in the system ($\mathbf{J}_{\mathrm{Ped}}.\mathbf{E}_{\mathrm{I}} > 0$); the magnetic flux tubes are the highly conducting wires connecting the magnetospheric EMF generator and the ionospheric load.

When the total depolarization current is equal to the total polarization current a steady state is reached and the polarization electric field **E** reaches an asymptotic maximum value. The plasma interchange velocity is then maximum, and the rate of change of the kinetic energy of the plasmoid tends to zero. The rate of loss of gravitational potential energy by the plasma element is then exactly balanced by the rate of energy dissipation through Joule heating in the 'walls' (or in the resistive layers of the ionosphere). When the integrated Pedersen conductivity along the magnetic field lines is large, the maximum interchange velocity is small. This is the case in the dayside local time sector where the E-region ionisation density is largest and where the value of Σ_p can be as high as 10 siemens. On the contrary, in the nightside region where the integrated conductivity is reduced to values below 0.5 siemens, the maximum interchange velocity can be quite significant.

We refer the reader to the work of Lemaire (1985) for a detailed discussion and derivation of the expression for the maximum interchange velocity, which is given here without demonstration:

$$V_{P,max} = \gamma \Delta n m_p g_{eff} V_0 L^3 / \Sigma_p B_e^2 S_0 \tag{5.17}$$

where Δn is the excess (or deficit) density in the equatorial plasma element with respect to the background plasma density model: $n_{bg}(L)$. The plasmoid is assumed to be located along the field line L; $V_0(L)$ and $S_0(L)$ are, respectively, the volume and ionospheric cross section of a magnetic flux tube of 1 weber. These quantities can be determined as a function of L in the case of a dipole magnetic field (see Lemaire, 1976a); γ is the fraction of that flux tube volume occupied by the plasma element, B_e is the equatorial magnetic field intensity ($B_e = 0.31 \times 10^{-4}$ T), g_{eff} is the effective gravitational acceleration taking into account the reduction due to centrifugal effects. A similar expression was obtained by Brice (1973) to determine the motion of artificial plasma clouds injected into the magnetosphere.

Interchange motion driven by grad-B and curvature drifts

Richmond (1973) calculated the interchange velocity due to the gradient-B and curvature drifts. The contribution of this additional drift will be discussed in more detail below, as well as the recent extension of Richmond's work by Huang, Wolf and Hill (1990). Southwood and Kivelson (1987, 1989) reviewed the question of interchange motion from two complementary points of view, closely following the formulation developed by Northrop and Teller (1960). In their comprehensive study they review also the early attempts of Gold (1959), Sonnerup and Laird (1963), Nakada et al. (1965) and Melrose (1967) to describe interchange motion in the magnetosphere based on local MHD transport equations. In these earlier formulations of interchange instability, the effect of the resistive ionosphere is ignored. Southwood and Kivelson (1989) give a mathematical formulation of the plasma interchange theory that is described above from a kinetic point of view. They confirm that the growth rate of interchange instability is inversely proportional to the integrated Pedersen conductivity, i.e. that the instability is not only determined by a local stability criterion, but that its growth rate depends also on non-local boundary conditions in the distant ionosphere.

Like bubbles of hot or cool air. . .

When the number densities inside and outside the plasma element are equal, $\Delta n = 0$. In this case the values of V_p, dV_p/dt, and $V_{p,max}$ given by equations (5.15),

(5.16) and (5.17) are all equal to zero. Under these circumstances there is no plasma interchange motion due to the gravitational force. When the background density is larger than the plasma density inside, Δn is negative and the interchange velocity is opposite to the gravitational force. Like a bubble of hot air, such a plasma density depression moves away from the Earth.

Order of magnitude calculations

It is useful to give some order of magnitude for $V_{p,max}$ in the magnetosphere. The value of the integrated Pedersen conductivity has a minimum of the order of 0.2 siemens in the post-midnight local time sector (Gurevitch et al., 1976). Consider a plasma element at $L = 4$ with an equatorial density of $200 \, cm^{-3}$, embedded in a background plasma density of $300 \, cm^{-3}$, so that $\Delta n = -100 \, cm^{-3}$. For $g_{eff} = 0.61 \, m/s^2$, $V = 4.6 \times 10^{19} \, cm^3/Wb$, $S_0 = 2.8 \times 10^8 \, cm^2/Wb$, $B_e = 3 \times 10^{-5} \, T$, and $\gamma = 1$, one obtains $V_{p,max} = 52 \, m/s = 0.03 R_E/h$. This value for the maximum interchange velocity is relatively small compared to the corotation velocity, which is equal to $1.8 \times 10^3 \, m/s$ or $1 R_E/h$ at $L = 4$. Nevertheless, it will be argued later that this small interchange velocity plays a key role in eroding the plasmasphere.

In the dayside local time sector, values up to 8 siemens are given by Brekke and Hall (1988) for the integrated Pedersen conductivity. This value is 40 times larger than during nighttime. As a matter of consequence the maximum plasma interchange velocity is 40 times smaller on the dayside (i.e. $0.00075 R_E/h$). This implies that on the dayside any plasma element is convected essentially parallel to the equipotential surface, as in the ideal MHD approximation. However, this fails to be the case in the nightside sector, where Σ_p is significantly reduced and where the growth rate of the plasma interchange instability is therefore considerably enhanced.

5.4.3 Thermally driven interchange instability

As indicated above and illustrated in Fig. 5.15d, the gradient-B and curvature drifts contribute to drive plasma interchange motions when the perpendicular and parallel temperature of the particles are not small, unlike in the case when the plasma is assumed to be cold, i.e. when $\beta = nkT\mu_0/B^2 \ll 1$. As indicated above, these additional magnetic drifts are proportional to the energy of the particles. The rate of change of Δx is then determined by the gradient-B and curvature drifts.

$$d(\Delta \mathbf{x})/dt = \Sigma_i m_i (v_\parallel^2 + v_\perp^2/2)/(Z_i eBR_c)\mathbf{e}_\Phi \qquad (5.18)$$

where R_c is the radius of curvature of the magnetic field lines. The mean parallel and perpendicular kinetic energies in (5.18) are related to the parallel and perpendicular temperatures by

$$<mv_\parallel^2/2> = kT_\parallel/2 \qquad\qquad (5.19)$$

$$<mv_\perp^2/2> = kT_\perp \qquad\qquad (5.20)$$

The rate of change of the electric polarization $\mathbf{P}$ (i.e. $ne\Delta x$), is now proportional to the difference of kinetic pressure between the inside and outside of the plasma elements. When the kinetic pressures are the same inside and outside the plasmoids no interchange motion is driven by the gradient-B or curvature drifts. In other words, when the kinetic pressure is the same inside and outside the plasma element the fluxes of positive and negative charges drifting toward and away from the surfaces of the plasmoid are the same. Surface charges do not accumulate at the surface of the plasma elements, the polarization current vanishes and, $V_{\mathrm{p,max}} = 0$. In this case, the plasmoids are floating in mechanical equilibrium within the background plasma like hot air balloons which have reached their equilibrium or stationary altitudes. In general, however, the kinetic pressure inside a plasmoid is different from that outside. It is the total pressure (kinetic plus magnetic pressures) that is balanced, across the surface of a plasmoid at equilibrium (Lemaire, 1985).

Evolution toward mechanical equilibrium

Total pressure balance across the surface of a plasma element is achieved when it is in mechanical equilibrium with the background plasma and fields. When the total pressure is higher inside than outside, the plasma element will expand adiabatically until its density and temperature are low enough to achieve global pressure balance. On the contrary, if the pressure inside the element is lower than outside, the volume of the plasmoid will shrink; its density will then increase as will its temperature due to adiabatic compression. Eventually, total pressure balance will be achieved over a characteristic time equal to the time required for an Alfvén wave to propagate across the plasmoid. In the plasmasphere this characteristic time is short compared to the time associated with the relatively slow interchange motion. When the equilibrium state is achieved there remains an unbalanced density which drives gravitational interchange motion, as well as a residual kinetic pressure difference which gives rise to thermally driven interchange motion in the presence of an inhomogeneous magnetic field. The hot air balloon analogy is rather illuminating in this respect: a balloon rises because the density of air inside the envelope is lower than the density outside. It does not rise due to the different temperatures inside and outside. Although there is a temperature difference, pressure balance across the envelope of the balloon tends to be achieved at any time. The pressure inside adjusts its value to stay precisely equal to the external atmospheric pressure at the corresponding altitude (the surface tension of the membrane of the balloon being disregarded in this analogy).

Results of MHD theories of interchange instabilities

Huang, Wolf and Hill (1990) consider the case in which the total pressure balance is locally perturbed and azimuthal pressure waves are generated. Their detailed mathematical formulation, basically confirms the results obtained two decades ago by Richmond (1973). Their main results are:

(i) that the special class of azimuthal kinetic pressure waves that they consider can become unstable for low-frequency waves whose wave vectors $\mathbf{k}$ are in the azimuthal direction ($\mathbf{k.r} = 0$ and $\mathbf{k.B} = 0$). It would be interesting to check that these conclusions hold for less restricted classes of plasma and field perturbations (waves) which would have non-zero wave vectors $\mathbf{k}$ in the radial direction as well as in the direction parallel to magnetic field lines ($\mathbf{k.r} \neq 0$ and $\mathbf{k.B} \neq 0$). Furthermore, the diamagnetism of warm plasma has to be taken into account in comprehensive calculations of the instability growth rates;

(ii) that the growth rate of unstable low-frequency waves at the plasmapause is controlled by the value of the integrated Pedersen conductivity. This was also noted by Richmond (1973) and Southwood and Kivelson (1989);

(iii) that for higher frequency azimuthal waves satisfying $\mathbf{k.B} = 0$ and $\mathbf{k.r} = 0$, the inertia of the magnetospheric particles limits the growth rate more than does the ionospheric conductivity;

(iv) by way of confirmation, that the energetic particles of the trapped ring current resist the formation of interchange ripples, because their normal configuration is highly stable against interchange motion (Nakada *et al.* 1965);

(v) that keV particles strongly inhibit the formation of long-wavelength ripples on the plasmapause; and

(vi) that the instability of electrostatic drift waves limits the lifetime of sharp plasmapause density gradients. A characteristic lifetime of 1/9 hour on the nightside where the integrated Pedersen conductivity has its smallest values has been calculated by Huang, Wolf and Hill (1990).

For the class of interchange motions considered in earlier studies, the volume and shape of interchanging magnetic flux tubes are unchanging during the process of interchange. For a discussion of this limitation in theories of inter-change motion, see Tserkonikov (1960), Newcomb (1961) and Cheng (1985).

5.4.4   Diamagnetic effects

There is a complementary effect which will now be addressed briefly. This effect is related to the diamagnetism of plasma density irregularities in a non-uniform external magnetic field like the geomagnetic field. Indeed, at the surface of a plasmoid the sum of magnetization and grad-B currents forms the net diamagnetic current density which perturbs locally the background magnetic field distribution. When the plasmoid is in pressure equilibrium the difference of magnetic pressure $\Delta B^2/2\mu_0$ across the surface of the plasma element balances the difference of kinetic pressure, Δp. The diamagnetic currents circulating at the surface and inside the plasmoid possess a dipole magnetic moment, $\mathbf{M_p}$. This dipole magnetic moment can either be parallel or anti-parallel to the external magnetic field direction, depending on the sign of Δp. For instance when the kinetic pressure inside the plasmoid is larger than outside, the magnetic field intensity is reduced inside. The magnetic moment $\mathbf{M_p}$ is then anti-parallel to $\mathbf{B}$, as illustrated in Fig. 5.15b. In this case $\mathbf{M_p}$ is parallel to the geomagnetic dipole moment, $\mathbf{M_E}$ and both magnetic dipoles tend to repel each other.

As a consequence of the repulsive magnetic force, $\mathbf{F_p} = -\mathbf{M_p} . \nabla B_{ext}$ between the Earth dipole $\mathbf{M_E}$ and $\mathbf{M_p}$, the plasma cloud tends to be accelerated away from the Earth as illustrated in Fig. 5.15c and d.

When the kinetic pressure inside the plasmoid is lower than outside, the diamagnetic currents and $\mathbf{M_p}$ are reversed. In this case, $\mathbf{M_p}$ and $\mathbf{M_E}$ are anti-parallel and the plasma hole tends to move closer to Earth. The diamagnetism of plasma irregularities which drives plasma interchange motion across an inhomogeneous magnetic field distribution is related to the thermally driven interchange discussed in Section 5.4.3.

5.4.5   Weak double-layers

There is an additional aspect whose importance has not always been fully appreciated. This is the effect of electrostatic double-layers (DL) on plasma interchange motion. The reason this effect is ignored is that in most models of plasma interchange instability, magnetic field lines are assumed to be electric equipotential lines. Furthermore, plasma elements are assumed to be filamentary structures which extend uniformly along magnetic field lines and preserve their filamentary shape during the interchange motion. But, a field-aligned plasma density distribution needs not be always uniform and a smoothly decreasing function of altitude as in the diffusive or exospheric models discussed in Sections 5.2.4. and 5.2.5. Indeed, 3-D plasma clouds (i.e. small-scale plasma density irregularities) can form in the magnetosphere. These 3-D plasmoids are not necessarily filamentary, but can have any shape, length and width, provided the former is larger than the Debye length.

To avoid loss of electrons from these clouds at a higher rate than their loss of ions, positive polarization charges accumulate on their surfaces. A localized thermoelectric field and potential difference are produced by these surface charges, as illustrated in Fig. 5.16a (Carlqvist, 1995). These double-layers are confined to the interface between the plasmoid and the background plasma.

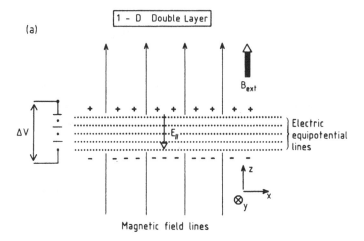

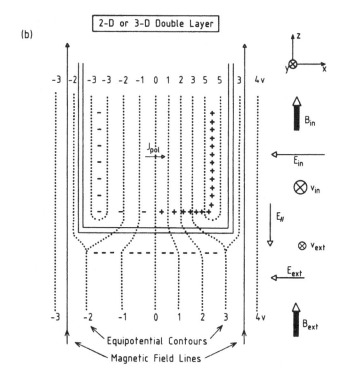

Figure 5.16 (a) Representation of a classical 1-D double layer in a uniform external magnetic field $\mathbf{B}_{ext}$. Thermo-electric charge separation produces a parallel electric field, $E_{\parallel}$, and a potential drop, ΔV, which are independent of x. (b) Illustration of electric equipotential contours in a double layer at the edge of a 2-D or 3-D plasmoid slipping across magnetic field lines with a differential convection velocity: $\mathbf{v}_{in} \neq \mathbf{v}_{ext}$. A uniform external magnetic field $\mathbf{B}_{ext}$ is assumed. The polarization charges on the vertical surfaces enhance the electric field intensity E_{in}. They increase the bulk velocity of the plasmoid ($\mathbf{v}_{in}$) with respect to the external plasma velocity: $v_{ext} = E_{ext}/B_{ext}$. The non-uniform distribution of surface charges on both sides of the horizontal double-layer produces a parallel electric field, $\mathbf{E}_{\parallel}$, whose intensity varies with x and/or y.

Their thickness can be much larger than one Debye length. In the diffusive equilibrium model discussed in Section 5.2.4, the Pannekoek–Rosseland electric potential difference extends over the whole magnetic field line.

The very existence of rather localized electrostatic double-layers is an additional feature which should be taken into account in the study of plasma interchange motion. Indeed, such microscopic double-layers decouple the plasma regions located on either side of the surface. Of course, such decoupling of two parts of a magnetic flux tube cannot, in principle, be accounted for in the framework of ideal MHD theory.

In 1-D double layer models the electrostatic potential jump ΔV is assumed to be the same at all places along the double-layer surface, as illustrated in Fig. 5.16a. On the other hand a 2-D or 3-D model of a double-layer is illustrated in Fig. 5.16b. In this more complex situation the electrostatic potential, ΔV, is a function of x and/or y as illustrated in this figure by the uneven and unparallel surface charge distribution on both sides of the double layer. The potential difference ΔV increases from left to right in the Ox direction, but it increases faster above the DL than below. The equipotential surfaces (dotted lines) 'cut' the vertical magnetic field lines (solid lines) at large angles within the DL region.

The external magnetic field is assumed to be uniform (no diamagnetic effect is considered here for the sake of simplicity) and it points upwards in the Oz direction; the convection velocity $\mathbf{v}_{ext} = (\mathbf{E}_{ext} \times \mathbf{B}_{ext})/B_{ext}^2$) is constant and points in the Oy direction. Inside the 3-D plasmoid the electric field intensity, $\mathbf{E}_{in}$, is larger than $\mathbf{E}_{ext}$. This is indicated by the closer spacing of the equipotential surfaces there. Therefore, the plasma inside the cloud moves faster than the plasma outside, i.e. $v_{in} > v_{ext}$. This figure illustrates how different parts of the same flux tube can be decoupled from each other. It indicates that the plasma they contain can move independently across magnetic field lines: i.e. that plasma is not necessarily 'frozen' in the magnetic field distribution in the presence of a microscopic double-layer.

In the 3-D double layer illustrated in Fig. 5.16b, the surface charges are not distributed uniformly along the double-layer surface. The larger accumulation of positive charges toward the right-hand side is due to the larger polarization current J_p inside the plasmoid. This polarization current supplies the polarization charges wherever they are needed to set up the polarization field $\mathbf{E}_{in}$ that is needed in order to conserve the momentum density $(nmv)_{in}$ of the moving plasmoid (see Schmidt, 1966; Lemaire, 1975, 1985). The same decoupling mechanism works at the surface of solar wind plasma density irregularities impulsively injected into the magnetosphere (see Lemaire and Roth, 1991).

The effect of these weak double-layers on the motion of interchanging plasma elements is difficult to model with large-scale hydromagnetic codes. This is due to the orders of magnitude difference in length scales involved. Nevertheless, these weak double-layers are important, since they permit decoupling of the motion of a 3-D magnetospheric plasma cloud (at high altitude) from the ionosphere (at low altitude). This inhibits the effect of the resistive ionosphere in limiting the maximum interchange velocity. When pairs of weak double-layers are present at the edges of plasma clouds, the plasmoid can drift across magnetic field lines with a velocity that can become much larger than the maximum interchange velocity given by equation (5.17). The maximum interchange velocity is then no longer determined by the value of the integrated Pedersen conductivity. The plasma cloud slips across magnetic field lines like any conducting spaceship orbiting in the magnetosphere. The Debye layer surrounding the spacecraft is similar to the DL envelopping the plasmoid. This additional effect needs to be addressed in more detail in future studies of the interchange instability. This recommendation applies also to the studies of interchange stability in the ionosphere as described by Ott (1978) and Ossakow and Chaturvedi (1978), as well as to numerical simulations of centrifugally driven interchange motion in the magnetosphere of Jupiter (Yang et al., 1992).

In future comprehensive studies of interchange motion in the plasmasphere, diamagnetism, the bending of magnetic field lines, changes in the volume and shape of plasma elements, and violation of the frozen-in magnetic field condition $\mathbf{E.B} = 0$ should definitely be taken into account.

5.5   Theories for the formation of the plasmapause

5.5.1   The plasmapause as the last closed equipotential (LCE)

The assumptions

The first theories proposed for the formation of the plasmapause were based on the ideal MHD approximation of plasma physics, which implies that magnetic field lines are equipotential lines ($\mathbf{E.B} = 0$), i.e. there are no field aligned electric potential drops or double layers. The magnetic field $\mathbf{B}$ was taken to be a centered dipole parallel to the Earth's rotation axis. The electric field was perpendicular to $\mathbf{B}$ and its distribution was assumed to be quasi-stationary. The electric and magnetic field distributions were therefore considered to be electrostatic and magnetostatic; electrodynamic effects resulting from variable B- and E-fields were regarded as unimportant or in any case were neglected.

The plasma was assumed to be cold ($T \simeq 0$; $\beta \simeq 0$), so that gradient-B/ curvature drifts, being proportional to the energy of the particles, could be neglected. The small gravitational drift was also neglected. Only the $\mathbf{E} \times \mathbf{B}/B^2$ drift was taken into account. Therefore, in early theories of the formation of the plasmapause, all plasma elements (density enhancements as well as density depressions) were forced to follow streamlines which were perpendicular to both the magnetic field and electric field directions, i.e. like single charged particles, any plasma blob had to drift parallel to the equipotential surfaces of the postulated external field. Theoretical and empirical electric field models were assumed to be a combination (superposition) of a corotational E-field of ionospheric origin and a solar wind driven convection E-field.

The former electric field was assumed to be represented by closed (almost circular) equipotential lines, while the E-field of solar wind origin was described by equipotential lines running along the magnetotail; this corresponds approximately to a uniform dawn-dusk electric field as illustrated in Fig. 5.14. In the outer regions of the magnetosphere the equipotentials lines were assumed to be directed back to the distant tail along the flanks of the magnetopause, as shown in Fig 5.17, taken from Nishida (1966). Note that in Brice's 'best estimate' E-field model (Fig. 5.14) the equipotential lines penetrate the daytime magnetopause instead of running back parallel to this surface. However, this major difference has no direct implication for convection in the inner magnetosphere, nor for the formation of the plasmapause.

The equipotential surfaces of Figs 5.14 or 5.17, and the cold plasma streamlines which are parallel to these surfaces can be divided into two categories: (i) those which pass into the magnetospheric tail (or penetrate the magnetopause); and (ii) those which do not pass into the tail (or which do not cross the magnetopause, as in Fig. 5.14). 'When magnetic field lines are carried to the tail and their ends become open, the plasma associated with them would diffuse along field lines toward the outer space. Since the temperature of the magnetospheric plasma is high, the rate of this diffusion is quite high, and tubes of force would be evacuated in less then several hours'. These sentences are quoted from Nishida's (1966) pioneering paper, which already contained the germs of what was to become somewhat later the polar wind flow (Banks and Holzer, 1968, Axford, 1968).

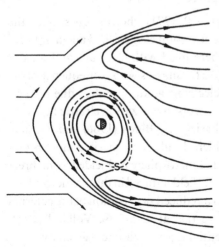

Figure 5.17 Equatorial cross-section of the equipotential surfaces corresponding to Nishida's magnetospheric electric field model. Equipotentials are closed inside the whistler knee (plasmapause). S is the 'stagnation point' where $E_\perp = 0$. According to the early MHD theory for the formation of the plasmapause the last closed equipotential (dashed) coincides with the location of the plasmapause (after Nishida, 1966).

As the convective system transports these field lines subsequently back from the tail, these field lines would be closed again, but the plasma supply from lower altitudes being a slow process, the plasma replenishment on these field lines would not proceed quickly. Hence the equilibrium plasma distribution would not be attained on these field lines before they are transported once again to magnetospheric tail and plasma starts to escape again.

Thus Nishida argued (p. 5670–71) that on those field lines which 'travel across the tail, the plasma density would be less than the value predicted on the basis of the diffusive equilibrium theory. However, for field lines that are never transported to the tail, plasma escape is always prevented by closed lines, and diffusive equilibrium would prevail. A sharp density gradient would appear at the boundary between these two types of field lines'. This boundary, shown by the dashed line in Fig 5.17, was identified as the plasmapause, i.e. the surface where Carpenter as well as Gringauz (see Chapter 1) had observed a sharp knee in the plasma density distribution.

This simple and appealing idea was then immediately adopted by Brice (1967) and by much of the space science community, although many alternative versions of the electrostatic electric field model were subsequently proposed by different authors. These electric field models have been presented in Section 4.10.2. The uniform dawn–dusk model of Kavanagh, Freeman and Chen (1968) (also sometimes called the 'tear drop' model because in this model the last closed equipotential (LCE) has the shape of a tear drop) has often been used instead of the original models drawn by Nishida and Brice. Later on, the uniform dawn–dusk E-field model was extended by Volland (1973, 1975) and Stern (1977) to include possible shielding effects. The aims of these improvements was to reduce the penetration of solar wind driven convection into the inner magnetosphere.

Sheilding of the inner magnetosphere

It was actually Block (1966) and Karlson (1970, 1971) who first questioned the validity of the simple superposition of the corotation electric field and a uniform dawn–dusk electric field in a magnetosphere filled with electrons and ions of non-zero energy and which have different inertial masses. As a consequence of the charge separation between the ions and electrons guiding centres due to their different inertial forces and different grad-B drifts, polarization electric fields are set up within the magnetosphere as emphasized by Block (1966) and Karlson (1971). Their pioneering contributions, which should have merited more attention, have led to the concept that the inner magnetosphere is shielded from direct penetration of the uniform dawn–dusk electric field induced in the magnetosphere by the solar wind flow. To account for such a reduced penetration of the external electric field deep into the magnetosphere (i.e. at $L \leq 5\text{–}6$), Volland (1973) postulated that the total electrostatic potential is approximately given by

$$\phi_E(L,\phi) = AL^\gamma \sin \phi - 91.5/L \text{ [in kV]} \tag{5.21}$$

where ϕ is the local time angle ($\phi = 0$ at midnight LT), and L is the equatorial distance (in R_E) of the magnetic field line. Magnetic field lines were 'of course' considered as equipotential lines ($E_\parallel = 0$). The free parameter γ is called the 'attenuation exponent' This parameter is generally adjusted to provide some amount of 'shielding'. Note that Volland's semi-empirical way of accounting for a reduction of the uniform dawn–dusk electric field intensity deep inside the plasmasphere is not based on a self-consistent physical theory of the coupled ionosphere–magnetosphere system. Furthermore, there is no connection between the *ad hoc* magnetospheric electric field distribution given by equation (5.21), and those obtained by Block (1966) and Karlson (1971) in their shielded electric field models. Work by various other groups (Vasyliunas, 1972; Jaggi and Wolf, 1973; Harel and Wolf, 1976; Harel *et al.*, 1981a;b) has confirmed the dominant role of plasmasheet and ring current pressure gradients on the magnetospheric electric field distribution.

Nevertheless, the electrostatic potential of Volland–Stern, given by equation (5.21), or its generalization by Volland (1978), are still rather popular. They are employed for their simplicity and quotability even in contemporary studies with different values of the attenuation exponent γ (Kurita and Hayakawa, 1985; del Pozo and Blanc, 1994; Soloviev *et al.*, 1989; Galperin *et al.*, 1995). When $\gamma = 1$ there is no shielding at all, as in the early models of Kavanagh, Freeman and Chen (1968), Grebowsky (1970), Chen and Wolf (1972) and others. A value of $\gamma = 2$ was often adopted in subsequent studies and simulations (Maynard and Chen, 1975; Galperin *et al.*, 1975, 1995; Kivelson, 1976; Southwood, 1977; Ejiri, 1978; Ejiri, Hoffman and Smith, 1978; Kivelson, Kaye and Southwood, 1979; Kaye and Kiveson, 1979; Kurita and Hayakawa, 1985; del Pozo and Blanc, 1994).

The second parameter (A) in equation (5.21) is adjusted to fit the positions of the LCE with observed positions of the plasmapause. So far there is no established physical theory relating the value of A to parameters of the solar-wind flow, which is the cause of this induced electric field component, nor to the level of geomagnetic activity (e.g. K_p). From observations of the plasmapause positions during disturbed geomagnetic conditions it was inferred that the parameter A should be an increasing function of the level of geomagnetic activity measured by the values of the index K_p, or some convenient combination of K_p during the previous 12 or 24 hours. Each of these electric field models was tailored for some specific application. They were not, however, deduced from *in situ* measurements like McIlwain's empirical E-field models (McIlwain, 1972, 1974), or from Incoherent Radar Scatter measurements (Blanc, 1978, 1983a; Fontaine *et al.*, 1986; Senior *et al.*, 1987, 1989) and other satellite measurements (Baumjohann, Haerendel and Melzer, 1985; Baumjohann and Haerendel, 1985).

Early dissentient views

The idea that the plasmapause could be identified with the last closed equipotential is 'simple' and quickly became very popular. Many scientists in the field have considered this explanation as 'obvious', and it has become one of the longest enduring paradigms in magnetospheric physics. However, already in 1966, Dungey (1967, p. 102) pointed out that it is rather surprising that 'Carpenter's knee should be so sharp if it is identified with the loop (i.e. the last closed equipotential) separating tubes of magnetic field circulating around the earth that never leave the magnetosphere, and those outside which have come from the tail'. Indeed, as argued by Dungey (1967), 'in reality the flow must be quite variable, being greatly enhanced during disturbed times, and the picture is then less simple. Some tubes of high density (i.e. from inside the plasmasphere) should sometimes be swept out on the day side, and some tubes, after entering from the tail, should enter the inner region and, after a few days, should have intermediate values of density. It then seems rather surprising that the knee should be so sharp, but the variable model would predict a patchy density in the region near the knee and this could be the true state'.

Furthermore, it could be added to Dungey's comment that, according to the MHD model, the plasmapause density gradient should be sharpest in the post dusk local time sector, i.e. closest to the 'stagnation point' where empty flux tubes from the tail split into two diverging streams of plasma. The tubes that are slowly convected along the immediate post-dusk LCE surface should be exposed to substantial refilling from below with ionospheric plasma. Furthermore, they should be subject to interchange instability à la Richmond (1973) or Huang, Wolf and Hill (1990); these mechanisms, which are claimed to have a time constant of 1/9 hour, should have enlarged the plasmapause region already by midnight local time. Observations indicate, however, that density gradients as steep as any yet reported can be observed in the midnight and post-midnight sectors. Unless the position of the stagnation point can be shifted from dusk into the midnight local time sector, it is difficult to explain such observations with the traditional theory.

Finally, according to the original theory, the formation of a plasmapause is the result of a steady state flow pattern where the plasmapause is described as the boundary between open and closed flux tubes. The knee should be sharpest when the electric field distribution is strictly stationary (i.e. time-independent). The observations indicate, on the contrary, that the formation of a plasmapause is not the consequence of a stationary convection pattern as initially postulated, but is on the contrary an as yet poorly understood consequence of unsteady and spatially structured flow activity.

The intensity of the dawn–dusk component of magnetospheric convection electric field models has generally been choosen a function of K_p to match empirical relationships between K_p and observed positions of the plas-mapause.The results deduced by Carpenter and Anderson (1992) from ISEE/SFR data and reported in Chapters 2 and 4 have shattered the twenty-year-old paradigm according to which the dawn–dusk asymmetry of the equatorial plasmapause and the equatorial distance of the 'stagnation point' may be used to infer the intensity of the 'uniform dawn–dusk' electric field component in the magnetosphere. Carpenter *et al.* (1993) discuss the underlying disagreement between observations of the dawn–dusk asymmetry and the role that asym-metry played in the early models as well as in subsequent theories for the formation of a plasmapause (see also Rycroft, 1974a).

Moldwin *et al.* (1994) using simultaneous observations of two geosyn-chronous spacecraft at different local times are also led to paint a different picture of the plasmaspheric morphology than that envisioned in the past where the plasmapause was assumed to coincide with the last closed equipotential.

The difficulties with the early steady state theory for plasmapause formation were not only noticed by Dungey (1967), but also by Grebowsky (1970, 1971) and Chen and Wolf (1972), who suggested then alternative ways to define the plasmapause position within the magnetosphere. Nevertheless even in modern textbooks authors often revert to the simple theory where the plasmapause coincides with the last closed equipotential of a stationary magnetospheric E-field distribution (e.g. Wolf, 1995).

5.5.2   Deformations of a last closed equipotential (LCE)

As already pointed out in the previous section, the definition of the plasmapause as coincident with a last closed equipotential is based on the presupposition of a steady-state magnetospheric electric field distribution. However, the solar wind plasma impinging on the magnetosphere is rarely stationary. Therefore, the magnetospheric convection E-field induced by the perpetually changing solar wind flow can rarely be in a steady state (see Lemaire and Roth, 1991). Overall enhancements of the solar wind momentum flux and fluctuations in the inter-planetary magnetic field (IMF) induce magnetospheric E-fields which perpet-ually change the instantaneous position of the LCE. The new positions of the LCE cannot be determined from its earlier position by integrating the equation of motion:

$$dr/dt = \mathbf{E} \times \mathbf{B}/B^2$$

Indeed, the points forming the LCE at some initial time, $t = 0$, drift to new positions which do not fall along the time dependent locations of the LCE at later instants of time (see Fig. 12 in Lemaire, 1987).

To overcome this basic difficulty, Grebowsky (1970) proposed to 'assume that a steady state configuration exists, corresponding to a dawn–dusk electric field of an initial magnitude. If this electric field component is everywhere suddenly increased in magnitude, then the plasma in the region between the initial plasmapause (i.e. the initial LCE) and the final steady state plasmapause will travel along the streamlines corresponding to the enhanced field, until it is lost from the magnetosphere at the magnetopause. By following the motion of the plasmapause plasma along the streamlines, the time dependent variation in the plasmapause position can be determined for this disturbance' (see p. 4332 in Grebowsky, 1970).

Figure 5.18 illustrates this evolution and the deformation of the initial LCE following an enhancement of the dawn–dusk E-field component. It can readily be seen that the most rapid changes in the plasmapause position occur near dusk, and that the final plasmapause configuration in the nightside sector is attained only 6 to 10 hours after the initial electric field increase. It takes 6–10 hours after the substorm onset to move the post-midnight plasmapause from its pre-storm position at midnight to its new equilibrium location closer to the Earth.

In Grebowsky's model plasma inside the initial LCE, corresponding to a low dawn–dusk E-field intensity, is carried from the dusk region in the sunward direction. The plasma thus entrained begins to escape from the magnetosphere

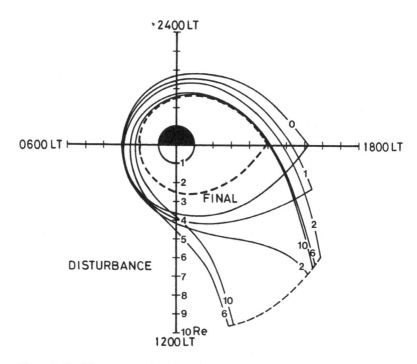

Figure 5.18   Plasmapause location plotted for selected times (0, 1, 2, 6 and 10 hours) after the dawn–dusk electric field component has increased suddenly from its initial steady state value of 0.28 mV/m to 0.58 mV/m (after Grebowsky, 1970).

along sunward directed drift paths. Figures 2.15a and b, corresponding to ISEE afternoon and duskside passes, clearly show narrow outlying features at $L \sim 4$ and $L \sim 6$. But one of the conclusions of Carpenter *et al.* (1993) was that MHD simulations were least effective in reproducing what is observed in the afternoon–dusk sectors. It is only in the 00–12 MLT sector that such MHD simulations do fairly well in terms of an apparent agreement with observations.

Grebowsky (1970) also computed the evolution of an initial LCE plasmapause position during quieting that he simulated by a sudden decrease in the dawn–dusk E-field intensity at an initial time t_0. He pointed out (p. 4333) 'that in the quieting process the ionosphere supplies plasma to the region between the initial and final plasmasphere boundaries. Because this region is characterized by closed streamlines, no loss of plasma occurs via transport out of the magnetosphere. Thus the corotation of the bulge should become obscured after a day or so by the increasing plasma density that arises from the transport of ionospherically produced plasma along magnetic field lines'.

The effect of a single sudden decrease of the dawn–dusk E-field intensity has also been studied by Chen and Wolf (1972) following similar steps. The effects of repeated, short-duration enhancements simulating a series of magnetically disturbed days was examined by Chen and Wolf (1972). The same method of simulation was also used by P. Corcuff (1978) and Corcuff *et al.* (1985) to determine the dynamical deformations of the plasmapause, which were then compared with satellites and whistler observations. Additional physics has been included in this kind of simulation by Spiro *et al.* (1981), who assumed a given electric potential distribution along a high-latitude boundary, and used a complex numerical code developed over several years at Rice University to calculate 'a self-consistent convection electric field' which fits the initial/boundary conditions. The evolution of the assumed pre-substorm plasmapause position was then determined using their calculated time-dependent E-field model. This model included shielding effects, the ring current and the effects of the ionosphere.

While limited by the necessity to select an arbitrary pre-storm plasmasphere configuration, this second generation of MHD models of the plasmapause had the merit of explaining a number of dynamic features such as rapid changes in plasmapause positions measured at widely spaced ground stations (Corcuff and Corcuff, 1982). They also appeared to account for certain observations, along Explorer 45 orbits, of regions of enhanced cold plasma separated from the main plasmasphere (Maynard and Chen, 1975).

Although this method of simulating the deformation of the plasmapause is rather appealing because of its simplicity, it suffers from difficulties that should not be overlooked in a comprehensive review. A first limitation of this method is the rather arbitrary choice of the model for the E-field distribution and of the initial/boundary conditions used. The postulated dependence of the E-field distribution on the value of the geomagnetic K_p index is an additional weakness; the dawn–dusk electric field intensity can be 'tuned' so that the stagnation point is at the 'right place' in the 1800 LT sector (or elsewhere!) for purposes of fitting the calculated positions of the LCE to various sets of whistler or satellite observations. Therefore, it is not surprising that the 'predicted' plasmapause positions fit the observations so well! Such 'a posteriori agreements' should not, however, be regarded as valid support for any of the 'ad hoc E-field models' used in the past, nor should they be regarded as ultimate support for the theory of formation of the plasmapause underlying the method with which they have been obtained. Perhaps this is why Chen, Grebowsky and Taylor (1975) noted (p. 970) 'that it would be more appealing from the physical standpoint to derive a relation between the dawn–dusk component of the convection electric field and K_p that is independent of plasmapause measurements'. Chen, Grebowsky and Taylor (1975) also point out (p. 970) that 'this model relies on measured plasmapause positions and suffers from the drawback that temporal phase differences (dependent upon local time) are known to exist between changes in the plasmapause location and changes in K_p (Chappell, 1972)'.

According to the MHD scenario, the 'new plasmapause', which is observed closer to Earth just after a substorm event, corresponds to the 'old' one which has been shifted sunwards in the nightside region and which has become sharper as a result of the 'compression of magnetic flux tubes'. Also as a result of this compression of the plasmasphere we would expect to observe higher density levels inside the nightside plasmasphere during and after a substorm, provided the density in the quiet plasmasphere falls off with L more slowly than the volume of flux tubes increases with L, as is in fact observed (e.g. Carpenter and Anderson, 1992). However, there is in fact a tendency for the density to decrease in the outer plasmasphere, in the aftermath of substorms. Indeed, as noted in Chapter 2 (Section 2.5.1), the thermal plasma concentration just inside the new plasmapause tends to be reduced by a factor of as much as three during some substorm events. The observations therefore suggest that physical processes other then those implicit in the standard MHD scenario are active in the vicinity of a newly formed plasmapause.

This early theory for the formation of the plasmapause has another drawback: the choice of the initial time (t_0) when MHD simulations start is a rather arbitrary choice, since there is no unique way to determine t_0. Indeed, in a K_p-dependent E-field distribution the shape and dimensions of the LCE, taken here as the initial plasmapause, will vary substantially as t_0 varies among conditions of a low, medium or high value of the K_p index. As a matter of consequence the subsequent positions of this boundary will depend on this arbitrarily chosen initial time t_0 (Lemaire, 1987).

In the second generation of time-dependent models, authors did not propose any physical mechanism for the formation of this 'initial plasmapause'; they only simulated how this surface would be displaced to other locations at later instants of time. It should also be pointed out that with non-steady E-field models, the plasmapause never does coincide with the last closed equipotential at any time t, except at the initial time $t = t_0$, when it was assumed (*a priori!*) to match a LCE. Therefore, we are tempted to conclude with Carpenter (1970) that 'the plasmapause is not generally coincident with an equipotential of the combined magnetosphere flow'.

To overstep these serious difficulties Chen and Wolf (1972), Chen and Grebowsky (1974) and Grebowsky, Tulunay and Chen (1974) introduced a new definition of the plasmapause which no longer coincides with the LCE. Nevertheless, the old concept of a plasmapause coincident with the last closed equipotential of the magnetospheric electric field still remains deep-seated within the backyard of the space science community, as can be seen in papers by Doe *et al.* (1992) and others, where the LCE concept is used, when 'steady state conditions' could be invoked.

5.5.3   A new approach and definition of the plasmapause; a third-generation model

The new definition of the plasmapause was based on the idea that some flux tubes drifting in a time-dependent/or a K_p-dependent E-field can have remained 'closed' for more than 6 days, while others typically at larger L-values, can have drifted in from the magnetotail during the preceding 6 days. The former flux tubes should be saturated, with diffusive equilibrium established; these tubes are confined within a region whose outer boundary can be computed at any time t by integrating a large number of drift paths backward in time from t to $t - 6$ days. The inner region where flux tubes have remained closed for at least 6 days is then identified as the plasmasphere; the boundary separating these tubes from those that were closed for only 5.99 days (or less) is considered to be the new working definition for the plasmapause.

Numerical method to determine the plasmapause position

This alternative definition of the plasmapause and the new procedure to calculate its position was first introduced by Chen and Wolf (1972). The equatorial plasma density depends on the time t' when a flux tube drifting in from the tail passed $L = 15$. If that flux tube was located beyond $15 R_E$ at times earlier than t', it is assumed that its equatorial density was $0.1 \, \mathrm{cm}^{-3}$ at time t'. Assuming a constant daytime refilling rate of $2\text{–}3 \times 10^8$ electrons $\mathrm{cm}^{-2} \mathrm{s}^{-1}$ and a smaller nightime depletion rate, Chen and Wolf (1972) computed the total tube content and equatorial density at time t by numerical integration between t' and t. This was done for a large number of magnetic flux tubes.

Some results of such demanding numerical calculations are shown in Fig. 5.19; the equatorial density profile is given versus L along a radial line in the dusk local time sector. The numbers along the density profile correspond to the filling times, in days, since t'. The assumed variation of the intensity of the uniform dawn–dusk electric field is illustrated in Fig. 5.20 as a function of universal time. In this case study the time between two repeated enhancements of the E-field intensity was 8 hours. The minimum and maximum electric field intensities were chosen to represent 'typical values of the equatorial electric field during substorms'.

In Fig. 5.19 all flux tubes at geocentric distances less than $3.9R_E$ have been circling the Earth for 6 days or more. However, flux tubes at $L > 3.9R_E$ have generally been circling the Earth for shorter periods of time, so that their degree of filling varies substantially with L. Flux tubes near $L = 4.1, 4.6,$ and 5.3 exhibit peaks in the equatorial density profile; they correspond to tubes forming attached plasmatails wrapped around the central core of the plasmasphere and shown in Fig. 5.21 at $t = 2400$ UT. These different plasmatail structures were formed as a consequence of successive E-field enhancements and the associated sunward surges of the duskside plasmasphere, as illustrated in Fig. 5.18.

The flux tubes inside the oldest plasmatails have been closed for more than 6 days; this is why their density is higher than most of the neighboring flux tubes. It can be seen from Figs 5.19 and 5.21 that three tails are found in the dusk meridian plane, at $t = 2400$ UT. Note that 7 hours earlier (i.e. at $t = 1700$ UT) only one tail would have been encountered along the radial direction at 1800 LT, while three tails would have been encountered then in the local noon sector.

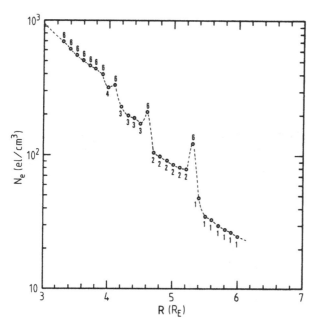

Figure 5.19 Profile of the equatorial electron density calculated for a gusty uniform dawn–dusk convection electric field whose time variation is displayed in Fig. 5.20. The closure times of flux tubes at different L-values in the dusk sector and at 2400 UT are indicated along the profile by the number of days of refilling since they crossed the drift shell: $L = 15$. The flux tubes below $L = 3.9$ and at $L = 4.1, 4.6,$ and 5.3 corresponding to peaks in the density profile have been closed for more than 6 days. These density peaks correspond also to plasmatails wrapped around the central core and shown in Fig. 5.21 (after Chen and Wolf, 1972).

The number of tails encountered at a fixed local time is a function of UT, and is a consequence of the corotation of the central core region and of the multiple tails that are wrapped around it and attached to the main plasmasphere. Each of the five tails in Fig. 5.21 were formed during the five enhancements of the dawn–dusk electric field intensity shown in Fig. 5.20.

Comparison with observations

The numerical calculations required to determine the shape of the plasmapause boundary all around the earth are extensive and demand large computer

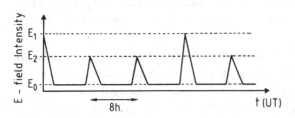

Figure 5.20 Assumed time variation of the uniform dawn dusk component of the convection electric field intensity used to obtain the results shown in the preceding and following figures (after Chen and Wolf, 1972)

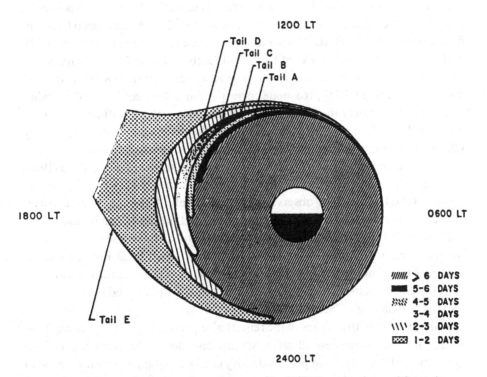

Figure 5.21 Plasmatail structure formed as a result of MHD convection and flux tube refilling in a time-dependent uniform dawn–dusk electric field. The gusty variation of the dawn–dusk electric field intensity is shown in Fig. 5.20. The plasmapause is defined as the limit between magnetic flux tubes which have refilled for more than 6 days and those which have refilled for less than 6 days. Different shadings are used for strands of tubes which have been closed for 5, 4, 3, 2 and 1 days (after Chen and Wolf, 1972).

resources. This new method of calculation has been used by Grebowsky, Tulunay and Chen (1974) to compute the positions of the plasmapause (at dawn and at dusk) every one or two hours UT during a period of 8 days between 23 and 27 May 1967 that included a substorm period. The calculated model plasmapause positions were then compared with a set of mid-latitude trough positions measured by the ARIEL-3 satellite. As usual, the K_p-index was taken to modulate the intensity of the uniform dawn–dusk E-field component. Instead of 6 days, these authors found it more economic to use a 'closure time' of 5 days. The L-values of the low altitude trough density minimum deduced from selected data sets of ARIEL-3 paralleled those of the calculated equatorial plasmapause; however, the agreement was better in the dusk region than in the dawn LT sector.

The same numerical method was used by Chen and Grebowsky (1974) to determine the position of plasmatails and to compare them with the positions of 'detached' plasma elements observed by OGO-5 outside the main plasmasphere. 'The satellite, according to the model, should have cut through regions with substantial variability in closure time and hence multiple plasmatails should have been observed. Consequently, the density should be characterized by many fluctuations'. But the OGO-5 observations did not reveal the expected complex multi-tail like structure. This led Chen and Grebowsky (1974) to say (p. 3853): 'Admittedly, the observation appears to be more complex than what this model predicts'. Some cases of approximate correspondance between the model predictions and the observations were in fact found by these authors, but they concluded (p. 3854) that 'whether or not all the detached events observed by OGO-5 can be reproduced by the simple model remains to be seen'.

Chen and Grebowsky (1978) and more recently Kurita and Hayakawa (1985) have used the same procedure to compute plasmapause positions and compare them to OGO 4 and S^3-A observations, respectively. Although, there is correspondance between some selected observations and the model predictions, these authors agree that this third generation model can hardly be used to explain all observations of the position of the plasmapause at all local time and at all universal times; nor can it be considered capable of explaining the position and width of all 'detached plasma' elements or 'attached plasmatails' observed by various satellites (see Chen and Grebowsky, 1978).

Carpenter's (1970) findings of the remarkable persistency of the plasmapheric bulge and the abruptness of its westward end do not support the lack of abruptness of the calculated plasma density gradient obtained with this brand of model. Similar conclusions have been drawn in the more recent paper of Carpenter et al. (1993). 'The foregoing line of interpretative work has provided and today still provides the basis for discussions of the plasmapause phenomenon' (Carpenter, 1995, personal communication).

Discussion

This third generation of MHD models 'eliminates the assumption of a steady state plasmapause configuration at some initial time and takes into account the partial filling of magnetic flux tubes by ionospheric plasma' (Grebowsky, Tulunay and Chen, 1974). It is an improvement upon the earlier ones discussed in the previous sections. But, it should be noted that the calculated plasmapause positions depend on the assumed 'closure times' (e.g. 6 or 5 days which are not sharply defined thresholds) as well as on a somewhat arbitrarily chosen distance in the tail (e.g. 15, 10 or $8R_E$) where magnetic flux tubes are supposed to 'enter' the magnetosphere. Indeed, changing these parameters in the computer program will change the instantaneous position of the plasmapause at any time t. Therefore, the position of the plasmapause depends on the choice of a series of parameters which are not determined in a unique way on solid physical grounds.

There is an additional limitation of this new approach. The large plasma density peaks illustrated in Figs 5.19 and 5.21 should be subject to slow plasma interchange motions in the gravitational and centrifugal potential field as described in Section 5.4. These plasmatails should fall back toward the central core of the plasmasphere when the gravitational pull dominates, while the gap between the tails and the main plasmasphere should increase when centrifugal effects become predominant. They should also be unstable with respect to MHD instabilities driven by pressure gradients as described by Richmond (1973). Since the value of the integrated Pedersen conductivity is not infinitely large these extended cold plasma tails should be disrupted in 1/9 hour because of the instabilities invoked by Vinas and Madden (1986) and Huang, Wolf and Hill (1990).

This remark leads us to a fourth generation of models for the formation of the plasmapause, proposed by Lemaire (1974, 1975). The grounds on which this theory are based are outlined in the next section. The theory is based on the effect of plasma interchange motion described in Section 5.4.2.

5.5.4   Plasmapause and LIT formation : a fourth-generation model

As was mentioned in Section 5.4.3, the interchange velocities driven by magnetization, gradient-B and curvature drifts vanish when the plasma is 'cold' or when the kinetic pressure is balanced across the surface of a plasma element. Furthermore, it has been shown that a plasma element with an excess mass density in a gravitational field (**g**) and magnetic field **B** will develop a polarization electric field whose intensity **E** is such that the $\mathbf{E} \times \mathbf{B}/B^2$ is equal to the free fall velocity in the gravitational field. This effect is always present whether the plasma is cold or not. The plasma is accelerated by the gravitational force until its velocity obtains a maximum value which is equal to $V_{\mathrm{p,max}}$ given by equation (5.17). As a matter of consequence, plasma density enhancements fall toward the Earth, while plasma density depressions tend to move upwards in the direction opposite to **g**, like a bubble of hot air or the ionospheric bubbles observed by Aggson *et al.* (1992) and others.

The Roche Limit or zero-parallel-force surface

The motion of plasma elements has components parallel and perpendicular to the direction of magnetic field. Due to the relatively large integrated Pedersen conducting the interchange motion of plasma elements is more strongly impeded transverse to magnetic field lines than in the direction parallel to **B**. Therefore, any plasma density depression will bubble up or down toward the surface where the parallel component of the gravitational force is equal to the parallel component of the pseudo-centrifugal force. This surface was compared to the 'Roche Limit' surface by Lemaire (1974). It has been called the Zero Parallel Force (ZPF) surface in Lemaire (1985). See Kopal (1989) for a definition of the Roche Limit in a binary system, and see Section 5.2.3 for a more detailed description of the ZPF surface. Beyond this surface any plasma density depression tends to move earthward, along magnetic field lines. Below the ZPF surface, where the total potential has a maximum any plasma density hole moves away from the Earth, and preferentially along magnetic field lines. In the case of a corotating protonosphere the ZPF surface has a circular equatorial cross-section; its equatorial radius is then equal to $L_c = 5.8 R_E$ (see Section 5.2.3).

The ripping off mechanism

Any plasma density enhancement with respect to a uniform background density located beyond the ZPF surface will move away from this surface as sketched in Fig. 5.22. All plasma density depressions formed on both sides of the ZPF tend to drift asymptotically toward this surface where they accumulate and form a trough in the field-aligned density profile. The plasma shell or blob (density enhancement) beyond the ZPF moves away from this surface and becomes detached from the plasma which is confined in the flux tube on the earthward side of this surface; the latter is trapped in the gravitational potential well while, on the other side, the former is accelerated toward the equatorial plane along all magnetic field lines for which $L > L_c$. This accelerated plasma moves upwards and accumulates in the equatorial potential well shown by the dashed lines in Fig. 5.1.

The plasma flowing up from both hemispheres converges toward the equatorial region where it is compressed. As a consequence of this compression the kinetic pressure in the equatorial region is enhanced in all flux tubes beyond $L = L_c$. This enhancement of the plasma pressure beyond the ZPF surface changes the hydrodynamic equilibrium in the direction transverse to the magnetic field lines. The excess of perpendicular kinetic pressure leads to cross-L interchange of the plasma away from this surface. The interchange motion is driven mainly by magnetic forces, as described in Section 5.4.3. The plasma bulges away from the Earth in the equatorial region of the magnetosphere as illustrated in Fig. 5.23. As a consequence of the excess kinetic pressure in the outermost layer of the plasmasphere the plasma is blown out away from the ZPF surface. All the plasma originally located beyond the ZPF surface along field lines for which $L > L_c$, is ripped off from the plasmasphere by the combined effect of field-aligned flow and cross-L interchange motion (Lemaire, 1997).

In other words, at the onset of substorm enhanced magnetospheric convection in the nightside sector, a potential well is formed in the equatorial region due to centrifugal effects. This produces converging field-aligned plasma flow toward the equatorial plane, where the kinetic pressure is increased. The resulting perpendicular pressure imbalance accelerates the compressed plasma away from the Earth, across magnetic field lines. The radial expansion enables the system to evacuate the plasma converging toward the equatorial plane, which otherwise would pile up there.

The cross-L expansion velocity cannot grow indefinitely, but it is limited to a maximum value which is inversely proportional to the value of integrated

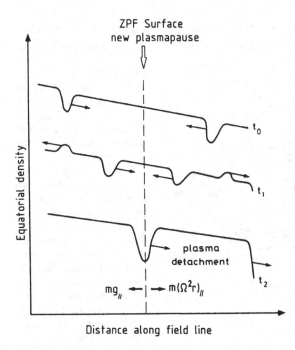

Figure 5.22 Density distribution showing two plasma holes drifting toward a common asymptotic trajectory determined by the balance between the field-aligned components of the gravitational force and the field-aligned component of the inertial (centrifugal) force. All plasma holes collect along this trajectory. As a consequence a trough develops there. The large plasma density enhancement formed beyond this trough drifts away from the main plasmasphere by plasma interchange motion. A sharp 'knee' in the equatorial density remains when the outer detached plasma shell has separated from the plasmasphere which is confined within the gravitational potential well. This figure illustrates therefore how a new plasmapause is formed by peeling off a shell of the plasmasphere by interchange motions (after Lemaire, 1985).

Pedersen conductivity. Since the integrated Pedersen conductivity is not equal to zero, the cross-L expansion cannot exceed some maximum value (see Section 5.4.2).

These transient field-aligned plasma flows and cross-L interchange motions are driven unstable only along magnetic field lines which traverse the ZPF surface, i.e. for which $L > L_c$. Indeed, as shown in Fig. 5.1, it is only along these field lines that a potential well is formed in the equatorial region.

The plasma along magnetic field lines closer to the Earth, for which L is smaller than L_c, remains stably trapped in the gravitational potential well. Therefore, the central plasmaspheric core, at $L < L_c$, remains trapped within the Earth gravitational field as long as a substorm associated convection electric field enhancement does not penetrate deeper into the magnetosphere.

Light-ion-trough formation

At lower altitude ionospheric plasma also flows upwards along all field lines corresponding to $L > L_c$: indeed, the potential barrier along these field lines is suddenly lowered due to the enhanced centrifugal effect. As a consequence, thermal plasma tends to flow up out of the topside ionosphere along all field lines for which $L > L_c$.

Since the heavy O^+ ions have a smaller density scale height than H^+ and He^+ ions, more light ions than heavy ones will be ripped off at high altitudes beyond

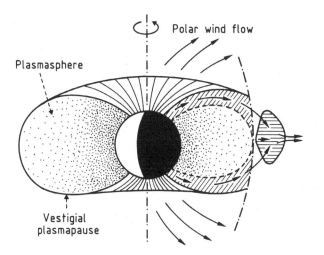

Figure 5.23 Illustration of the detachment of a fragment of the plasmasphere beyond the Zero-Parallel-Force surface (dashed-dotted line) where the parallel components of the pseudo-centrifugal force and gravitational force balance each other in the post-midnight local time sector. The field-aligned flow of ionization toward the equatorial plane, and the radial plasma interchange motion driven by the sudden enhancement of the eastward convection velocity carry a detached element of cold plasma away from the Earth toward the magnetopause (Lemaire, 1997).

the ZPF surface. As a matter of consequence more light ions (H^+ and He^+) are sucked out of the ionosphere than O^+ ions. This upward ionization flow produces a light-ion-trough (LIT) at low altitudes. This LIT starts to develop soon after the ZPF surface is forming closer to the Earth, due to enhanced eastward magnetospheric convection in the midnight sector. Therefore, troughs are forming in the latitudinal profiles of light ion density but not directly in the latitudinal profile of the O^+ density, except indirectly, via the charge exchange reaction (4.1).

The resulting polar wind like flux vanishes along magnetic field lines tangent to the ZPF surface, i.e. for $L = L_c$. For $L > L_c$, the intensity of the upward ion flux is an increasing function of $(L - L_c)$. Indeed, the centrifugal effects become increasingly important as L increases beyond L_c. Since the upward flow is smaller near $L = L_c$ than at larger values of L, the light ion densities are not significantly changed at the equatorward edge of the light-ion-trough. However, at larger L a significant fraction of the total flux tube content is ripped off; therefore much more plasma from low altitude is sucked out of the mid–latitude ionosphere to replace that which is 'falling' in the equatorial potential well. Thus the light ion density decreases faster at higher L that close to $L = L_c$.

According to this scenario the LIT starts to form gradually in the topside ionosphere soon after the ripping off mechanism operates at high altitude along the ZPF surface. But while a sharp density gradient may form at high altitude immediately after the onset of enhanced eastward magnetospheric convection, it takes hours for the light-ion-trough to develop at lower altitude in the mid-latitude ionosphere. The reasons are (i) the finite time for the signal carried by a pressure wave to propagate along field lines from the high altitude ZPF surface down into the ionosphere, (ii) the limiting upward diffusion flux of H^+ ions trough the O^+ ion background in the topside ionosphere, and (iii) the replacement of H^+ ion by the charge exchange reaction (4.1). Therefore, the latitudinal density gradient at low altitude is expected to be much more gradual (smoother) than that observed at high altitude below the ZPF surface. A sharp knee may be observed at high altitude in the plasma density distribution below the Roche limit while no corresponding gradient has had time to develop at the altitude of 500–1000 km along the field lines beyond $L = L_c$. Thus the lack of a clear one-to-one correspondence between low-altitude LIT signatures and high-altitude plasmapause ones, is not surprising within the frame of the scenario described above.

The midnight local time sector where the action takes place!

According to this scenario the ZPF surface is the place where small amplitude perturbations in a uniform plasma density distribution, become convectively unstable and have a growth rate which, for cross-L motion, is inversely proportional to the value of the integrated Pedersen conductivity (see Section 5.4.2).

Since the integrated Pedersen conductivity has a minimum value in the night-side region, it is in this LT sector that the growth rate for this instability is maximum, and that the maximum plasma interchange velocity reaches its largest value. It is therefore in the nightside sector that the ripped off plasma is most quickly evacuated by cross-L interchange motion.

Furthermore, according to current magnetospheric convection models (e.g. those of McIlwain, 1972, 1974, 1986; Blomberg and Marklund, 1988b), it is in the post-midnight sector that the magnetospheric convection velocity is maximum and exceeds the corotation speed, beyond $L = 4$ or 5. During strongly disturbed conditions (i.e. when K_p is large), the eastward angular velocity of cold plasma becomes significantly larger than the corotation velocity for $L > 3$. Under these conditions, the *parallel component* of the centrifugal force exceeds that of the gravitational force at an equatorial distance smaller than $5.8R_E$ (see Section 5.2.3). Note that the *radial (cross-L) component* of the centrifugal force exceeds that of the gravitational force beyond the zero radial force (ZRF) surface. This ZRF surface crosses the equatorial plane at L_0 which is larger than L_c. For corotating plasma $L_0 = 6.6$. For realistic magnetospheric E-field distributions the ZRF and ZPF surfaces are closest to the Earth between 0100 and 0400 LT. The sharpest plasmapause gradients have, indeed, been observed by Chappell, Harris and Sharp (1970) in the midnight and post-midnight sectors.

Lemaire's mechanism explains why the new plasmapause is found closer to Earth when the maximum value of K_p has been large during the 6 to 12 hours preceeding a particular observation of the plasmapause position (Carpenter, 1967). The observed anti-correlation between the plasmapause position and the maximum value of K_p during the 12 preceding hours is a key feature that any satisfactory physical theory for plasmapause formation must be able to account for.

Step-like plasma density profiles observed after a series of substorms of decreasing peak intensity can also be explained by Lemaire's peeling off mechanism. Indeed, in the case of successive substorm events of decreasing intensity, the deepest points of penetration of the ZRF and ZPF surfaces shift to larger values of L in the post–midnight sector at each new onset.

In early versions of this theory for the formation of the plasmapause, it was pointed out that the field-aligned distribution of the total gravitational-centrifugal potential has a well at high altitude with a minimum in the equatorial plane. This led Lemaire (1974) to argue that the plasmapause is the equipotential surface which is tangent to the ZPF (or Roche limit) surface. In subsequent work and model simulations Lemaire (1985, 1987) emphasized the unstable cross-L interchange motion of equatorial plasma element beyond the ZRF surface, but in his numerical simulations the converging field-aligned motion of plasma elements was not included. In Lemaire's (1997) most recent work both aspects are merged; the field-aligned flow which is followed by the compression of ripped off plasma elements, as well as their outward cross-L motion in the equatorial plane beyond the ZPF surface are included. These are complementary aspects which contribute (i) at high altitude to the formation of a plasmapause 'knee', and (ii) to the formation of a light-ion-trough in the mid-latitude ionosphere.

Following this scenario a flux tube is never completely emptied during an erosion event, an initial condition which is assumed in some time-dependent hydrodynamical refilling models or kinetic simulations.

5.5.5   Smoothing of the plasmapause density gradient

It has been shown by Horwitz (1983) that the observed broadening of the plasmapause region cannot be explained by Bohm diffusion or by wave–particle interactions. When corotating from the post-midnight local time sector to the dayside, broadening of the plasmapause region is expected due to the slight divergence of convection streamlines in the post-dawn LT sector prior to noon (e.g. in Figs 4.19 and 5.28). But there are other additional physical mechanisms which can contribute to smooth the sharp plasma density gradients at the plasmapause boundary or even deeper inside the plasmasphere. Lemaire (1987) has pointed out the role that Coulomb collisions play in smoothing density gradients in the plasmasphere. He considered that the effect of Coulomb collisions had suffered unwarranted neglect in earlier plasmaspheric studies and especially in considerations of the smoothing of density gradients.

Huang, Wolf and Hill (1990) studied the role of plasma interchange driven by azimuthal pressure gradients in broadening the plasmapause region. Furthermore, gravitationally driven plasma interchange also contributes to broaden the plasmapause region soon after the plasmapause has been formed. This is illus-

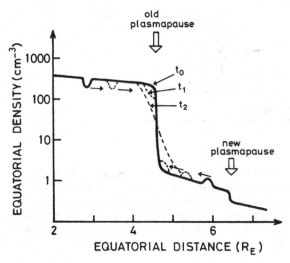

Figure 5.24   Illustration of a mechanism responsible for the broadening of the plasmapause region. The steep plasmapause density gradient formed during a substorm in the post-midnight sector broadens by plasma interchange motion when it corotates to the dayside local time sector (after Lemaire, 1985).

trated in Fig. 5.24 where plasma density enhancements formed in the intermediate region are shown to drift inward toward the old (vestigial) plasmapause. By this mechanism plasma is transported to the outer edge of the old plasmapause, thus increasing the density in magnetic flux tubes located just beyond the sharp initial density gradient. A plasma hole formed earthward of the old plasmapause tends to drift upward until it reaches the inner edge of the density knee. It will dissolve there, where its density becomes equal to the background density. This contributes to decreasing the equatorial density in magnetic flux tubes at the inner edge of the vestigial plasmapause. The broken lines in Fig. 5.24 illustrate the evolution of the equatorial density distribution as a function of time. It shows the broadening of the vestigial plasmapause region associated with the decreasing of the steepness of the density gradient.

The characteristic time constant for this non-adiabatic smoothing of sharp density gradients depends in this case of the integrated Pedersen conductivity (Σ_p) in the ionosphere. Indeed, we have shown in Section 5.4.2 that the value of Σ_p determines the maximum plasma interchange velocity and, consequently, the related broadening time (t_B). When the value of Σ_p is small the maximum interchange velocity due to gravity is enhanced; the time constant for t_B is of the order of 30 h for $\Sigma_p = 0.3$ siemens.

It can therefore be concluded that the sharp density gradient formed in the post-midnight region at the onset of a substorm event broadens not only because of the divergence of the plasma streamlines in the dayside local time sector, but also via additional non-adiabatic processes, like plasma interchange

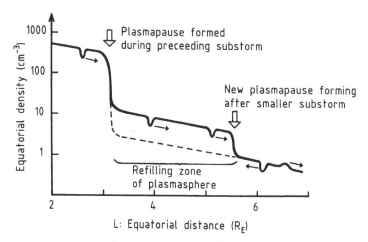

Figure 5.25   Illustration of the formation of 'multiple plasmapauses' after an extended period of high geomagnetic activity. The gradual refilling of a depleted region between the old position of a vestigial plasmapause and a new one is shown by 'steps' in the equatorial plasma density distribution. Only the outermost density 'knee' determines the position of the actual (new) plasmapause region (after Lemaire, 1985).

motion, this in addition to the effect of Coulomb collisions which should not be overlooked in future studies.

5.5.6   Multiple plasmapause formation

Figure 5.25 illustrates the effect of plasmasphere refilling on the equatorial density distribution. The intermediate region between an old vestigial plasmapause and a new one formed during a less severe substorm event refills gradually at a rate which is nearly independent of L. This is why the slope of the equatorial density (on a log scale) versus L remains nearly constant in time and $n(L,t)$ is approximately proportional to L^{-4}, during the refilling process.

When magnetospheric convection remains stationary for several days in a row, the equatorial density should return to the saturated level. If there should be no net outward transport of plasma across drift shells, the saturation density level would necessarily correspond to the hydrostatic-diffusive equilibrium density distribution discussed in Section 5.2.4. But it has been found experimentally that this state of hydrostatic equilibrium is never attained, even after 8 days of prolonged quieting. This is one of the reasons which led Lemaire and Schunk (1992, 1994) to suggest that the whole plasmasphere is slowly expanding like the equatorial region of the solar corona. They call this continuous outward flux of ionization the 'plasmaspheric wind'. Detailed studies of mid-latitude convection by the whistler method, or by radar would be required to confirm such a slow expansion of the plasmasphere.

5.5.7   Determination of the plasmapause position in the post-midnight sector

Observations show that the eastward convection velocity is generally larger than the corotation velocity in the post-midnight local time sector beyond $L = 3 - 4$, and that the eastward plasma drift is enhanced there at the onset of magnetic substorms. This characteristic feature is built in most empirical electric field models based on satellite observations and radar observations. It is illustrated in Figs 4.19 and 5.27.

Using the analytical electric field model E3H of McIlwain (1974) which is illustrated in Figs 5.26 and 5.27, one can determine the position where the inertial (centrifugal) force, $m(\mathbf{u}.\nabla)\mathbf{u}$, is balanced by the gravitational force, $m\mathbf{g}$. The locus of points where the radial components of these forces balance each other corresponds to the Zero Radial Force (ZRF) surface. The locus of points where the parallel components of these forces balance each other is located slightly more earthward along the ZPF surface. The equatorial cross-section of the ZPF coincide with that of the dipole field lines $L = L_c$, where L_c is given by equation (5.6). It is equal to 5.78 for corotating plasma. The equatorial

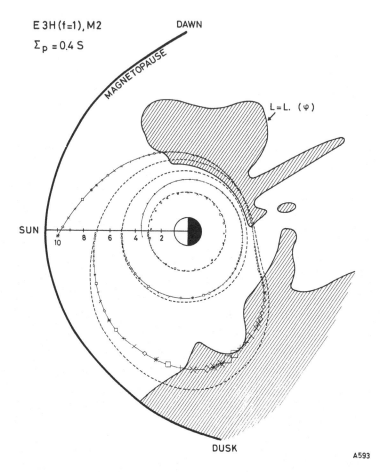

Figure 5.26 Equatorial cross-section of the the Zero Radial Force surface corresponding to the E3H electric and M2 magnetic field models of McIlwain (1974). The solid line forming the border of the shaded area corresponds to the Zero Radial Force (ZRF) surface where the radial component of the centrifugal pseudo-force becomes larger than the gravitational force. Inside the shaded region cross-L plasma interchange motion is directed away from the Earth for all cold plasma density enhancements. The dashed closed curves are equipotential lines of the E3H field. These curves represent streamlines of plasma elements when the effect of interchange motion is ignored. The trajectories of three equatorial plasma density enhancements for which $\Delta n/n = 0.4$ are shown by series of symbols. The size of these symbols is proportional to the equatorial cross-sections of the drifting plasma elements. The type of symbols has been changed every new UT hour. The local time and latitudinal distribution of the integrated Pedersen conductivity adopted in these simulations is that of Gurevitch *et al.* (1976). The nightside value of the integrated Pedersen conductivity is 0.5 S, while on the dayside it reaches a maximum value of 11 S. Note that the equatorial plasma density is not assumed to be enhanced by converging equatorward field-aligned motions. Plasma elements starting respectively at $L = 3$ and 5 (at noon LT) do not penetrate through the ZRF surface: they spiral inward. The outermost plasma density enhancement spirals aways from the Earth as soon as it penetrates the shaded region beyond the ZRF surface (after Lemaire, 1985).

cross-section of the ZRF surface is at $L = L_0(3/2)^{1/3}L_c = 1.14L_c$, i.e. at $6.6R_E$ in the case of corotation.

The wavy solid lines around the shaded area in Fig. 5.26 represent the equatorial section of the ZRF surface when the electric field and magnetic field distributions are respectively given by the E3H and M2 models of McIlwain (1974). Plasma density enhancements drifting in the dayside region (where the value of the integrated Pedersen conductivity is high) closely follow the equipotential lines (dashed lines) of the large-scale electric field. But as soon as these equatorial density enhancements penetrate into the nightside region where the Pedersen conductivity in the E-region is significantly reduced, the plasma interchange velocity is enhanced in the direction perpendicular to magnetic field lines. The drift path then departs significantly from the broken equipotential lines. The innermost density enhancements tend to spiral toward the Earth, falling down in the gravitational potential well, while the outermost ones tend to drift irreversibly away from the Earth as soon as they penetrate inside the shaded area beyond the ZRF. On the contrary, equatorial density depressions (plasma holes) tend to move toward the ZRF surface, provided of course that the background density decreases fast enough with L (approximately as L^{-4}) to prevent the density inside the moving and contracting element to become equal to the external (background) density. This property is used in Lemaire's (1974; 1985) numerical codes to locate the equatorial distance, L_0, of deepest penetration of the ZRF surface and, L_c that of the ZPF surface which is at a slightly lower L according to equation (5.6).

The solid line in Fig. 5.27 running across the nightside magnetosphere from the dawn to dusk corresponds to the equatorial cross-section of the ZPF surface for the E3H convection electric field model (Lemaire, 1976b). The broken lines correspond to the electric equipotential contours. The equipotential line labeled -10kV is tangent to the ZPF surface in the post-midnight sector at $L = 4.5$. According to Lemaire the hatched area inside this equipotential line corresponds to the equatorial cross-section of the plasmapause for this electric field model.

Since the E3H model did not have a stagnation point, the last closed equipotential (LCE) corresponding to the Alfvén layer for zero energy electrons or ions extends here to the magnetopause. This LCE coincides here with the inner edge of the hatched area extending to the magnetopause. Of course, in other E-field models like E5D, the presence of a stagnation point at dusk drastically changes the shape and position of the LCE as well as of the ZRF and ZPF surfaces in this local time sector. In the post-midnight sector, however, the equatorial cross-sections of these surfaces are nearly the same for the E3H and E5D empirical models.

The ZRF and ZPF surfaces can be determined for any other electric field model, for instance, the Volland–Stern models. In the Volland–Stern models the intensity of the electric field is symmetric with respect to the dawn–dusk meridian plane. Since the observed day–night asymmetry and post-midnight enhancement of the convection electric field are not present in the Volland–Stern models, but are present in the E3H empirical model, Lemaire (1975, 1985) preferred to use the latter model in his numerical simulations, despite the absence of a stagnation point in the dusk local time sector. Arguing that the distribution of the electric field in the dusk region has no direct effect on physical processes taking place far away in the post-midnight sector, he considers the formation of a new plasmapause boundary by centrifugaly driven plasma inter-

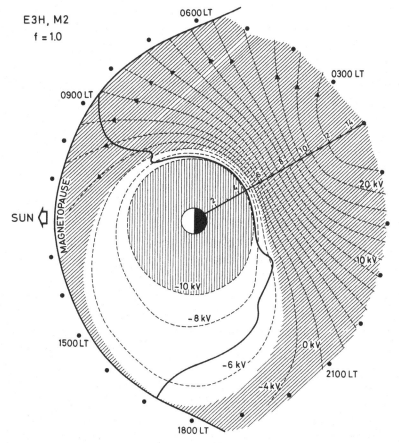

Figure 5.27   Equatorial cross-sections of the electrostatic potentials surfaces corresponding to McIlwain's (1974) electric field model E3H are shown by broken lines. A similar figure was given in Lemaire (1974) for another earlier electric field model. The wavy solid line across the nightside magnetosphere corresponds to the equatorial cross-section of the ZPF surface along which the field-aligned component of the centripetal force is equal to the gravitational force. The central hatched area tangent to the ZPF surface nearly coincides with the equatorial cross-section of the plasmasphere for this model electric field (after Lemaire, 1976b).

change in the midnight sector as a separate phenomenon from the erosion events which can also possibly take place near a stagnation point in the dusk sector.

McIlwain (1986) published a new empirical electric field model E5D as an improvement or an update of his E3 and E3H older versions. The equipotential lines of the E5D electric field model are illustrated in Fig. 5.28 for $K_p = 0$ and for $K_p = 6$. The analytic representation of the E5D electric field distribution is much simpler than the expansion describing the earlier E3 and E3H versions. While the E3H model contained 120 terms which were determined for $K_p = 1$–2, the newer E5D model contains only five independent parameters which are functions of K_p. Furthermore, unlike the teardrop model or the Volland–Stern family of semi-empirical models, the E5D model is deduced from dynamic particle spectra measured with the geosynchronous satellites ATS-5 and 6 in the region of the magnetosphere close to where the plasmapause is located. For all these reasons we wish to recommend the E5D electric field model as a useful reference for future studies. It would anyway be useful to compare the E5D analytical model with those models deduced from other types of measurements (e.g. from available incoherent scatter radar data, DE-2, Viking and FREJA satellites), as Baumjohann and Haerendel (1985) did for GEOS electric field measurements.

5.5.8 Detached plasma elements

According to Lemaire's interpretation, the plasmapause is expected to form in the post-midnight LT sector in non-symmetric E-field distributions like those of the E3H and E5D models. With the Volland–Stern symmetric models the new plasmapause should be formed near dawn, but not at dusk as in the theory of the last closed equipotential. Lemaire (1974) considered that the equatorial position of the plasmapause at later local times could be traced by the streamline which is tangent to the ZPF surface in the post-midnight sector. Indeed, once the sharp gradient is formed somewhere between 2300 and 0500 LT, it will be convected toward the dayside, in a region which is convectively stable with respect to field-aligned as well as transverse interchange motion. On the dayside, the integrated Pedersen conductivity is enhanced by more than an order of magnitude; there the newly formed plasmapause tends to move along equipotential surfaces of the external magnetospheric electric field. This is why in Fig. 5.27 the plasmapause cross-section is assumed to coincide almost with an equipotential line of the steady state electric field distribution.

The plasma shell or fragment which is detached from the plasmasphere in the post-midnight region moves slowly outwards. OGO-5 as well as ISEE have observed such detached plasma elements or blobs in the post-midnight sector, very close to the plasmapause (Chen and Grebowsky, 1978; Kowalkowski and

Lemaire, 1979). One of these detached plasma element observed in the post-midnight sector is shown in Fig. 4.16. These observations can hardly be explained by erosion at a stagnation point in the dusk sector. They can easily be explained by ripping off and plasma interchange mechanisms similar to those discussed above.

Of course, many detached plasma elements are also observed on the dayside and especially in the afternoon LT sector (Chappell, 1974; Chen and Grebowsky, 1978; see also Section 2.5.2 and 4.9). It is generally assumed that these dayside density enhancements are tail-like structures stretched out from the duskside plasmasphere by sunward convection surges associated with substorm events. However, one cannot exclude the possibility that some of the plasma elements seen between 0600 and 1700 LT have been detached near midnight, but did not yet reach the magnetopause. They could well have been convected along an outward–inward drift path into the afternoon LT sector. Of course, the plasma stretching mechanism proposed by Grebowsky (1970) and Chen and Wolf (1972) and illustrated in Figs 5.18 and 5.21 possibly contributes to the formation of attached plasmatails. See Carpenter *et al.* (1993) for a recent and forthright discussion of both possibilities. Both mechanisms could well operate simultaneously, one in the post-midnight sector and the other in the dusk sector.

5.5.9 Substorm associated peeling off events

According to the physical mechanism described in Section 5.5.4 for the formation of a new plasmapause, a sharp knee is formed in the cold plasma distribution when the angular rotational speed of the plasmasphere (or part of it) is suddenly enhanced at the onset of a magnetic storm. Let us assume that before this enhancement occurs the field-aligned plasma distribution is, for instance, in diffusive equilibrium and corotating with the angular velocity of the underlying ionosphere, as shown by the upper solid line in Fig. 5.5. In this case, the field-aligned component of the pressure gradient force is balanced by all other forces including the gravitational and inertial (centrifugal) forces. Let us now consider that part of the plasmasphere acquires suddenly an additional eastward bulk velocity. This can occur when a hot plasma element from the plasmasheet is impulsively injected deeper into the magnetosphere. As a result of the velocity shear and of the different temperatures at the interface between the cold unperturbed plasma and the injected hot plasma, an oblique electrostatic double layer forms along the surface of the hot plasma cloud, as described in Section 5.4.5.

Furthermore, as a consequence of enhanced azimuthal convection velocity the centrifugal forces are enhanced inside the volume of the hot plasma cloud, but not outside. As a consequence of the larger centrifugal force the equilibrium of the field-aligned flux of momentum density is perturbed; the cold plasma is accelerated away from the Earth along magnetic field lines and compressed as already described in Section 5.5.4. Due to the resulting enhancement of plasma pressure in the equatorial region the pressure balance equilibrium is also perturbed in the direction perpendicular to the magnetic field lines. As a matter of consequence, plasma interchange motion will carry the ripped off plasma element away from the Earth, leaving behind a sharp density gradient at the surface of the unperturbed region of the plasmasphere where the convection E-field has not been enhanced.

Note that in this section the ZPF surface is replaced by the interface between two different plasma regions : one being filled with cold corotating plasma, the other being an admixture of cold and hot plasma drifting more rapidly in the sunward (eastward) direction. The distribution of the convection velocity across this interface is then discontinuous, unlike in standard convection electric field models.

When the convection electric field distribution is sheared and the plasma in the upper part of magnetic flux tubes slips with respect to its low-altitude part, the surface of velocity shears plays the same role as the ZPF surface. Under these circumstances, instead of identifying the plasmapause with the equipotential line tangent to the ZPF surface, as in Fig. 5.27, it is then appropriate to identify the plasmapause with the equipotential surface of the convection electric field which is tangent to the volume of the intruding plasma cloud. The deeper hot plasma clouds are injected into the nightside plasmatrough and/or plasmasphere, the closer to the Earth the new plasmapause may be formed.

5.5.10   Search for an ideal time-dependent electric field model

The basic difference between the earlier theories for formation of the plasmapause and Lemaire's is that it is unstable plasma interchange, rather then laminar MHD plasma flows diverging at a stagnation point, that forms new (sharp) plasmapause density gradients. Of course, both mechanisms can be at work simultaneously in different LT sectors.

Much resistance to this fourth-generation model has come from the fact that the numerical simulations published by Lemaire (1974, 1975, 1976b, 1985) were based on a magnetospheric E-field model (E3H) which had no stagnation point near dusk. Indeed, since the publication of the original papers of Nishida and Brice in 1966–67, all magnetospheric electric field models had to have a point of singularity somewhere, preferably near dusk. A second reason is that in his early theory Lemaire (1975, 1976b) considered interchange motions in 'cold' plasma, neglecting therefore non-zero temperature effects; but the cold plasma approximation was also assumed in ideal MHD theories.

Of course, there are other weaknesses in Lemaire's dynamical simulations:

e.g. the E3H model had to be scaled to represent electric field distributions outside the range of K_p for which this E-field model was designed originally (i.e. for $K_p \simeq$ 1–2). The scaling factor (f) introduced by McIlwain (1974) was arbitrarily adjusted by Lemaire (1976b, 1985) in order to match the theoretical position of the ZPF surface with plasmapause positions determined experimentally in the post-midnight sector for other levels of geomagnetic activity. The empirical relation $f = f(K_p)$, was chosen in an *ad hoc* manner to fit the observed plasmapause positions given by the empirical relationship (4.13), deduced by Carpenter and Park (1973). This procedure is no less questionable than the one used by Grebowsky (1970) and by Chen and Wolf (1972) and others to modulate the intensity of the dawn–dusk electric field component as a function of K_p. While such procedures have been and still remain common practice, however, common practice is not a warranty that these procedures are necessarily realistic ones or 'best estimates'. Future modelers should resist this practice and avoid arguing, *a posteriori*, that the good agreement of their *ad hoc* model predictions with a given set of observations proves the validity of these models or simulations.

Without any comprehensive and independently determined electric field model based on reliable measurements, modelers in the 1970s were led to tailor their own E-field models mostly so as to fit their specific applications. Fortunately, thanks to the continued efforts of experimentalists engaged in routine ground-based observational campaigns and difficult measurements in space, more comprehensive K_p-dependent or solar-wind-dependent electric and magnetic field models for the magnetosphere are now becoming available. Thanks also to the efforts of interdisciplinary groups of scientists analyzing the large amount of data available, empirical models have been assembled and made available to all. They constitute road maps for future magnetospheric explorers.

For instance, the IGRF, the Olsen and Pfitzer (1974; 1977) and the Tsyganenko (1987, 1989) magnetic field models as well as McIlwain's (1986) E5D electric field model illustrated in Fig. 5.28 are good examples of models that are needed to test and improve magnetospheric theories and processes; the global models currently developed by IAGA, COSPAR, ISTP, and GEM are needed as inputs for numerical simulations and modeling of the formation and deformation of the plasmapause.

To be most useful these empirical field models must be portable; this implies that they should preferably be given in terms of mathematical expressions or in matrix format instead of graphs or output listings of complex computer programs. This is precisely one additional advantage of the E5D model. It has a simpler analytic expression than the E3H model, making it easier to implement in terms of computer codes. The following calculations and the simulations displayed in Fig. 5.29 are based on the E5D model. The results are qualitatively similar to those obtained earlier with E3H and shown in figs. 5.26 and 5.27,

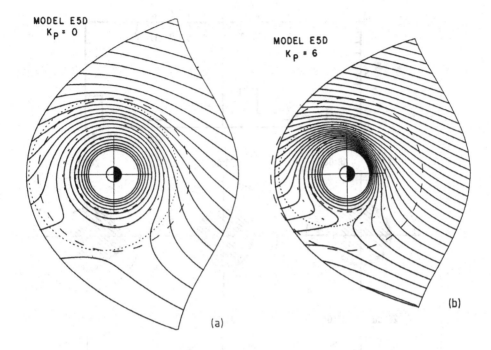

Figure 5.28   Equatorial cross-section of the electric equipotential surfaces corresponding to McIlwain's E5D model (a) for $K_p = 0$ (left panel) and (b) for $K_p = 6$ (right panel). K_p-dependent analytical expressions describe this comprehensive electric field model (after McIlwain, 1986).

except in the dusk LT sector, where the presence of the stagnation point drastically changes the plasma streamline pattern.

It may be argued that the stationary E5D electric field model is also questionable and that, in certain cases, it underestimates the eastward convection velocity in the post-midnight LT sector. This is a possibility: under highly variable conditions the peak values of the electric field intensity exceed the corresponding average values used by McIlwain (1986) to build the large-scale quasi-stationary model E5D. There is also the possibility that smooth and quasi-stationary electric field models, including the E3H and E5D models, are not adequate to approximate the actual time-dependent magnetospheric electric field distribution during disturbed geomagnetic conditions. Why should the electric field distribution be quasi-stationary and 'electrostatic' when the magnetic field distribution in the magnetosphere is almost continuously changing with time? Why should the magnetospheric electric field distribution not have significant 'induction' and 'parallel' components? Furthermore, why should it remain smooth during a substorm, instead of developing discontinuous Alfvén layers and tangential discontinuities? These time-dependent kinetic features are

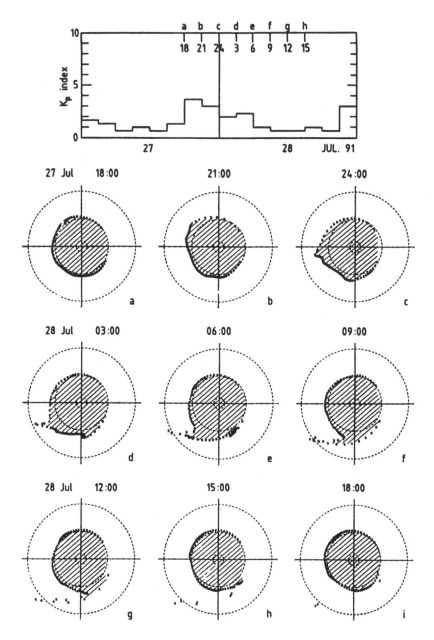

Figure 5.29 Equatorial cross-section of the plasmapause positions based on Lemaire's theory for the formation of the plasmapause by plasma interchange instability. McIlwain's (1986) K_p-dependent electric field model, E5D, has been employed in this simulation. The successive snapshots illustrate the deformation of the plasmapause every next 3 hours of Universal Time for a period of 2 days between 27 and 29 July 1991. A substorm event occurred during this interval. The development of a dayside bulge and of a plasma tail like structure by differential convection velocity is illustrated. The dashed circles correspond to $L = 5$ and 10.

not included in current empirical or semi-empirical models like the E3H or E5D models, nor in any earlier Volland–Stern type models for the magnetospheric electric field distributions.

The results of Afonin et al. (1996), illustrated in Fig. 4.13 and discussed in Section 4.8.5, tend to support the existence of such a sharp boundary in the nightside plasmatrough. An Alfvén layer like plasma boundary surges earthward during the main phase of a geomagnetic storm. This sharp ionization front erodes the subauroral electron temperature enhancement (SETE) which is associated with the high altitude plasmapause (see Fig. 4.12). These challenging observations as well as those of subauroral ion drift (SAID) by Galperin et al. (1974) and Anderson et al. (1991, 1993) lead us to infer that during disturbed conditions, sharp discontinuities or Alfvén layers develop in the nightside magnetosphere. Their presence in the plasmatrough and the plasmapause region invalidates, of course, the commonly used concept of a smoothly varying electric field intensity across the plasmapause surface (Karlson, 1971).

Therefore, if the results of current numerical results do not always fit the observations, we should not necessarily conclude that the physical theory or the numerical code used are questionable, but one should perhaps first question the realism of the large scale electric field model employed as an input in these numerical simulations (see Section 4.10.3).

5.5.11 Application for improved empirical electric field models

Instead of smooth and quasi-stationary electrostatic models like the Volland–Stern, the E3H or E5D models, electrodynamic models would be needed for future, more comprehensive magnetospheric studies and numerical simulations. The electric field variations induced by a substorm event do not necessarily occur simultaneously and in phase everywhere around the world. These field variations may be out of phase in different local time sectors. In current empirical models, including the E5D model, it is postulated that the electric field variations are changing everywhere in phase. This weakness plagues not only MHD theories for the formation of the plasmapause but all others as well. However, until comprehensive time dependent electric and magnetic field models are available we are obliged to make the best use of simple (portable) models like the E5D models. This is what is attempted in the following section to illustrate how the positions of the plasmapause can be calculated for this particular E-field model.

5.5.12 Dynamical model calculation of the plasmapause positions

As a result of the continuously changing E-field distribution, the plasmapause position cannot simply be identified with a last equipotential surface nor with an equipotential tangent to the ZPF surface or to an intruding plasmasheet cloud. When the E-field intensity changes rapidly with time, a more pragmatic approach is probably needed to trace the position of plasmapause as a function of LT and UT. Indeed, the deepest point of penetration of the ZPF surface is then changing all the time, as well as the position of the stagnation point in the K_p-dependent electric field model E5D.

To determine the minimum equatorial distance of the ZRF and ZPF surfaces under such variable conditions, Lemaire (1985, 1986) proceeds as follows. He calculates the drift paths of plasma holes to determine the position of the ZRF surface in the post-midnight sector; these drift paths invariably tend toward the position of the ZRF surface. Initially, the equatorial plasma holes are released at $L = 6.6$ and $\phi = 2300$ LT. Their drift paths are calculated by numerical integration of the equation of motion in the variable external electric field and magnetic field. The $\mathbf{E} \times \mathbf{B}/B^2$ drift velocity and the maximum interchange velocity are taken into account in the equation of motion of the plasma blob. A reduced Pedersen conductivity can be introduced to enhance the rate of convergence of the drift paths toward the ZRF surface. The minimum radial distance L_0 of the calculated drift paths of plasma holes corresponds to the point of deepest penetration of the ZRF surface in the post-midnight sector; according to Lemaire's theory the new plasmapause is forming at $L_c = (2/3)^{1/3} L_0$ which is the minimum equatorial distance of the ZPF surface when the magnetic field line distribution corresponds to a dipole. Once this point of space is determined in the equatorial plane, the next step is to calculate the streamline of background plasma density elements, i.e. plasma elements whose density is equal to the background density. The drift paths of background plasma elements released at the point of deepest penetration of the ZPF surface are then used to trace the positions of the new plasmapause around the Earth in other local time sectors. The drift paths of the plasma density elements are calculated by integrating the equation of motion; a standard Runge–Kutta method is used for this purpose.

The trajectory of a background plasma element coincides with the equipotential tangent to the ZPF when the E-field distribution is stationary as initially proposed by Lemaire (1974). But for time dependent (or K_p-dependent) E-field models, this drift path experiences deformations. In his latest model calculations Lemaire (1994) determines the successive positions of a series of plasma elements released every 20 minutes UT near midnight, at the point of formation of the plasmapause. Over a period of 24 hours, 72 plasma elements are released and used to trace the position of the new plasmapause at later local times. When the local time of a particular plasma element exceeds a limit, ϕ_{max}, it is abandoned to save computer storage. Only those plasma elements whose local times are smaller than ϕ_{max} and larger than ϕ_{min} (the LT where they are released in the post-midnight sector) are tracked in consecutive runs. Those elements which reach the magnetopause are also eliminated. The instantaneous position (i.e. the equatorial distance, $L_i(t)$ and local time, $\phi(t)$) of each plasma element that has survived are stored every fixed interval of time, Δt. The positions of the surviving elements are displayed in plots like those shown in Fig. 5.29. In this simulation ϕ_{max} was set equal to 2100 LT. The instantaneous positions of the plasma elements forming the outer boundary of the plasmasphere are shown every 3 hours of universal time for the period between 27 July 1991 at 1800 UT and 28 July at 1800 UT. As indicated in the top panel of Fig. 5.29, the geomagnetic activity index K_p gradually increased during this period up to a maximum value of $K_p = 4-$ during the evening of July 27, and then decreased to lower values on the following day.

The simulations presented in Fig. 5.29 show that the plasmasphere was almost circular and featureless at 1800 UT on 27 July at the end of an extended period of reduced geomagnetic activity (see panel a). During this quiet period the refilling mechanism could have increased the equatorial density inside and outside the plasmapause so that no sharp knee may have existed at the location indicated in panel a. Panel b shows a snapshot 3 hours later at 2100 UT, just after the dawn–dusk component of the electric field had increased to a maximum intensity at about 19:00 UT. The plasmapause is now forming at a lower L-shell in the post-midnight sector. Furthermore, a bulge is developing around $\phi = 10:00$ LT in the pre-noon sector. This bulge is formed as a consequence of the enhanced sunward convection velocity associated with the enhanced dawn-dusk E-field component.

At 2400 UT (panel c) the level of activity was decreasing, the ZRF and ZPF surfaces were receding to larger radial distances and the dayside bulge has corotated into the afternoon LT sector. Three hours later this bulge has developed into a tail-like structure with a well defined westward edge near 1400 LT. In the following panels (d–i) it can be seen how this plasmatail became more elongated. This stretching is due to the decrease of the eastward convection velocity as a function of L; i.e. the differential angular rotation velocity in the dayside sector. The tip of the tail-like structure moves radially outward along the open equipotential lines of the E5D electric field distribution (see Fig. 5.28). This tail-like structure does not corotate: it drifts with a smaller angular velocity; it constitutes a sort of stagnant plasma region extending up to the magnetopause.

This plasma region may be related to the 'detached plasma regions' observed in the afternoon local time sector by Chappell, Harris and Sharp (1970b). If the scenario illustrated in Fig. 5.29 is correct, such plasma regions are not truly detached from the main body of the plasmasphere but their density is significantly reduced due to their expansion as they stretch out to large L. Similar attached plasma-tails have been obtained in earlier MHD simulations illustrated in Figs 5.18 and 5.21. However, these latter plasma-tails were then formed in a rather different way, by stretching plasma originally in the vicinity of the duskside 'stagnation point' and carrying it into the afternoon local time sector under the influence of the enhanced sunward convection. The most recent simulations of this kind have been presented by Weiss *et al.* (1997) and are illustrated on the Web at [http://space.rice.edu/IMAGE/psph/ca/p1.gif].

Although this mechanism cannot be excluded, however, the outputs of these simulations depend significantly on the electric field model adopted, as well as on the intitial shape and position of the plasmapause at the time $t = 0$. As already pointed out in Section 5.5.2, this initial position is not uniquely determined nor based on any specified physical mechanism for the formation of the plasmapause.

In contrast, in the scenario illustrated in Fig. 5.29, the tail-like structure is the residue of a bulge formed in the pre-noon sector by substorm enhanced (sunward) convection in this LT sector; subsequently, due to the sheared convection velocity at large radial distances in the dayside magnetosphere, this dayside bulge is stretching as a consequence of the slower azimuthal convection velocities at larger radial distances than closer to the Earth. Note that similar tail-like structures have been obtained by Lemaire (1985; see Fig. 28) with a Kp-dependent electric field model, despite the absence of a 'stagnation point' at 1800 LT in this E-field model.

The sequence of events illustrated in Fig. 5.29 corresponds well to whistler observations reported by Carpenter (1970) and Ho and Carpenter (1974). They found evidence of tail-like features that move in the direction of the Earth's rotation during quieting. The westward edge of the bulge tends to become sharper due to the differential convection velocity in the dayside. It marks the edge of the stretched plasmasphere extension moving across the afternoon sector into the nightside sector with an azimuthal velocity smaller than corotation.

From the last panels, it can be seen that after the episode of enhanced magnetospheric convection, the plasmasphere relaxes again to a more circular shape with a slight dawn–dusk asymmetry. There is also a permanent noon-midnight asymmetry which results from the noon–midnight asymmetry built in the E5D electric field model (see Figs 5.28a and b). A quiet time noon–midnight asymmetry has been observed by Gringauz and his colleagues, as discussed in Chapters 3 and 4.

5.5.13   Contemporary modeling efforts

Instead of using the point of deepest penetration of the ZPF surface in the post-midnight plasmasphere to determine the plasmapause position, a more pragmatic approach has been proposed by Soloviev *et al.* (1989) to determine, the position of the plasmapause from the observations of its projection in the nightside LT sector. It had already been suggested earlier by Galperin *et al.*

(1975) that the positions in the equatorial plane where new plasmapauses are formed correspond to an Alfvén layer. Note that the Alfvén layers for zero energy protons and electrons coincide with the last closed equipotential (LCE). The projection in the ionosphere of the Alfvén layer of zero energy protons and electrons can be determined by the observed position of subauroral ion drift (SAID) (Galperin et al., 1973b; 1974; Smiddy et al., 1977). The ionospheric projection of this layer which is assumed to coincide with the high altitude plasmapause is also identified by Galperin et al. (1977) and Sauvaud et al. (1983) with the location where 1–2 keV soft plasmasheet electrons are precipitated and detected in the ionosphere. The equatorward edge of the region of precipitation of these soft electrons is called by them the 'soft electron boundary' (SEB) The average position of this boundary is a function of local time and depends on the level of geomagnetic activity during the 3 (or 12) previous hours.

According to Galperin et al. (1977) and Valchuk et al. (1986), the K_p-dependent position of the SEB is a function of ϕ, the magnetic local time. From a statistical point of view the mean invariant latitude Λ_0^{SEB} of the SEB is given by

$$90° - \Lambda_0^{SEB} = 18.53 + 1.25\underline{K}_p + 0.18\underline{K}_p^2 + 2.84 + 1.24\underline{K}_p$$
$$- 0.076\underline{K}_p^2(\phi/6 - 3)$$
for 18 MLT $\leq \phi \leq$ 24 MLT $\qquad\qquad$ (5.22a)

$$90° - \Lambda_0^{SEB} = 24.354 + 2.719\underline{K}_p - 0.203\underline{K}_p^2 + 0.753\underline{K}_p\phi/6$$
$$- 6.304(\phi/6 - 0.5)^2 - 1.698\phi/6$$
for 00 MLT $< \phi <$ 0600 MLT $\qquad\qquad$ (5.22b)

where $\underline{K}_p$ is an index of geomagnetic activity deduced from the values of K_p for the two previous 3-h intervals. This corresponds to the time needed for the inner edge of the plasmasheet (assumed to be an Alfvén layer) to propagate earthward to the point of deepest penetration into the nightside plasmasphere. Note that this is about the time it took the 'sharp ledge' or 'wall' in the density profile shown in Fig. 4.13, to propagate from $\Lambda^0 = 64°$ to $60°$, where the erosion of the plasmasphere started to take place during the geomagnetic storm of 1–2 December 1977.

Following the idea that Alfvén layers for zero energy particles determine the position of new equatorial plasmapauses, Soloviev et al. (1989) and Galperin et al. (1995, 1997) used the position of the SEB to determine the free parameters A and γ of the Volland–Stern electric field model given by equation (5.21) (see section 5.5.1). The values of A and γ are adjusted in order to fit the last closed equipotential (LCE) of this E-field with the position of the SEB at all local times between $\phi = 21$ MLT and 06MLT. Assuming that $\gamma = 2$ is a satisfactory fit value for the 'attenuation exponent' or 'shielding effect', the intensity of the dawn-dusk electric field, which is proportional to $A(\underline{K}_p)$, is determined from the K_p dependence of $\Lambda_0^{SEB}(\underline{K}_p,\phi)$ given by equation (5.22). But, since this SEB surface does not fit the LCE of a standard Volland–Stern model where $A(\underline{K}_p)$ is a constant independent of local time, Soloviev et al. (1989) and Galperin et al. (1995) force such a fit, for a given MLT, by imposing that A becomes a function of ϕ: $A(\underline{K}_p, \phi)$. Once the dependence of A on K_p and ϕ is obtained empirically, the spatial distribution of this modified Volland–Stern electric field model is uniquely determined for a particular instant of time, t. The universal time-dependence of this semi-empirical global magnetospheric electric field is then also fully determined since $\underline{K}_p$ can be obtained from the series of K_p-values preceeding the time t. Note that in this model the intensity of the convection electric field is again a smooth function of the radial distance and does not feature the discontinous spatial gradient which is expected across physical Alfvén layers. Therefore, in our understanding this electric field model does not include the large velocity shears observed equatorward of the SEB by Galperin et al. (1973a; 1975) and others (Sivtseva et al., 1983, 1984). Furthermore, no theoretical justification is given to adapt the modified Volland–Stern E-field model. The authors tried other electric fields, but the final results were rather similar, while the computation time was longer (Galperin, 1997, personal communication).

Empty flux tubes, convected in this time-dependent E-field model are refilling slowly with thermal plasma escaping from the ionosphere as described in Section 5.2.6. The expression for the upward ionization flux is taken from Krinberg and Tashchilin (1982). In the time-dependent plasmasphere model of Soloviev et al. (1989) the equatorial plasma density in a flux tube is given by

$$n_{eq}(L \geq L_{min}, t) = N_\infty(1 - e^{-t/\tau}) \tag{5.23}$$

where $\tau = \tau^* L^4$ is a refilling time scale characterizing the time needed to reach the saturation density N_∞. According to Krinberg and Tashchilin (1982), $\tau^* = 0.2$ days and $N_\infty = 4 \times 10^3\,cm^{-3}$; L_{min} refers to the minimum equatorial distance in the plasmasphere where the flux tube has been carried since it is tracked backwards in time. For $L < L_{min}$ Soloviev et al. (1989) assume that

$$n_{eq}(L < L_{min}, \phi) = N_0 L^{-\alpha} \tag{5.24}$$

with $\alpha = 3.5 + 0.5 \sin \phi$; $N_0(\phi)$ is determined by the matching condition: $n_{eq}(L = 2, \phi) = N_\infty$ adopted from Moore et al. (1987).

On closed drift trajectories (i.e. inside the plasmasphere), the flux tubes fill up since the last magnetic storm. Open drift trajectories are earmarked by tracing the drift path backwards in time. Any flux tube which stayed beyond the SEB for more than 3 hours before time t is assumed to be on an open drift path. This flux tube is then refilling at a constant rate since $t' = t - 3\,h$, according to equation (5.23). In other words, the accumulated plasma density is 'reset to zero' at time t' once the flux tube has remained more than 3 h outside the plasmasphere, i.e. outside the LCE of the modified Volland–Stern model. Another assumption made in this time-dependent model of the plasmasphere is that pitch angle diffusion of the upflowing ionospheric particles is strong enough to establish rapidly an isotropic ion velocity distribution function up to the equatorial plane.

In the Soloviev *et al.* (1989) simulations the magnetic field distribution is approximated by a centered dipole, as in most other model simulations. The cross-section $S(L)$ of the drifting flux tubes is such that their magnetic flux is conserved during the $\mathbf{E} \times \mathbf{B}/B^2$ drift motion, while their volume $V(L)$ changes with L. Due to the expansion or contraction of the flux tubes their equatorial density decreases or increases as a consequence of the conservation of the total accumulated mass.

Figure 5.30 illustrates a time sequence of equatorial plasma density profile in

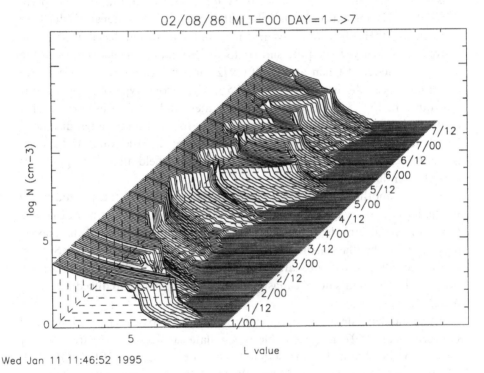

02/08/86 MLT=00 DAY=1−>7

Wed Jan 11 11:46:52 1995

Figure 5.30 Equatorial density profiles obtained by Galperin *et al.* (1995, 1997) with their time-dependent model of the plasmasphere. The density distribution (vertical scale) is given as a function of the equatorial distance L in the midnight local time sector ($\phi = 0$). The stack diagram corresponds to a sequence of profiles corresponding to a period of time of seven days starting 8 February 1986. The time-dependent electric field distribution in the magnetosphere was simulated by a modified Volland–Stern semi-empirical model. The amplitude of the shielded dawn–dusk electric field is assumed to depend on the level of geomagnetic activity (K_p) and on local time so that the last closed equipotential matches the average positions of the 'soft electron boundary' (SEB) for all values of K_p and for all local times ϕ between 21 MLT and 06 MLT (courtesy of Galperin and Torkar, personal communication, 1994).

the midnight local time sector ($\phi = 0$) calculated by Galperin *et al.* (1995, 1997) for seven successive days between 8 and 15 February 1986. The horizontal axis in this stack diagram corresponds to L while the vertical axis is proportional to the logarithm of the equatorial density. Note the existence of steplike density profiles, with sharp density drops at newly formed plasmapauses. Note also the inward and outward shifts of the plasmapause 'knee' which are in phase with the variation of the geomagnetic activity level $\underline{K}_p$. Even detached plasmatails located outside the midnight plasmasphere are obtained with this time-dependent simulation of the plasmasphere. In their simulations two distinct types of plasmatails are obtained. The first type is due to the gradual increase of the dawn–dusk E-field component near 1800 MLT around the 'stagnation point'. This is not apparent in Fig. 5.30, but an example is shown in Fig. 3d of Galperin *et al.* (1997). The formation of such plasmatails with an excess density is due here to particular MHD flow patterns arising from the distorted distribution of the convection velocity. It is quite similar to the formation of plasmatails in the dynamical model of Chen and Wolf (1972) or Kurita and Hayakawa (1985) which have been discussed in Section 5.5.3. The second type of plasmatails is illustrated in Fig. 5.30. The formation of these tails is a consequence of a relatively long residence time of particular flux tubes at large radial distances. For these flux tubes the accumulation time is longer, and the equatorial density is enhanced. This is a potential mechanism to produce field-aligned irregularities and whistler ducts.

There is another characteristic feature observed inside the plasmasphere which does not show up in the ideal MHD simulation of Galperin *et al.* (1995): this is the significant decrease by as much as a factor of three of the plasma density inside the plasmasphere after the onset of some substorms (see Section 2.5.1). According to the simulation illustrated in Fig. 5.30 the density distribution within the nightside plasmasphere at $L < L_{min}$ is not reduced when $\underline{K}_p$ increases.

The time-dependent plasmaspheric model of Soloviev *et al.* (1989) and Galperin *et al.* (1997) has been able to simulate satisfactorily electron density measurements obtained along selected nightside passes of ISEE-1, GEOS-1, EXOS-B and DE-1 between 23 and 06 MLT (see Soloviev *et al.*, 1989). The authors point out, however, that the agreement between their model predictions and the satellite observations is poor in about 20% of the cases considered and deteriorates at local times before 21 MLT. This is surprising in view of their assumption that the plasmapause coincides with the last closed equipotential extending all the way up from the stagnation point at 18 LT to 21 MLT where the fit is unsatisfactory. They attribute this disagreement to the simplicity of the Volland–Stern model which they are using for the global electric field distribution.

The physical mechanism which erodes the plasmapause along magnetic field lines corresponding to the low-energy electron precipitation boundary (SEB), is not specified in the paper of Soloviev *et al.* (1989). But their numerical model has many interesting features and advantages compared to earlier ideal MHD models.

To check the E-field models described above as well as the theories for the formation of the plasmapause discussed in the present Chapter, simultaneous electric field measurements and cold plasma observations would be needed between $L = 3$ and $L = 8$ both in the post-midnight sector and in the duskside region. In addition to values of the plasma density, simultaneous measurements of the temperatures, and of the bulk velocity of the different ions species are needed at all energies but especially between 0.1 and 10 eV. Global imaging of the plasmasphere using new techniques of observations like those recently proposed by various groups of scientists in the USA, will be required to reach the most comprehensive understanding of the physical mechanisms leading to the formation of the plasmapause, the plasmatrough, the plasma density irregularities outside and inside the plasmasphere, the light-ion-trough (LIT), the subauroral electron temperature enhancement (SETE), the subauroral ion drift events (SAID) and polarization jets (PJ) which are a special variety of SAIDs.

Epilogue

Although we are aware that the theories reviewed in Chapter 5 neither address nor explain all aspects of currently available observations, the authors hope nevertheless that the Chapter has described the state of the art in this field of investigation. We have outlined, in a historical perspective, the successive steps in modelling the plasmasphere and its outer boundary, the plasmapause. As has happened in many other fields of investigation, the first model is the best known of all, but several successive generations of models were proposed later on to improve or replace the initial picture. The successive improvements on the preceding work have been outlined, as well as the limitations of each of the successive models. The authors will be rewarded for their efforts in writing this last Chapter and the whole book if it succeeds in setting the stage for the generation of future theories, if it stimulates new ideas and if it produces a revival of interest for the plasmasphere and plasmapause region.

The history of the discovery of the plasmapause by Gringauz and Carpenter, respectively, in the former Soviet Union and United States, has been reported in Chapter 1. Those who worked in this field in the 1960s will remember some of the episodes, but probably will have discovered other aspects of the story that are not published anywhere else.

The main experimental techniques used for the study of the plasmasphere and its outer boundary are whistler wave observations and *in situ* satellite measurements. Both have been reviewed in Chapters 2 and 3.

In Chapter 4, we have tried to give a unified picture of what has been learned from all these experimental measurements combined. This attempt to put all available pieces of the puzzle together contains unavoidable repetitions, and is probably not fully comprehensive. Some of the interpretations presented here may be found to be incorrect in a few years, indeed, many pieces of the puzzle are still missing and should be discovered in future missions to the magnetosphere. Radio sounding and global imaging of the magnetosphere in EUV or with other new techniques like the observation of energetic neutral atoms (ENA) are the tools of the future to achieve these challenging and important goals.

If our book should stimulate discussions and give additional momentum to this field of research, we would then consider worthwhile our efforts and the time we have devoted to this monograph.

For the young scientists and perhaps for our grandchildren, these chapters should offer an opportunity to browse through three decades of existing experimental discoveries, and to obtain a fairly comprehensive list of references where they may find more detailed information concerning the plasmasphere and plasmapause.

J.F.L., K.I.G., D.L.C. and V.B.

References

Abe, T., Oyama, K. I., Amemiya, H., Watanabe, S., Okuzawa, T. & Schlegel, K. (1990). Measurements of temperature and velocity distribution of thermal electrons by Akebono (EXOS-D) satellite: Experimental setup and preliminary results. *Journal Geomagn. Geoelectricity*, **42**, 537–54.

Abe, T., Oyama, K. I., Watanabe, S. & Fukuniski, H. (1993). Characteristic features of electron temperature and density variations in field-aligned current regions. *Journal Geophysical Research*, **98**, 11257–66.

Afonin, V. V., Bassolo, V. S., Smilauer, J. & Lemaire, J. F. (1997). Motion and erosion of the nightside plasmapause region and of the associated subauroral electron temperature enhancement: COSMOS-900 observations. *Journal of Geophysical Research*, **102**, 2093–103.

Afonin, V. V., Grechnev, K. V., Ershova, V. A., Roste, O. Z., Smirnova, N. F., Smilauer, J. & Shulchishin, Yu. A. (1994). Ion composition and temperature of ionosphere at maximum of 22 solar activity cycle from satellite Intercosmos-24) (project Active). *Cosmic Research*, **32**, 82–94.

Aggson, T. L., Maynard, N. C., Hanson, W. B. & Saba, J. L. (1992). Electric field observations of equatorial bubbles. *Journal of Geophysical Research*, **97**, 2997–3009.

Alfvén, H. (1967). Partial corotation of a magnetized plasma. *Icarus*, **7**, 387–93.

Alfvén, H. (1976). On frozen-in field lines and field-line reconnection. *Journal of Geophysical Research*, **81**, 4019–21.

Alfvén, H. & Arrhenius, G. (1976). Evolution of the solar system. NASA-SP-345, 288–94.

Alfvén, H. & Fälthammar, C.-G. (1963). *Cosmical Electrodynamics, Fundamental Principles*, 228 pp., Clarendon Press, Oxford.

Allcock, G. M. (1959). The electron density distribution in the outer ionosphere derived from whistler data. *Journal of Atmospheric and Terrestrial Physics*, **14**, 185–99.

Almeida, O. G. (1973). Protonospheric columnar electron content determination, I, Analysis. *Journal of Atmospheric and Terrestrial Physics*, **35**, 1657–75.

Anderson, D. N., Decker, D. T. & Valladares, C. E. (1996). Global theoretical ionospheric model (GTIM). *Solar Terrestrial Energy Program: Handbook of Ionospheric Models*, ed. R. W. Schunk, CASS-USU Logan, UT 84322-4405, pp. 133–52.

Anderson, D. N., Forbes, J. M. & Codrescu, M. (1989). A fully analytic low- and mid-latitude ionospheric model. *Journal of Geophysical Research*, **94**, 1520–4.

Anderson, R. R. (1987). A comprehensive survey of

the electron number density in the earth's magnetosphere as deduced from the ISEE/PWI data. Univ. of Iowa, preprint A687-162.

Anderson, R. R. (1994). CRRES plasma wave observations during quiet times, during geomagnetic disturbances, and during chemical releases. In *Dusty Plasma, Noise and Chaos in Space and in the Laboratory*, ed. H. Kikuchi, New York.

Anderson, R. R. & Gurnett, D. A. (1973). Plasma wave observations near the plasmapause with the S3-A satellite. *Journal of Geophysical Research*, **78**, 4756–64.

Anderson, P. C., Hanson, W. B., Heelis, R. A., Craven, J. D., Baker, D. N. & Frank, L. A. (1993). A proposed production model of rapid subauroral ion drifts and their relation to substorm evolution. *Journal of Geophysical Research*, **98**, 6069–78.

Anderson, P. C., Heelis, R. A. & Hanson, W. B. (1991). The ionospheric signatures of rapid subauroral ion drifts. *Journal of Geophysical Research*, **96**, 5785–92.

Anderson, R. R. & Kurth, W. S. (1989). Discrete electromagnetic emissions in planetary magnetospheres. In *Plasma Waves and Instabilities at Comets and in Magnetospheres*, eds. B. T. Tsurutani and H. Oya, pp. 81–117, Washington, DC, American Geophysical Union.

André, M & Chang, T. (1992). Ion heating perpendicular to the magnetic field, in Physics of Space Plasma, SPI Conference Proceedings and reprints series, no 12, 35–71.

Andrews, M. K., Knox, F. B., & Thomson, N. R. (1978). Magnetospheric electric fields and protonospheric coupling fluxes inferred from simultaneous phase and group path measurements on whistler-mode signals. *Planetary and Space Science*, **26**, 171–83.

Angerami, J. J. (1966). Whistler study of the distribution of thermal plasma in the magnetosphere. PhD thesis, Tech. Rep. 3412-7, Stanford Univ., Stanford, CA.

Angerami, J. J. (1970). Whistler duct properties deduced from VLF observations made with the OGO 3 satellite near the magnetic equator. *Journal of Geophysical Research*, **75**, 6115–35.

Angerami , J. J. & Carpenter, D. L. (1966). Whistler studies of the plasmapause in the magnetosphere, 2, Electron density and total tube content near the knee in magnetospheric ionization. *Journal of Geophysical Research*, **71**, 711–25.

Angerami, J. J. & Thomas, J. O. (1964). Studies of planetary atmospheres, 1. The distribution of electrons and ions in the earth's exosphere. *Journal of Geophysical Research*, **69**, 4537–60.

Arnoldy, R. L., Dragoon, K., Cahill, L. J. Jr., Mende, S. B. & Rosenberg, T. J. (1982). Detailed correlations of magnetic field and riometer observations at $L = 4.2$ with pulsating aurora. *Journal of Geophysical Research*, **87**, 10 449–56.

Axford, W. I. (1968). The polar wind and the terrestrial helium budget. *Journal of Geophysical Research*, **73**, 6855–9.

Axford, W. I. & Hines, C. O. (1961). A unifying theory of high-latitude geophysical phenomena and geomagnetic storms. *Canadian Journal of Physics*, **39**, 1433–64.

Bailey, G. J. (1980). The topside ionosphere above Arecibo at equinox during sunspot maximum. *Planetary Space Science*, **28**, 47–59.

Bailey, G. J. (1983). The effect of a meridional ExB drift on the thermal plasma at $L = 1.4$. *Planetary and Space Science*, **31**, 389–409.

Bailey, G. J., Moffett, R. J. & Murphy, J. A. (1977). Theoretical study of night-time field-aligned O^+ and H^+ fluxes in a midlatitude magnetic field tube at equinox under sunspot maximum conditions. *Journal of Atmospheric Terrestrial Physics*, **39**, 105–10.

Bailey, G. J. & Sellek, R. (1990). A mathematical model of the Earth's plasmasphere and its application to a study of He^+ at $L = 3$. *Annales Geophysicae*, **8**, 171–90.

Bailey, G. J., Simmons, P. A. & Moffett, R. J. (1987). Topside and interhemispheric ion flows in the mid-latitude plasmasphere. *Journal of Atmospheric and Terrestrial Physics*, **49**, 503–19.

Balsiger, H., Eberhardt, P. Geiss, J., Ghielmetti, A., Walker, H. P., Young, D. T., Loidl, H. & Rosenbauer, H. (1976). A satellite-borne ion mass spectrometer for the energy range 0 to 16 keV. *Space Science Instrumentation*, **2**, 499–521.

Banks, P. M. & Holzer, T. E. (1968). The polar wind. *Journal of Geophysical Research*, **73**, 6846–54.

Banks, P. M., Nagy, A. F. & Axford, W. I. (1971). Dynamical behavior of thermal protons in mid-latitude ionosphere and magnetosphere. *Planetary and Space Science*, **19**, 1053–67.

Barakat, A. R. & Barghouthi, I. A. (1994a). The

effect of wave-particle interactions on the polar wind O^+. *Geophysical Research Letters*, **21**, 2279–2282.

Barakat, A. R. & Barghouthi, I. A. (1994b). The effect of wave-particle interactions on the polar wind: preliminary results. *Planetary and Space Science*, **42**, 987–992.

Barakat, A. R., Barghouthi, I. A. & Schunk, R. W. (1995a). Double-hump H^+ velocity distribution in the polar wind. *Geophysical Research Letters*, **22**, 1857–60.

Barakat, A. R. & Lemaire, J. (1990). Monte Carlo study of the espace of a minor species. *Physical Review*, **A42**(6), 3291–302.

Barakat, A. R. & Schunk, R. W. (1982). Transport equations for multicomponent anisotropic space plasmas: A review. *Plasma Physics*, **24**, 389–98.

Barakat, A. R., Schunk, R. W., Barghouti, I. A. & Lemaire, J. (1990). Monte-Carlo study of the transition from collision-dominated to collisionless polar wind flow in *Physics of Space Plasmas*, eds T. Chang, G. B. Crew & J. R. Jaspers, pp. 431–7, Scientific Publisher, Cambridge, MA.

Barakat, A. R., Thiemann, H. & Schunk, E. W. (1995b). Comparison of macroscopic PIC and semikinetic models of the polar wind, submitted to *Journal of Geophysical Research*.

Barbier, D. (1958). L'activité aurorale aux basses latitudes. *Annales de Géophysique*, **14**, 334–55.

Barfield, J. N., Burch, J. L. & Williams, D. J. (1975). Substorm associated reconfiguration of the dusk side equatorial magnetosphere : a possible source mechanism for isolated plasma regions. *Journal of Geophysical Research*, **80**, 47–55.

Barfield, J. N. & McPherron, R. L. (1972). Investigation of interaction between Pc1 and 2 and Pc5 micropulsations at the synchronous orbit during magnetic storms. *Journal of Geophysical Research*, **77**, 4707–19.

Barghouthi, I. A. (1994). The effects of wave-particle interactions on H^+ and O^+ outflow at high latitude: a comparative study, submitted to *Journal of Geophysical Research*.

Barghouthi, I. A., Barakat, A. R. & Persoon, A. M. (1994). Effects of altitude-dependent wave particle interactions on the polar wind plasma, submitted to *Annales Geophysicae*.

Barghouthi, I. A., Barakat, A. R. & Schunk, R. W. (1993). Monte Carlo study of the transition region in the polar wind: an improved collision model. *Journal of Geophysical Research*, **98**,

17583–91.

Barghouthi, I. A., Barakat, A. R., Schunk, R. W. & Lemaire, J. (1990). H^+ outflow in the polar wind : A Monte Carlo simulation. AGU Fall Meeting, San Francisco, California. *EOS*, **71**, 1493.

Barkhausen, H. (1919). Zwei mit Hilfe der neuen Verstärker entdeckte Erscheinungen. *Physikalische Zeitschrift*, **20**(1919), 401–3.

Barrington, R. E., Belrose, J. S. & Keeley, D. A. (1963). Very-low-frequency noise bands observed by the Alouette 1 satellite. *Journal of Geophysical Research*, **68**, 6539–42.

Bartels, J. (1932). Terrestrial-magnetic activity and its relations to solar phenomena. *Terrestrial Magnetism Atmospheric Electricity*, **37**, 1.

Bauer, S. J. (1962). On the structure of the topside ionosphere. *Journal of Atmospheric Science*, **19**, 276–8.

Bauer, S. J. (1963). Helium ion belt in the upper atmosphere. *Nature*, **197**, 36–7.

Bauer, S. J. (1966). Chemical processes involving helium ions and the behavior of atomic nitrogen ions in the upper atmosphere. *Journal of Geophysical Research*, **71**, 1508–11.

Bauer, S. J. (1970). Satellite measurements of the cold plasma in the magnetosphere. In *Progress in Radio Science, 1966–1969*, **1**, eds G. M. Brown, N. D. Clarence & M. J. Rycroft, pp. 159–170, Brussels: International Union of Radio Science.

Baugher, C. R., Chappell, C. R., Horwitz, J. L., Shelley, E. G. & Young, D. T. (1980). Initial thermal plasma observations from ISEE 1. *Geophysical Research Letters*, **7**, 657–60.

Baumjohann, W., Haerendel, G. (1985). Magnetospheric convection observed between 0600 and 2100 LT: solar wind and IMF dependence. *Journal of Geophysical Research*, **90**, 6370–8.

Baumjohann, W., Haerendel, G. & Melzner, F. (1985). Magnetospheric convection observed between 0600 and 2100 LT: variations with K_p. *Journal of Geophysical Research*, **90**, 393–8.

Beghin, C. Pandey, R. & Roux, D. (1985). North–South asymmetry in quasi-monochromatic plasma density irregularities observed in night-time equatorial F-region. *Advances in Space Research*, **5**(4) 209–12.

Behr, A. M. & Siedentopf H. (1953). Untersuchungen über Zodiakallicht und

Gegenschein nach lichtelektrischen Messungen auf dem Jungfraujoch. *Zeitschrift Astrophysik*, **32**, 19.

Bell, T. F., Helliwell, R. A. & Hudson, M. K. (1991). Lower hybrid waves excited through linear mode coupling and the heating of ions in the auroral and subauroral ionosphere. *Journal of Geophysical Research*, **96**, 11 379–88.

Bell, T. F., Helliwell, R. A., Inan, U. S. & Lauben, D. S. (1993). The heating of suprathermal ions above thunderstorm cells. *Geophysical Research Letters*, **20**, 1991–4.

Bell, T. F. & Ngo, H. D. (1988). Electrostatic waves stimulated by coherent VLF signals propagating in and near the inner radiation belt. *Journal of Geophysical Research*, **93**, 2599–618.

Bell, T. F. & Ngo, H. D. (1990). Electrostatic lower hybrid waves excited by electromagnetic whistler mode waves scattering from planar magnetic-field-aligned plasma density irregularities. *Journal of Geophysical Research*, **95**, 149–72.

Berchem, J. & Etcheto, J. (1981). Experimental study of magnetospheric convection. *Advances in Space Research*, **1**(1), 179–84.

Bernhardt, P. A. (1979). Theory and analysis of the 'super whistler'. *Journal of Geophysical Research*, **84**, 5131–42.

Berning, W. W. (1951). Charge densities in the ionosphere from radio Doppler data. *Journal of Meteorology*, **8**, 175–9.

Bezrukikh, V. V. (1970). Results of charged particle observations aboard ELECTRON-2 and ELECTRON-4 satellites. *Kosmicheskie Issledovaniya*, **2**, 271–7.

Bezrukikh, V. V. & Gringauz, K. I. (1965). Issledovaniya kosmicheskogo prostranstva, Nauka, Moscow, **177**.

Bezrukikh, V. V. & Gringauz, K. I. (1976). The hot zone in the outer plasmasphere of the earth. *Journal of Atmospheric and Terrestrial Physics*, **38**, 1085–91.

Bezrukikh, V. V. & Gringauz, K. I. (1995). Experiments on cold plasma studies in the Earth's magnetosphere from INTERBALL satellites, in Interball, mission and payload, **214–7**, CNES-IKI-RSA publication.

Bhonsle, R. V., Da Rosa, A. V. & Garriott, O. K. (1965). Measurements of the total electron content and the equivalent slab thickness of the midlatitude ionosphere. *Journal Research of the National Bureau of Standards*, **69D**, 929–37.

Biermann, L. (1957). Solar corpuscular radiation and the interplanetary gas. *The Observatory*, **77**, 109.

Bilitza, D. (1990). International Reference Ionosphere 1990. NSSDC 90-22, Greenbelt, Md.

Bilitza, D., Rawer, K., Bossy, L. & Gulyaeva, T. (1993). International Reference Ionosphere – Past, present and future. II. Plasma temperatures, ion composition and ion drift. *Advances in Space Research*, **13**(3), 15–23.

Binsack, J. H. (1967). Plasmapause observations with the M.I.T. experiment on IMP2. *Journal of Geophysical Research*, **72**, 5231–37.

Blanc, M. (1978). Midlatitude convection electric fields and their relation to ring current development. *Geophysical Research Letters*, **5**, 203–6.

Blanc, M. (1983a). Magnetospheric convection effects at midlatitudes, 1. Saint-Santin observations. *Journal of Geophysical Research*, **88**, 211–23.

Blanc, M. (1983b). Magnetospheric convection effects at mid-latitudes 3. Theoretical derivation of the disturbance convection pattern in the plasmasphere. *Journal of Geophysical Research*, **88**, 235–51.

Blanc, M. & Amayenc, P. (1976). Contribution of incoherent scatter radars to the study of middle and low latitude ionospheric electric fields. In *Atmospheric Physics from Spacelab*, ed. J. J. Burger, pp. 61–90, Dordrecht, Holland, D. Reidel Publishing Company.

Blanc, M., Amayenc, P., Bauer, P. & Taieb, C. (1977). Electric field induced drifts from the French incoherent scatter facility. *Journal of Geophysical Research*, **82**, 87–97.

Blanc, M. & Richmond, A. D. (1980). The ionospheric disturbance dynamo. *Journal of Geophysical Research*, **85**, 1669–86.

Block, L. P. (1966). On the distribution of electric fields in the magnetosphere. *Journal of Geophysical Research*, **71**, 855–64.

Block, L. P., & Carpenter, D. L. (1974). Derivation of magnetospheric electric fields from whistler data in a dynamic geomagnetic field. *Journal of Geophysical Research*, **79**, 2783–89.

Block, L. P., Fälthammar, C.-G., Lindqvist, P.-A., Marklund, G. T., Mozer, F. S., Pedersen, A., Potemra, T. A. & Zanetti, L. J. (1987). Electric field measurements on Viking : first results. *Geophysical Research Letters*, **14**, 435–8.

Blomberg, L. G. & Marklund, G. T. (1988a). The

influence of conductivities consistent with field-aligned currents on high-latitude convection patterns, Report TRITA-EPP-88-02, The Royal Inst. of Technol., Stockholm, Sweden.

Blomberg, L. G. & Marklund, G. T. (1988b). A numerical model of ionospheric convection derived from field-aligned currents and the corresponding conductivity. Report TRITA-EPP-88-02, The Royal Inst. of Technol., Stockholm, Sweden.

Boardsen, S. A., Gallagher, D. L., Gurnett, D. A., Peterson, W. K. & Green, J. L (1992). Funnel-shaped, low frequency equatorial waves. *Journal of Geophysical Research*, **97**, 14967–76.

Boskova, J., Jiricek, F., Smilauer, J. & Triska, P. (1993a). VLF plasmaspheric emissions – 'Wave' signatures of inner magnetosphere boundaries. *Journal of Atmospheric and Terrestrial Physics*, **55**, 1789–1795.

Boskova, J., Jiricek, F., Smilauer J., Triska, P., Afonin, V. V. & Istomin, V. G. (1993b). Plasmaspheric refilling phenomena observed by the Intercosmos 24 satellite. *Journal of Atmospheric and Terrestrial Physics*, **55**, 1595–1603.

Bosqued, J.-M. (1987). AUREOL–3 results on ion precipitation. *Physica Scripta*, **18**, 158–166.

Bostick, W. H. (1956). Experimental study of ionized matter projected across a magnetic field. *Physical Review*, **104**, 292–9.

Boström, R., Gustafsson, G., Holback, B., Holmgren, G., Koskinen, H. & Kintner, P. (1988). Characteristics of solitary waves and weak double layers in the magnetospheric plasma. IRF Preprint 105, Swedish Inst. of Space Physics, Uppsala, Sweden, 1988.

Boström, R., Koskinen, H. & Holback, B. (1987). Low frequency waves and small scale solitary structures observed by Viking, Proc. 21st ESLAB Symp., 22–25 June 1987, Bolkesjo, Norway. ESA SP-275, 185–92.

Bouriot, M., Tixier, M. & Corcuff, Y. (1967). Etude de l'ionisation magnetosphérique entre 1.9 et 2.6 rayons géocentriques au moyen des sifflements radioélectriques, reçus à Poitiers au cours d'un cycle solaire. *Annales de Geophysique*, **23**, 527–34.

Brace, L. H., Chappell, C. R., Chandler, M. O., Comfort, R. H., Horwitz, J. L. & Hoegy, W. R. (1988). F-region electron temperature signatures of the plasmapause based on

Dynamics Explorer 1 and 2 measurements. *Journal of Geophysical Research*, **93**, 1896–908.

Brace, L. H., Maier, E. J., Hoffman, J. H., Whitteker, J. & Shepherd, G. G. (1974). Deformation of the nightside plasmasphere and ionosphere during the August 1972 geomagnetic storm. *Journal of Geophysical Research*, **79**, 5211–8.

Brace, L. H., Reddy, B. M. & Mayr, H. G. (1967). Global behavior of the ionosphere at 1000-kilometer altitude. *Journal of Geophysical Research*, **72**, 265–83.

Brace, L. H. & Theis, R. F. (1974). The behavior of the plasmapause at mid-latitudes: ISIS–1 Langmuir probe measurements. *Journal of Geophysical Research*, **79**, 1871–84.

Brekke, A. & Hall, C. (1988). Auroral ionospheric quiet summer time conductances. *Annales Geophysicae*, **6**, 361–75.

Brice, N. M. & Smith, R. L. (1965). Lower hybrid resonance emissions. *Journal of Geophysical Research*, **70**, 71–80.

Brice, N. M. (1967). Bulk motion of the magnetosphere. *Journal of Geophysical Research*, **72**, 5193–211.

Brice, N. M. (1973). Differential drift of plasma clouds in the magnetosphere (unpublished preprint).

Brinton, H. C., Pickett, R. A. & Taylor, H. A. (1968). Thermal ion structure of the plasmasphere. *Planetary and Space Science*, **16**, 899–909.

Brittain, R., Kintner, P. M., Kelley, M. C., Siren, J. C. & Carpenter, D. L. (1983). Standing wave patterns in VLF hiss. *Journal of Geophysical Research*, **88**, 7059–64.

Bullough K., & Sagredo, J. L. (1970). Longitudinal structure in the plasmapause : VLF goniometer observations of knee-whistlers. *Nature*, **225**, 1038–9.

Burch, J. L., and co-investigators (1995). Imager for Magnetopause-to-Aurora Global Exploration (IMAGE). http://image.gsfc.nasa.gov/

Burgess, J. M. (1969). Flow equations for composite gases, Academic, Orlando, Fla.

Burgess, W. C. & Inan, U. S. (1993). The role of ducted whistlers in the precipitation loss and equilibrium flux of radiation belt electrons. *Journal of Geophysical Research*, **98**, 15 643–65.

Burtis, W. J. & Helliwell, R. A. (1976). Magnetospheric chorus: occurrence patterns and normalized frequency. *Planetary and Space*

Science, **24**, 1007–24.

Burton, E. T. & Boardman, E. M. (1933). Audio-frequency atmospherics. *Proceedings of the Institute of Radio Engineers*, **21**, 1476–94. Also published in *Bell System Technical Journal*, **12**, 498–516.

Carlqvist, P. (1995). Multi-component double layers and selective acceleration of charged particles. *Journal of Geophysical Research*, **100**, 205–12.

Carpenter, D. L. (1960). Identification of Whistler Sources on Visual Records and a Method of Routine Whistler Analysis. Stanford Electronics Laboratories Technical Report, 5, contract AF 18(603)-126. Stanford, California: Stanford University.

Carpenter, D. L. (1962a). New experimental evidence of the effect of magnetic storms on the magnetosphere. *Journal of Geophysical Research*, **67**, 135–45.

Carpenter, D. L. (1962b). Electron-density variations in the magnetosphere deduced from whistler data. *Journal of Geophysical Research*, **67**, 3345–60.

Carpenter, D. L. (1962c). The Magnetosphere During Magnetic Storms; a Whistler Analysis. Ph D thesis. Stanford, California: Stanford University.

Carpenter, D. L. (1963a). Whistler measurements of electron density and magnetic field strength in the remote magnetosphere. *Journal of Geophysical Research*, **68**, 3727–30.

Carpenter, D. L. (1963b). Whistler evidence of a 'knee' in the magnetospheric ionization density profile. *Journal of Geophysical Research*, **68**, 1675–82.

Carpenter, D. L. (1965). Whistler measurements of the equatorial profile of magnetospheric electron density, in *Progress in Radio Science 1960–1963, III, The Ionosphere*, ed. G. M. Brown, Elsevier, Amsterdam, pp. 76–91.

Carpenter, D. L. (1966). Whistler studies of the plasmapause in the magnetosphere, I. Temporal variations in the position of the knee and some evidence on plasma motions near the knee. *Journal of Geophysical Research*, **71**, 693–709.

Carpenter, D. L. (1967). Relations between the dawn minimum in the equatorial radius of the plasmapause and Dst, Kp and the local K at Byrd station. *Journal of Geophysical Research*, **72**, 2969–71.

Carpenter, D. L. (1968). Ducted whistler mode propagation in the magnetosphere; a half-gyrofrequency upper intensity cut off and some associated wave growth phenomena. *Journal of Geophysical Research*, **73**, 2919–28.

Carpenter, D. L. (1970). Whistler evidence of the dynamic behavior of the duskside bulge in the plasmasphere. *Journal of Geophysical Research*, **75**, 3837–47.

Carpenter, D. L. (1971). OGO 2 and 4 VLF observations of the asymmetric plasmapause near the time of SAR arc events. *Journal of Geophysical Research*, **76**, 3644–50.

Carpenter, D. L. (1978a). New whistler evidence of a dynamo origin of electric fields in the quiet plasmasphere. *Journal of Geophysical Research*, **83**, 1558–64.

Carpenter, D. L. (1978b). Whistler and VLF noises propagating just outside the plasmapause. *Journal of Geophysical Research*, **83**, 45–57.

Carpenter, D. L. (1981). A study of the outer limits of ducted whistler propagation in the magnetosphere. *Journal of Geophysical Research*, **86**, 839–45.

Carpenter, D. L. (1983). Some aspects of plasmapause probing by whistlers. *Radio Science*, **18**, 917–25.

Carpenter, D. L. (1986). Whistler studies of the plasmasphere shape and dynamics. *Advances in Space Research*, **6**(3), 5–19.

Carpenter, D. L. (1995). Earth's plasmasphere awaits rediscovery, *EOS, Transactions American Geophysical Union*, **76**(9), 89–92.

Carpenter, D. L. & Akasofu, S.-I. (1972). Two substorm studies of relations between westward electric fields in the outer plasmasphere, auroral activity, and geomagnetic perturbations. *Journal of Geophysical Research*, **77**, 6854–63.

Carpenter, D. L. & Anderson, R. R. (1992). An ISEE/whistler model of equatorial electron density in the magnetosphere. *Journal of Geophysical Research*, **97**, 1097–108.

Carpenter, D. L., Anderson, R. R., Bell, T. F. & Miller, T. R. (1981). A comparison of equatorial electron densities measured by whistlers and by a satellite radio technique. *Geophysical Research Letters*, **8**, 1107–10.

Carpenter, D. L. & Chappell, C. R. (1973). Satellite studies of magnetospheric substorms on August 15, 1968, 3, Some features of magnetospheric convection. *Journal of Geophysical Research*, **78**, 3062–7.

Carpenter, D. L., Foster, J. C., Rosenberg, T. J. &

Lanzerotti, L. J. (1975). A subauroral and mid-latitude view of substorm activity. *Journal of Geophysical Research*, **80**, 4279–86.

Carpenter, D. L., Giles, B. G., Chappell, C. R., Décréau, P. M. E., Anderson, R. R., Persoon, A. M., Smith, A. J., Corcuff, Y. & Canu, P. (1993). Plasmasphere dynamics in the duskside bulge sector : a new look at an old topic. *Journal of Geophysical Research*, **98**, 19 243–71.

Carpenter, D. L. & Labelle J. W. (1982). A study of whistlers correlated with bursts of electron precipitation near $L = 2$. *Journal of Geophysical Research*, **87**, 4427–34.

Carpenter, D. L. & Miller, T. R. (1983). Rare ground based observations of Siple VLF transmitter signals outside the plasmapause. *Journal of Geophysical Research*, **88**, 10 227–32.

Carpenter, D. L. & Orville, R. E. (1989). The excitation of active whistler mode signal paths in the magnetosphere by lightning: Two case studies. *Journal of Geophysical Research*, **94**, 8886–94.

Carpenter, D. L. & Park, C. G. (1973). On what ionosphere workers should know about the plasmapause-plasmasphere. *Reviews of Geophysics and Space Physics*, **11**, 133–54.

Carpenter, D. L., Park, C. G., Arens, J. F. & Williams, D. J. (1971). Position of the plasmapause during a stormtime increase in trapped energetic ($E > 280$ keV) electrons. *Journal of Geophysical Research*, **76**, 4669–73.

Carpenter, D. L., Park, C. G. & Miller, T. R. (1979). A model of substorm electric fields in the plasmasphere based on whistler data. *Journal of Geophysical Research*, **84**, 6559–63.

Carpenter, D. L., Park, C. G., Taylor, H. A., Jr. & Brinton, H. C. (1969). Multi-experiment detection of the plasmapause from EOGO satellites and Antarctic ground stations. *Journal of Geophysical Research*, **74**, 1837–47.

Carpenter, D. L. & Seely, N. T. (1976). Cross-L plasma drifts in the outer plasmasphere: quiet time patterns and some substorm effects. *Journal of Geophysical Research*, **81**, 2728–36.

Carpenter, D. L. & Smith, R. L. (1964). Whistler measurements of electron density in the magnetosphere. *Reviews of Geophysics*, **2**, 415–41.

Carpenter, D. L., Smith, A. J., Giles, B. L., Chappell, C. R. & Décréau, P. M. E. (1992). A case study of plasma structure in the dusk sector associated with enhanced magnetospheric convection. *Journal of Geophysical Research*, **97**, 1157–66.

Carpenter, D. L. & Stone, K. (1967). Direct detection by a whistler method of the magnetospheric electric field associated with a polar substorm. *Planetary and Space Science*, **15**, 395–7.

Carpenter, D. L., Stone, K., Siren, J. C. & Crystal, T. L. (1972). Magnetospheric electric fields deduced from drifting whistler paths. *Journal of Geophysical Research*, **77**, 2819–34.

Carpenter, D. L. & Sulic, D. M., (1988). Ducted whistler propagation outside the plasmapause. *Journal of Geophysical Research*, **93**, 9731–42.

Carpenter, D. L., Walter, F., Barrington, R. E. & McEwen, D. J. (1968). Alouette 1 and 2 observations of abrupt changes in whistler rate and of VLF noise variations at the plasmapause: A satellite-ground study. *Journal of Geophysical Research*, **73**, 2929–40.

Carpenter, G. B. & Colin, L. (1963). On a remarkable correlation between whistler-mode propagation and high-frequency North scatter. *Journal of Geophysical Research*, **68**, 5649–57.

Caudal, G. & Blanc, M. (1988). Magnetospheric convection during quiet or moderately disturbed times. *Reviews of Geophysics*, **26**, 809–22.

Chan, K.-W. & Holzer, R. E. (1976). ELF hiss associated with plasma density enhancements in the outer magnetosphere. *Journal of Geophysical Research*, **81**, 2267–74.

Chandler, M. O., Kozyra, J. U., Horwitz, J. L., Comfort, R. H., Peterson, W. K. & Brace, L. H. (1988). Modelling of thermal plasma in the outer plasmasphere : a magnetospheric heat source. In *Modelling Magnetospheric Plasma*, Geophys. Monogr. Ser. vol. 44, eds T. E. Moore and J. H. Waite Jr., AGU, Washington DC.

Chandler, M. O., Ponthieu, J. J., Cravens, T. E., Nagy, A. F. & Richards, P. G. (1987). Model calculations of minor ion populations in the plasmasphere. *Journal of Geophysical Research*, **92**, 5885–95.

Chandrasekhar, S. (1960). *Plasma Physics*, **217 pp.**, University of Chicago Press, Chicago.

Chang, T., Crew, G. B., Hershkowitz, N., Jasperse, J. R., Retterer, J. M. & Winningham, J. D. (1986). Transverse acceleration of oxygen ions by electromagnetic cyclotron resonance with broad band left-hand-polarized waves. *Geophysical Research Letters*, **13**, 636–9.

Chapman, S. (1957). Notes on the solar corona and the terrestrial atmosphere. *Smithsonian*

Contributions to Astrophysics, **2**, 1.

Chappell, C. R. (1972). Recent satellite measurements of the morphology and dynamics of the plasmasphere. *Reviews Geophysics and Space Physics*, **10**, 951–79.

Chappell, C. R. (1974). Detached plasma regions in the magnetosphere. *Journal of Geophysical Research*, **79**, 1861–70.

Chappell, C. R. (1982). Initial observations of thermal plasma composition and energetics from Dynamics Explorer-1. *Geophysical Research Letters*, **9**, 929–32.

Chappell, C. R., Fields, S. A., Baugher, C. R., Hoffman, J. H., Hanson, W. B., Wright, W. W. & Hammack, H. D. (1981). The retarding ion mass spectrometer on Dynamics Explorer-A. *Space Science Instruments*, **5**, 477–91.

Chappell, C. R., Harris, K. K. & Sharp, G. W. (1970a). A study of the influence of magnetic activity on the location of the plasmapause as measured by OGO 5. *Journal of Geophysical Research*, **75**, 50–6.

Chappell, C. R., Harris, K. K. & Sharp, G. W. (1970b). The morphology of the bulge region of the plasmasphere. *Journal of Geophysical Research*, **75**, 3848–61.

Chappell, C. R., Harris, K. K. & Sharp, G. W. (1971a). The dayside of the plasmasphere. *Journal of Geophysical Research*, **76**, 7632–47.

Chappell, C. R., Harris, K. K. & Sharp, G. W. (1971b). OGO 5 measurements of the plasmasphere during observations of stable auroral red arcs. *Journal of Geophysical Research*, **76**, 2357–65.

Chappell, C. R., Olsen, R. C., Green, J. L., Johnson, J. F. E. & Waite, J. H., Jr. (1982). The discovery of nitrogen ions in the Earth's magnetosphere. *Geophysical Research Letters*, **9**, 937–40.

Chen, A. J. & Grebowsky, J. M. (1974). Plasma tail interpretations of pronounced detached plasma regions measured by OGO 5. *Journal of Geophysical Research*, **79**, 3851–5.

Chen, A. J., Grebowsky, J. M. & Taylor, H. A., Jr. (1975). Dynamics of mid-latitude light ion trough and plasma tails. *Journal of Geophysical Research*, **80**, 968–76.

Chen, A. J. & Grebowsky, J. M. (1978). Dynamical interpretation of observed plasmasphere deformations. *Planetary and Space Science*, **26**, 661–72.

Chen, A. J. & Wolf, R. A. (1972). Effects on the plasmasphere of time-varying convection electric field. *Planetary and Space Science*, **20**, 483–509.

Chen, L. & Hasegawa, A. (1974). A theory of long-period magnetic pulsations, **2**, Impulse excitation of surface eigenmode. *Journal of Geophysical Research*, **79**, 1033–7.

Cheng, A. F. (1985). Magnetospheric interchange instability. *Journal of Geophysical Research*, **90**, 9900–4.

Chernov, A. A., Khazanov, G. V. & Tanygin, S. V. (1990). Modeling of electron thermal fluxes into the ionosphere during the excitation of ion cyclotron waves by the ring current. *Annales Geophysicae*, **8**(12), **825–7**.

Chiu, Y. T., Luhmann, J. G., Ching, B. K. & Boucher, D. J. Jr. (1979). An equilibrium model of plasmaspheric composition and density. *Journal of Geophysical Research*, **84**, 909–16.

Church, S. R. & Thorne, R. M. (1983). On the origin of plasmaspheric hiss: ray path integrated amplification. *Journal of Geophysical Research*, **88**, 7941–57.

Clilverd, M. A., Smith, A. J. & Thompson, N. R. (1991). The annual variation in quiet time plasmaspheric electron density, determined from whistler mode group delays. *Planetary and Space Science*, **39**, 1059–67.

Clilverd, M. A., Smith, A. J. & Thomson, N. R. (1992). The effects of ionospheric horizontal electron density gradients on whistler mode signals. *Journal of Atmospheric and Terrestrial Physics*, **54**, 1061–74.

Cole, K. D. (1965). Stable auroral red arcs, sinks for energy of Dst main phase. *Journal of Geophysical Research*, **70**, 1689–706.

Cole, K. D. (1970a). Magnetospheric processes leading to mid-latitude auroras. *Annales de Géophysique*, **26**, 187–93.

Cole, K. D. (1970b). Relationship of geomagnetic fluctuations to other magnetospheric phenomena. *Journal of Geophysical Research*, **75**, 4216–4223.

Cole, K. D. (1971). Formation of field-aligned irregularities in the magnetosphere. *Journal of Atmospheric and Terrestrial Physics*, **33**, 741–50.

Cole, K. D. (1975). Coulomb collisions of ring current particles – indirect source of heat for the ionosphere, NASA document GSFC X-621-75-108.

Comfort, R. H. (1986). Plasmasphere thermal structure as measured by ISEE-1 and DE-1. *Advances in Space Research*, **6**(3), 31–40.

Comfort, R. H. (1996). Thermal structure of the plasmasphere. *Advances in Space Research*, **17**(10), 175–84.

Comfort, R. H., Baugher, C. R. & Chappell, C. R. (1982). Use of the thin sheath approximation for obtaining ion temperatures from the ISEE-1 limited aperture RPA. *Journal of Geophysical Research*, **87**, 5109–23.

Comfort, R. H. & Horwitz, J. L. (1981). Low-energy pitch-angle distributions observed on the dayside at geosynchronous altitudes. *Journal of Geophysical Research*, **86**, 1621–7.

Comfort, R. H., Richards, P. G., Craven, P. D. & Chandler, M. O. (1996). Problems in simulating ion temperatures in low density flux tubes, in press for AGU Monograph. *Coupling of Micro- and Mesoscale Processes in Space Plasma Transport*, eds. J. L. Horwitz and N. Singh.

Comfort, R. H., Waite, Jr., J. H. & Chappell, C. R. (1985). Thermal ion temperatures from the retarding ion mass spectrometer on DE-1. *Journal of Geophysical Research*, **90**, 3475–86.

Corcuff, P. (1977). Méthodes d'analyse des sifflements électroniques: 1 – Application à des sifflements théoriques. *Annales de Géophysique*, **33**, 443–54.

Corcuff, P. (1978). Contribution à l'étude de la convection magnétosphérique et de la dynamique de la plasmasphere. Thèse de doctorat d'état, Université de Poitiers.

Corcuff, P., Corcuff, Y., Carpenter, D. L., Chappell, C. R., Vigneron, J. & Kleimenova, N. (1972). La plasmasphère en période de recouvrement magnétique. Etude combinée des données des satellites OGO 4, OGO 5 et des sifflements reçus au sol. *Annales de Géophysique*, **28**, 679–95.

Corcuff, P., Corcuff, Y. & Tarcsai, G. (1977). Méthodes d'analyse des sifflements électroniques: 2 – Application à des sifflements observés au sol. *Annales de Géophysique*, **33**, 455–9.

Corcuff, Y. (1961). Variation de la dispersion des sifflements radioélectriques au cours des orages magnétiques. *Annales de Géophysique*, **17**, 374–7.

Corcuff, Y. (1962). La dispersion des sifflements radioélectriques au cours des orages magnétiques; ses variations nocturnes, annuelle et semi-annuelle en périodes calmes. *Annales de Géophysique*, **18**, 334–40.

Corcuff, Y. (1965). Etude de la Magnétosphère au Moyen des Sifflements Radioélectriques. Thèse

de doctorat d'état. Poitiers: University of Poitiers.

Corcuff, Y. (1975). Probing the plasmapause by whistlers. *Annales de Géophysique*, **31**, 53–67.

Corcuff, Y. & Corcuff, P. (1982). Structure et dynamique de la plasmapause-plasmasphère les 6 et 14 juillet 1977. Etude à l'aide des données de sifflements reçus au sol et de données des satellites ISIS et GEOS-1. *Annales de Géophysique*, **38**, 1–24.

Corcuff, Y., Corcuff, P. & Lemaire, J. (1985). Dynamical plasmapause positions during the July 29–31, 1977, storm period : a comparison of observations and time-dependent model calculations. *Annales Geophysicae*, **3**, 569–79.

Corcuff, Y. & Delaroche, M. (1964). Augmentation du gradient d'ionisation dans la proche magnétosphère en période de forte activité magnétique. *Comptes-Rendus de l'Académie des Sciences*, **258**, 650–3.

Cordier, S. (1994). Hyperbolicity of Grad's extension of hydrodynamic models for ionospheric plasma. *Mathematical Models in Applied Sciences*, **4/5**, 625–45; 647–67.

Cornwall, J. M., Coroniti, F. V. & Thorne, R. M. (1970). Turbulent loss of ring current protons. *Journal of Geophysical Research*, **75**, 4699–709.

Cornwall, J. M., Coroniti, F. V. & Thorne, R. M. (1971). Unified theory of SAR-arc formation at the plasmapause. *Journal of Geophysical Research*, **76**, 4428–45.

Curtis, S. A. (1985). Equatorial trapped plasmasphere ion distributions and transverse stochastic acceleration. *Journal of Geophysical Research*, **90**, 1765–70.

Daniell, G. J. (1986). Analytic properties of the whistler dispersion function. *Journal of Atmospheric and Terrestrial Physics*, **48**, 271–5.

Daniell, R. E., Brown, L. D., Anderson, D. N., Fox, M. W., Doherty, P. H., Decker, D. T., Sojka, J. J. & Schunk, R. W. (1995). PIM: A global ionospheric parameterization based on first principles models. *Radio Science*, **30**, 1499–510.

Daniell, R. E., Brown, L. D., Anderson, D. N., Whalen, J. A., Sojka, J. J. & Schunk, R. W. (1990). *Proceedings STP Symposium* in Australia (preprint).

Davies, K. (1980). Recent progress in satellite radio beacon studies with particular emphasis on the ATS–6 radio beacon experiment. *Space Science Reviews*, **25**, 357–430.

Davies, K., Anderson, D. N., Paul, A. K.,

Degenhardt, W., Hartmann, G. K & Leitinger, R. (1979). Night-time increases in total electron content observed with the ATS6 radio beacon. *Journal of Geophysical Research*, **84**, 1536–42.

Davies, K., Degenhardt, W., Hartmann, G. K. & Leitinger, R. (1980). Comparison of the total electron content made with the ATS-6 radio beacon over the U. S. and Europe. *Journal of Atmospheric and Terrestrial Physics*, **42**, 411–16.

Davies, K., Hartmann, G. K. & Leitinger, R. (1977). A comparison of several methods of estimating the columnar electron content of the plasmasphere. *Journal of Atmospheric and Terrestrial Physics*, **39**, 571–80.

de la Beaujardière, O., Wickwar, V. B. & King, J. H. (1986). Sondrestrom radar observations of the effect of the IMF B_y component on polar cap convection. In *Solar Wind-Magnetosphere Coupling*, eds Y. Kamide and J. A. Slavin, TERRAPUB, Tokyo, pp. 495–505.

Décréau, P. M. E., Beghin, C. & Parrot, M. (1978b). Electron density and temperature, as measured by the mutual impedance experiment on board GEOS-1. *Space Science Reviews*, **22**, 581–95.

Décréau, P. M. E., Beghin, C. & Parrot, M. (1982). Global characteristics of the cold plasma in the equatorial plasmapause region as deduced from the GEOS 1 mutual impedance probe. *Journal of Geophysical Research*, **87**, 695–712.

Décréau, P. M. E., Beghin, C. & Parrot, M. (1984). Contribution of mutual impedance experiments to the understanding of magnetospheric processes. In *Proceedings of the Conference on Achievements of the IMS, 26–28 June 1984, Graz, Austria*, ESA SP-217, pp. 705–7.

Décréau, P. M. E., Carpenter, D. L., Chappell, C. R., Comfort, R. H., Green, J. L., Olsen, R. C. & Waite J. H. (1986a). Latitudinal plasma distribution in the dusk plasmaspheric bulge : refilling phase and quasi-equilibrium state. *Journal of Geophysical Research*, **91**, 6929–43 and 7147.

Décréau, P. M. E., Etcheto, J., Knott, R., Pedersen, K., Wrenn, G. L. and Young, D. T. (1978a). Multi-experiment determination of plasma density and temperature. *Space Science Reviews*, **22**, 633–45.

Décréau, P. M. E., Lemaire, J., Chappell, C. R. & Waite, J. H. (1986b). Nightside plasmapause positions observed by DE-1 as a function of geomagnetic indices : Comparison with whistler observations and model calculations. *Advances in Space Research*, **6**(3), 209–14.

Degenhardt, M., Hartmann, G. K. & Leitinger, R. (1977). Effects of a magnetic storm on the plasmasphere electron content. *Journal of Atmospheric and Terrestrial Physics*, **39**, 1435–40.

Dejnakarintra, M. & Park, C. G. (1974). Lightning-induced electric fields in the ionosphere. *Journal of Geophysical Research*, **79**, 1903–10.

Delcourt, D. C., Chappell, C. R., Moore, T. E. & Waite, J. H. Jr. (1989). A three-dimensional numerical model of ionospheric plasma in the magnetosphere. *Journal of Geophysical Research*, **94**, 11893–920.

del Pozo, C. F. & Blanc, M. (1994). Analytical self-consistent model of the large-scale convection electric field. *Journal of Geophysical Research*, **99**, 4053–68.

Demars, H. G. & Schunk, R. W. (1986). Solutions to bi-maxwellian transport equations for SAR-arc conditions. *Planetary and Space Science*, **34**, 1335–48.

Demars, H. G. & Schunk, R. W. (1987a). Comparison of solutions to bi-maxwellian and maxwellian transport equations for subsonic flows. *Journal of Geophysical Research*, **92**, 5969–90.

Demars, H. G. & Schunk, R. W. (1987b). Temperature anisotropies in the terrestrial ionosphere and plasmasphere. *Reviews of Geophysics*, **25**, 1659–79.

Demars, H. G. & Schunk, R. W. (1991). Comparison of semi-kinetic and generalized transport models of the polar wind. *Geophysical Research Letters*, **18**, 713–6.

Denby, M., Bullough, K., Alexander, P. D. & Rycroft M. J. (1980). Observational and theoretical studies of a cross meridian refraction of VLF waves in the ionosphere and magnetosphere. *Journal of Atmospheric and Terrestrial Physics*, **42**, 51–60.

Dessler, A. J. & Cloutier, P. A. (1969). Discussion of letter by Peter M. Banks and Thomas E. Holzer, The polar wind. *Journal of Geophysical Research*, **74**, 3730–3.

Doe, R. A, Moldwin, M. B. & Mendillo, M. (1992). Plasmapause morphology determined from an empirical ionospheric convection model. *Journal of Geophysical Research*, **97**, 1151–6.

Doolittle, J. H. & Carpenter, D. L. (1983). Photometric evidence of electron precipitation by first hop whistlers. *Geophysical Research*

Letters, **10**, 611–4.

Dowden, R. L., McKay, A. D., Amon, L. E. S., Koons, H. C. & Dazey, M. H. (1978). Linear and nonlinear amplification in the magnetosphere during a 6.6-kHz transmission. *Journal of Geophysical Research*, **83**, 169–81.

Draganov, A. B., Inan, U. S., Sonwalkar, V. S. & Bell, T. F. (1992). Magnetospherically reflected whistlers as a source of plasmaspheric hiss. *Geophysical Research Letters*, **19**, 233–6.

Dunckel, N. & Helliwell, R. A. (1969). Whistler mode emissions on the OGO-1 satellite. *Journal of Geophysical Research*, **74**, 6371–85.

Dunckel, N., Ficklin, B., Rorden, L. & Helliwell, R. A. (1970). Low-frequency noise observed in the distant magnetosphere with OGO 1. *Journal of Geophysical Research*, **75**, 1854–62.

Dungey, J. W. (1954). Electrodynamics of the outer atmosphere. Pennsylvania State University Report No. 69 and 57.

Dungey, J. W. (1955a). Electrodynamics of the outer atmosphere. In Report of the Physical Society Conference on the Physics of the Ionosphere, Physical Society, London, p. 229.

Dungey, J. W. (1955b). The physics of the ionosphere. The Physical Society, London, 21.

Dungey, J. W. (1967). The theory of the quiet magnetosphere, in *Proceedings of the 1966 Symposium on Solar-Terrestrial Physics*, Belgrade, eds King, J. W. and Newman, W. S., pp. 91–106.

Eckersley, T. L. (1935). Musical atmospherics. *Nature*, **135**, 104–5.

Edgar, B. C. (1976). The upper- and lower-frequency cutoffs of magnetospherically reflected whisters. *Journal of Geophysical Research*, **81**, 205–11.

Ejiri, M. (1978). Trajectory traces of charged particles in the magnetosphere. *Journal of Geophysical Research*, **83**, 4798–810.

Ejiri, M., Hoffman, R. A. & Smith, P. H. (1978). The convection electric field model for the magnetosphere based on EXPLORER 45 observations. *Journal of Geophysical Research*, **83**, 4811–15.

Engebretson, M. J., Cahill, L. J. Jr., Arnoldy, R. L., Anderson, B. J., Rosenberg, T. J., Carpenter, D. L., Inan, U. S. & Eather, R. H. (1991). The role of the ionosphere in coupling upstream ULF wave power into the dayside magnetosphere. *Journal of Geophysical Research*, **96**, 1527–42.

Engebretson, M. J., Zanetti, L. J., Potemra, T. A. &

Acuna, M. H. (1986). Harmonically structured ULF pulsations observed by the AMPTE/CCE magnetic field experiment. *Geophysical Research Letters*, **13**, 905–8.

Etcheto, J. & Bloch, J. J. (1978). Plasma density measurements from the GEOS-1 relaxation sounder. *Space Science Reviews*, **22**, 597–610.

Evans, J. V. (1972). Measurements of horizontal drifts in the E and F regions at Millstone Hill. *Journal of Geophysical Research*, **73**, 2341–52.

Evans, J. V. & Holt, J. M. (1978). Night-time proton fluxes at Millstone Hill. *Planetary and Space Science*, **26**, 727–44.

Eviatar, A., Lenchek, A. M. & Singer S. F. (1964). Distribution of density in an ion-exosphere of a nonrotating planet. *Physics of Fluids*, **7**, 1775–9.

Fahleson, U., Fälthammar, C.-G., Pedersen, A., Knott, K., Brommundt, G., Schumann, G., Haerendel, G. & Rieger, E. (1971). Simultaneous electric field measurements made in the auroral ionosphere by using three independent techniques. *Radio Science*, **6**, 233–45.

Fahr, H. J. and Shizgal, B. (1983). Modern exospheric theories and their observational relevance. *Reviews of Geophysics and Space Physics*, **21**, 75–124.

Fairfield, D. H. & Vinas, A. F. (1984). The inner edge of the plasma sheet and the diffuse aurora. *Journal of Geophysical Research*, **89**, 841–54.

Fälthammar, C.-G., Block, L. P., Lindqvist, P.-A., Marklund, G. T., Pedersen, A. & Mozer, F. S. (1987). Preliminary results from the D.C. electric field experiment on Viking. *Annales Geophysicae*, **5A**, 171–5.

Farrugia, C. J., Geiss, J., Young, D. T. and Balsiger, M. (1988). GEOS-1 observations of low-energy ions in the earth's plasmasphere: a study on composition, and temperature and density structure under quiet geomagnetic conditions. *Advances in Space Research*, **8**(8), 25–33.

Farrugia, C. J., Young, D. T., Geiss, J. & Balsiger, M. (1989). The composition, temperature and density structure of cold ions in the quiet terrestrial plasmasphere : GEOS-1 results. *Journal of Geophysical Research*, **94**, 11 865–91.

Filippov, V. M., Shestakova, L. V. & Gal'perin, Yu. I. (1984). Belt of fast ion drift in the subauroral F-region and its manifestation in the structure of the high-latitude ionosphere. *Kosmicheskie Issledovanya*, **22**, 557–64.

Fok, M.-C., Kozyra, J. U., Nagy, A. F., Rasmussen, C. E. & Khazanov, G. V. (1993). Decay of equatorial ring current ions and associated aeronomical consequences. *Journal of Geophysical Research*, **98**, 19381–93.

Fok, M.-C., Moore, T. E., Kozyra, J. U., Ho, G. C. & Hamilton, D. C. (1995). Three-dimensional ring current decay model. *Journal of Geophysical Research*, **100**, 9619–32.

Fontaine, D. & Blanc, M. (1983). A theoretical approach to the morphology and the dynamics of diffuse auroral zones. *Journal of Geophysical Research*, **88**, 7171–7184.

Fontaine, D., Blanc, M., Reinhart, L. & Glowinski, R. (1985). Numerical simulations of the magnetospheric convection including the effects of electron precipitation. *Journal of Geophysical Research*, **90**, 8343–60.

Fontaine, D., Perraut, S., Alcaydé, D., Caudal, G. & Higel, B. (1986). Large-scale structures of the convection inferred from coordinated measurements by EISCAT and GEOS 2. *Journal of Atmospheric and Terrestrial Physics*, **48**, 973–86.

Förster, M. & Jakowski, N. (1988). The night-time winter anomaly (NWA) effect in the american sector as a consequence of interhemispheric ionospheric coupling. *PAGEOPH*, **127**, 447–71.

Förster, M., Jakowski, N., Best, A. & Shmilauer, Y. (1992). Plasmaspheric response to the geomagnetic storm period March 20–23, 1990, observed by the ACTIVNY (MAGION–2) satellite. *Canadian Journal of Physics*, **70**, 569–74.

Foster, J. C. (1983). An empirical electric field model derived from Chatanika radar data. *Journal of Geophysical Research*, **88**, 981–7.

Foster, J. C. (1984). Ionospheric signatures of magnetospheric convection. *Journal of Geophysical Research*, **89**, 855–65.

Foster, J. C., Holt, J. M., Musgrove, R. G. & Evans, D. S. (1986a). Ionospheric convection associated with discrete levels of particle precipitation. *Geophysical Research Letters*, **13**, 656–9.

Foster, J. C., Holt, J. M., Musgrove, R. G. & Evans, D. S. (1986b). Solar wind dependencies of high-latitude convection and precipitation. in *Solar Wind-Magnetosphere Coupling*, eds Y. Kamide and J. A. Slavin, TERRAPUB, Tokyo, pp. 477–94.

Foster, J. C., Park, C. G., Brace, L. H., Burrows, J. R., Hoffman, J. H., Maier, E. J. & Whitteker, J. H. (1978). Plasmapause signatures in the ionosphere and magnetosphere. *Journal of Geophysical Research*, **83**, 1175–82.

Fraser-Smith, A. C. (1987). Centered and eccentric geomagnetic dipoles and their poles, 1600–1985. *Reviews of Geophysics*, **25**, 1–16.

Fraser-Smith, A. C. (1993). ULF magnetic fields generated by electrical storms and their significance to geomagnetic pulsation generation. *Geophysical Research Letters*, **20**, 467–70.

Freeman, R., Norman, K. & Willmore, A. P. (1970). Electron density measurements in the thermal plasma of the magnetosphere using a Langmuir probe. In *International satellite observations*, eds V. Manno and D. E. Page, Reidel, Dordrecht, pp. 524–34.

Freeman, M. P., Southwood, D. J., Lester, M., Yeoman, T. K. & Reeves, G. D. (1992). Substorm-associated radar auroral surges. *Journal of Geophysical Research*, **97**, 12 173–85.

Fridman, M. & Lemaire, J. (1980). Relationship between auroral electrons fluxes and field-aligned electric potential difference. *Journal of Geophysical Research*, **85**, 664–70.

Fuller-Rowell, T. J., Rees, D., Quegan, S., Moffett, R. J. & Bailey, G. J. (1987). Interactions between neutral thermospheric composition and the polar ionosphere using a coupled ionosphere-thermosphere model. *Journal of Geophysical Research*, **92**, 7744–8.

Fuller-Rowell, T. J., Rees, D., Quegan, S., Moffett, R. J., Codrescu, M. V. & Millward, G. H. (1996). A coupled thermosphere-ionosphere model. *Solar Terrestrial Energy Program: Handbook of Ionospheric Models*, ed. R. W. Schunk, CASS-USU, Logan, UT 84322-4405, pp. 217–38.

Gail, W. B. & Carpenter, D. L. (1984). Whistler induced suppression of VLF noise. *Journal of Geophysical Research*, **89**, 1015–22.

Gallagher, D. L. & Craven, P. D. (1988). Initial development of a new empirical model of the Earth's inner magnetosphere for density, temperature, and composition. In *Modelling Magnetospheric Plasma*, Geophysical Monograph Series, vol. 44, eds T. E. Moore and J. H. Waite, Jr., p. 61, Washington, DC, American Geophysical Union.

Gallagher, D. L., Craven, P. D. & Comfort, R. H. (1988). An empirical model of the earth's plasmasphere. *Advances in Space Research*, **8**(8), 15–24.

Galperin, Yu. I., Crasnier, J., Lissakov, Yu. V., Nikolayenko, L. M., Sinitsyn, V. M., Sauvaud,

J.-A. & Khalipov, V. L. (1977). The diffuse auroral zone. I. A model for the equatorial boundary of the diffuse surge zone of auroral electrons in the evening and midnight sectors. *Kosmicheskie Issledovanya*, **15**, 421–34.

Galperin, Yu. I., Khalipov, V. L. & Filippov, V. M. (1986). Signature of rapid subauroral ion drifts in the high-latitude ionosphere structure. *Annales Geophysicae*, **4**, 145–54.

Galperin, Yu. I., Ponomarev, Y. N. & Zosimova, A. G. (1973a). Direct measurements of drift rate of ions in the upper atmosphere during a magnetic storm 1. Problems of method and some results of measurements during magnetically quiet period. *Kosmicheskie Issledovanya*, **11**, 273–83.

Galperin, Yu. I., Ponomarev, V. N. & Zosimova, A. G. (1973b). Direct measurements of ion drift velocity in the upper ionosphere during a magnetic storm, 2. Experiment quiet time. *Kosmicheskie Issledovanya*, **11**, 284–96.

Galperin, Yu. I., Ponomarev, V. N. & Zosimova, A. G. (1974). Plasma convection in polar ionosphere. *Annales de Géophysique*, **30**, 1–7.

Galperin, Yu. I., Ponomarev, V. N., Ponomarev, Yu. N. & Zosimova, A. G. (1975). Plasma convection in the evening sector of the magnetosphere and the nature of the plasmapause. *Kosmicheskie Issledovanya*, **18**, 669–86.

Galperin, Yu. I., Soloviev, V. S., Torkar, K. & Foster, J. C. (1995). Predicting the plasmaspheric density radial profiles for the INTERBALL space mission (preprint).

Galperin, Y. I., Soloviev, V. S., Torkar, K., Foster, J. C. & Veselov, M. V. (1997). Predicting plasmaspheric radial density profiles. *Journal of Geophysical Research*, **102**, 2079–91.

Ganguli, S. B. (1996). The polar wind. *Reviews of Geophysics*, **34**, 311–48.

Ganguli, S. B., Mitchell, H. G. Jr. & Palmadesso, P. J. (1987). Behavior of ionized plasma in the high latitude topside ionosphere. *Planetary and Space Science*, **35**, 703–13.

Ganguli, S. B. & Palmadesso, P. J. (1987). Plasma transport in the auroral return current region. *Journal of Geophysical Research*, **92**, 8673–90.

Gary, D. E. & Hurford, G. J. (1989). Solar radio burst spectral observations, particle acceleration, and wave-particle interactions. In *Solar System Plasma Processes*, eds J. H. Waite, Jr., J. L. Burch and R. L. Moore, p. 237, Washington, DC, American Geophysical Union.

Gary, S. P., Moldwin, M. B., Thomsen, M. F., Winske, D. & McComas, D. J. (1994). Hot proton anisotropies and cool proton temperatures in the outer magnetosphere. *Journal of Geophysical Research*, **99**, 23, 603–15.

Geisler, J. E. (1967). On the limiting daytime flux of ionization into the protonosphere. *Journal of Geophysical Research*, **72**, 81–5.

Geisler, J. E. & Bowhill, S. A. (1965). The relation between the dispersion of whistlers and the electron temperature in the protonosphere. *Journal of Atmospheric and Terrestrial Physics*, **27**, 122–5.

Geiss, J., Balsiger, H., Eberhardt, P., Walker, H. P., Weber, L., Young, D. T. & Rosenbauer, H. (1978). Dynamics of magnetospheric ion composition as observed by GEOS mass spectrometer. *Space Science Reviews*, **22**, 537–66.

Geiss, J. & Young, D. T. (1981). Production and transport of O^{++} in the ionosphere and plasmasphere. *Journal of Geophysical Research*, **86**, 4739–50.

Giles, B. L., Chappell, C. R., Moore, T. E., Comfort R. H. & Waite, J. H. Jr. (1994). Statistical survey of pitch angle distributions in core (0–50 eV) ions from Dynamics Explorer 1: Outflow in the auroral zone, polar cap, and cusp. *Journal of Geophysical Research*, **99**, 17 483–501.

Glass, N. W., Wolcott, J. H., Miller, L. W. & Robertson, M. M. (1970). Local time behavior of the alignment and position of a stable auroral red arc. *Journal of Geophysical Research*, **75**, 2579–82.

Glassmeier, K. H. (1994). Geomagnetic pulsations. In *Handbook of Atmospheric Electrodynamics*, Vol. II, ed. H. Volland, Boca Raton, Florida, CRC Press.

Gold, T. (1959). Motions in the magnetosphere of the earth. *Journal of Geophysical Research*, **61**, 1219–24.

Gombosi, T. I. & Nagy, A. F. (1988). Time dependent polar wind modeling. *Advances in Space Research*, **8**(8), 59–68.

Gombosi, T. I. & Rasmussen, C. E. (1991). Transport of gyration-dominated space plasmas of thermal origin. 1. Generalized transport equations. *Journal of Geophysical Research*, **96**, 7759–78.

Gonzales, C. A., Kelley, M. C., Carpenter, D. L., Miller, T. R. & Wand, R. H. (1980).

Simultaneous measurements of ionospheric and magnetospheric electric fields in the outer plasmasphere. *Geophysical Research Letters*, **7**, 517–20.

Gorbachev, O. A., Khazanov, G. V., Gamayunov & Krivorutsky, E. N. (1992). A theoretical model for the ring current interaction with the Earth's plasmasphere. *Planetary and Space Science*, **40**, 859–72.

Gorbachev, O. A., Konikov, Yu. V. & Khazanov, G. V. (1988). Quasilinear heating of electrons in the earth's plasmasphere. *Pure and Applied Geophysics*, **127**, 545–59.

Grebowsky, J. M. (1970). Model study of plasmapause motion. *Journal of Geophysical Research*, **75**, 4329–33.

Grebowsky, J. M. (1971). Time dependent plasmapause motion. *Journal of Geophysical Research*, **76**, 6193–7.

Grebowsky, J. M. (1972). Model development of supersonic trough wind with shocks. *Planetary Space Science*, **20**, 1923–34.

Grebowsky, J. M. & Chen, A. J. (1976). Effects on the plasmasphere of irregular electric fields. *Planetary and Space Science*, **24**, 689–96.

Grebowsky, J. M., Hoegy, W. R. & Chen, T. C. (1990). Solar maximum-minimum extremes in the summer noontime polar cap F region ion composition: the measurements. *Journal of Atmospheric and Terrestrial Physics*, **95**, 12269–76.

Grebowsky, J. M., Hoegy, W. R. & Chen, T. C. (1993). High latitude field aligned light ion flows in the topside ionosphere deduced from ion composition and plasma temperature. *Journal of Atmospheric and Terrestrial Physics*, **55**, 1605–17.

Grebowsky, J. M., Hoffman, J. H. & Maynard, N. C. (1978). Ionospheric and magnetospheric 'plasmapauses'. *Planetary and Space Science*, **26**, 651–60.

Grebowsky, J. M., Maynard, N. C., Tulunay, Y. K. & Lanzerotti, L. J. (1976). Coincident observations of ionospheric troughs and the equatorial plasmapause. *Planetary and Space Science*, **24**, 1177–85.

Grebowsky, J. M., Tulunay, Y. & Chen, A. J. (1974). Temporal variations in the dawn and dusk midlatitude trough and plasmapause position. *Planetary and Space Science*, **22**, 1089–99.

Green, J. L., Benson, R. F., Fung, S. F., Smith, M. F., Calvert, W., Carpenter, D. L., Gallagher, D.

L., Reiff, P. H., Reinisch, B. W. & Taylor, W. W. L. (1994). Magnetospheric Radio Sounding (proposal to NASA).

Green, J. L., Waite, J. H., Jr., Chappell, C. R., Chandler, M. O., Doupnik, J. R., Richards, P. G., Heelis, R., Shawnan, S. D. & Brace, L. H. (1986). Observations of ionospheric magnetospheric coupling: DE and Chatanika coincidences. *Journal of Geophysical Research*, **91**, 5803–15.

Greenspan, M. E., Burke, W. J., Rich, F. J., Hughes, W. J. & Heelis, R. A. (1994). DMSP F8 observations of the mid-latitude and low-latitude topside ionosphere near solar minimum. *Journal of Geophysical Research*, **99**, 3817–26.

Greenstadt, E. W., McPherron, R. L., Anderson, R. R. & Scarf, F. L. (1986). A storm time, Pc 5 event observed in the outer magnetosphere by ISEE 1 and 2: wave properties. *Journal of Geophysical Research*, **91**, 13398–410.

Greenstadt, E. W., McPherron, R. L. & Takahashi, K. (1980). Solar wind control of daytime, midperiod geomagnetic pulsations. *Journal of Geomagnetism and Geoelectricity*, **32**, 89–110.

Gringauz, K. I. (1958). Rocket measurements of electron density in the ionosphere by means VHF-dispersion interferometer. *Doklady Akademiya Nauk. SSSR*, **120**, 1234.

Gringauz, K. I. (1961a). Some results of experiments in interplanetary space by means of charged particle traps on soviet space probes. *Space Research II*, 539–53.

Gringauz, K. I. (1961b). The structure of the Earth's ionized gas envelope based on local charged particle concentrations measured in the USSR. *Space Research II*, 574–92.

Gringauz, K. I. (1963). The structure of the ionized gas envelope of Earth from direct measurements in the USSR of local charged particle concentrations. *Planetary and Space Science*, **11**, 281–96.

Gringauz, K. I. (1965). Some results of USSR experiments in the ionosphere and interplanetary space. In *Progress in Radio Science 1960–1963*, III, The Ionosphere, ed. G. M. Brown, Elsevier, Amsterdam, pp. 65–75.

Gringauz, K. I. (1983). Plasmasphere and its interaction with the ring current. *Space Science Reviews*, **34**, 245–57.

Gringauz, K. I. (1985). Structure and properties of the Earth's plasmasphere. *Advances in Space Resarch*, **5**(4), 391–400.

Gringauz, K. I. & Bezrukikh,V. V. (1976).
Asymmetry of the Earth's Plasmasphere in the
direction noon-midnight from PROGNOZ-1
and PROGNOZ-2 data. *Journal of
Atmospheric and Terrestrial Physics*, **38**, 1071–6.

Gringauz, K. I. & Bezrukikh, V. V. (1977).
Plasmasphere of the earth (review).
Geomagnetizm i Aeronomiya, **17**, 784–803.

Gringauz, K. I., Bezrukikh, V. V., Musatov, L. S.,
Rybchinsky, R. E. & Sheronova, S. M. (1964).
Measurements made in the earth's
magnetosphere by means of charged particle
traps aboard the Mars-1 probe. *Space
Research*, **4**, 621–6.

Gringauz, K. I., Bezrukikh, V. V. & Ozerov, V. D.
(1961). The results of measurements of the
concentration of positive ions in the
atmosphere by means of ion traps aboard the
third SPUTNIK, in *Artificial Earth Satellites*, **6**,
63–100.

Gringauz, K. I., Bezrukikh, V. V., Ozerov, V. D. &
Rybchinsky, R. E. (1960a). The study of the
interplanetary ionized gas, high-energy
electrons and corpuscular radiation of the Sun,
employing three-electrode charged particle
traps on the second Soviet space rocket.
Doklady Akademiya Nauk SSSR, **131**, 1302–4;
translated in (1960) *Soviet Physics Doklady*, **5**,
361–4; published again in (1962) *Planetary and
Space Science*, **9**, 103–7.

Gringauz, K. I., Kurt, V. G., Moroz, V. I. &
Shklovsky, I. S. (1960b). Ionized gas and fast
electrons near the earth and in the
interplanetary space. *Doklady Akademiya Nauk.
SSSR*, **132**, 1062–5.

Gringauz, K. I., Kurt, V. G., Moroz, V. I. &
Shklovsky, I. S. (1960c). Results of observations
of charged particles up to R = 100 000 km with
the aid of charged particle traps on Soviet
cosmic rockets. *Astronomicheskii. Zhurnal*, **37**,
716–35; translated in 1961, *Soviet Astronomy A.
J.*, **4**, 680–95.

Gringauz, K. I. & Zelikman, M. C. (1957). A
measurement of the positive ion density along
the orbit of an artifical earth satellite. *Uspekhi
Fizicheskikh Nauk*, **63**, 239–52.

Guiter, S. M. & Gombosi, T. I. (1990). The role of
high-speed plasma flows in plasmaspheric
refilling. *Journal of Geophysical Research*, **95**,
10427.

Guiter, S. M., Gombosi, T. I. & Rasmussen, C. E.
(1995a). Two-stream modeling of plasmaspheric
refilling. *Journal of Geophysical Research*, **100**,

9519–26.

Guiter, S. M., Rasmussen, C. E., Gombosi, I. I.,
Sojka, J. J. & Schunk, R. W. (1995b). What is
the source of observed annual variations in the
plasmaspheric density? *Journal of Geophysical
Research*, **100**, 8013–20.

Gurevich, A. V., Krylov, A. L. & Tsedilina, Ye. Ye.
(1976). Electric fields in the earth's
magnetosphere and ionosphere. *Space Science
Reviews*, **19**, 59–160.

Gurnett, D. A. (1974). The earth as a radio source:
terrestrial kilometric radiation. *Journal of
Geophysical Research*, **79**, 4227–38.

Gurnett, D. A. (1976). Plasma wave interactions
with energetic ions near the magnetic equator.
Journal of Geophysical Research, **81**, 2765–70.

Gurnett, D. A., Anderson, R. R., Scarf, F. L.,
Fredricks, R. W. & Smith, E. J. (1979). Initial
results from the ISEE-1 and -2 plasma wave
investigation. *Space Science Reviews*, **23**,
103–22.

Gurnett D. A. & Frank L. A. (1974). Thermal and
suprathermal plasma densities in the outer
magnetosphere. *Journal of Geophysical
Research*, **79**, 2355–61.

Gurnett, D. A. & Inan, U. S. (1988). Plasma wave
observations with the Dynamics Explorer 1
spacecraft. *Reviews of Geophysics*, **26**, 285–316.

Gurnett, D. A. & Scarf, F. L. (1967). Summary
report on session on new developments. In
Progress in Radio Science, 1963–1966, pp.
1106–7. Brussels: International Union of Radio
Science.

Gurnett D. A. & Shaw R. R. (1973).
Electromagnetic radiation trapped in the
magnetosphere above the plasma frequency.
Journal of Geophysical Research, **78**, 8136–49.

Guthart, H. (1965). An anisotropic electron
velocity distribution for the cyclotron
absorption of whistlers and VLF emissions.
Radio Science, **69D**, 1403–15.

Hamar, D., Tarcsai, Gy., Lichtenberger, J., Smith,
A. J. & Yearby, K. H. (1990). Fine structure of
whistlers recorded digitally at Halley,
Antarctica. *Journal of Atmospheric and
Terrestrial Physics*, **52**, 801–10.

Hanson, W. B. (1962). Upper atmosphere helium
ions. *Journal of Geophysical Research*, **67**,
183–8.

Hanson, W. B. (1964). Dynamic diffusion process in
the exosphere. In *Electron Density Distribution
in the Ionosphere and Exosphere*, ed. E. Thrane,

pp. 361–370, Amsterdam, North-Holland.

Hanson, W. B., Heelis, R. A. & Power, R. A., Lippincott, C. R., Zuccaro, D. R., Holt, B. J., Harmon, L. H. & Sanatani, S. (1981). The retarding potential mass spectrometer for the Dynamics Explorer-B. *Space Science Instrumentation*, **5**, 503–10.

Hanson, W. B. & Ortenburger, I. B. (1961). The coupling between the protonosphere and the normal F region. *Journal of Geophysical Research*, **66**, 1425–35.

Hanson, W. B. & Patterson, T. N. L. (1963). Diurnal variation of the hydrogen concentration in the exosphere. *Planetary and Space Science*, **11**, 1035–52.

Hanson, W. B. & Patterson, T. N. L. (1964). The maintenance of the night-time F-layer. *Planetary and Space Science*, **12**, 979–97.

Hanson, W. B., Patterson, T. N. L. & Degaonkar S. S. (1963). Some deductions from a measurement of the hydrogen ion distribution in the high atmosphere. *Journal of Geophysical Research*, **68**, 6203–5.

Harel, M. & Wolf, R. A. (1976). Convection, in *Physics of Solar Planetary Environments*, Vol. 2, ed. D. J. Williams, AGU, Washington DC, pp. 617–29.

Harel, M., Wolf, R. A., Reiff, P. H., Spiro, R. W., Burke, W. J., Rich, F. J. & Smiddy, M. (1981a). Quantitative simulation of a magnetospheric substorm. 1. Model logic and overview. *Journal of Geophysical Research*, **86**, 2217–41.

Harel, M., Wolf, R. A., Spiro, R. W., Reiff, P. H., Chen, C.-K., Burke, W. J., Rich, F. J. & Smiddy, M. (1981b). Quantitative simulation of a magnetospheric substorm. 2. Comparison with observations. *Journal of Geophysical Research*, **86**, 2242–60.

Harris, I. & Priester, W. (1962). Theoretical models for the solar-cycle variation of the upper atmosphere. *Journal of Geophysical Research*, **67**, 4585–91.

Harris, K. K., Sharp, G. W. & Chappell, C. R. (1970). Observations of the plasmapause from OGO-5. *Journal of Geophysical Research*, **75**, 219–24.

Hartle, R. E. (1969). Ion-exosphere with variable conditions at the baropause. *Physics of Fluids*, **12**, 455–62.

Hasegawa, A. (1971). Drift wave instability at the plasmapause. *Journal of Geophysical Research*, **76**, 5361–4.

Hayakawa, M. (1994). Whistlers. In *Handbook of Atmospheric Electrodynamics*, Vol. II, ed. H. Volland, Boca Raton, Florida, CRC Press.

Hayakawa , M. & Sazhin, S. S. (1992). Mid-latitude and plasmaspheric hiss: a review. *Planetary and Space Science*, **40**, 1325–38.

Heelis, R. A., Bailey, G. J., Sellek, R., Moffett, R. J. & Jenkins, B. (1993). Field-aligned drifts in subauroral ion drift events. *Journal of Geophysical Research*, **98**, 21 493–9.

Heelis, R. A. & Coley, W. R. (1992). East–West ion drifts at mid-latitudes observed by Dynamics Explorer 2. *Journal of Geophysical Research*, **97**, 19 461–69.

Heelis, R. A., Coley, W. R., Loranc, M. & Hairston, M. R. (1992). Three-dimensional ionospheric plasma circulation. *Journal of Geophysical Research*, **97**, 13 903–10.

Heelis, R. A., Hanson, W. B. & Bailey, G. J. (1990). Distributions of He^+ at middle and equatorial latitudes during solar maximum. *Journal of Geophysical Research*, **95**, 10 313–20.

Heelis, R. A., Hanson, W. B., Lippincott, C. R., Zuccaro, D. R., Harmon, L. H., Holt, B. J., Doherty, J. E. & Power, R. A. (1981a). The ion drift meter for Dynamics Explorer-B. *Space Science Instrumentation*, **5**, 511–21.

Heelis, R. A., Murphy, J. A. & Hanson, W. B. (1981b). A feature of the behavior of He^+ in the nightside high-latitude ionosphere during equinox. *Journal of Geophysical Research*, **86**, 59–64.

Helliwell, R. A. (1961). Exospheric electron density variations deduced from whistlers. *Annales de Géophysique*, **17**, 76–81.

Helliwell, R. A. (1963). Coupling between the ionosphere and the earth-ionosphere waveguide at very low frequencies. In *Proceedings of the International Conference on the Ionosphere, London, July 1962*, pp. 452–460. Dorking, England, Bartholomew Press.

Helliwell, R. A. (1965). *Whistlers and Related Ionospheric Phenomena*, Stanford, CA, Stanford University Press.

Helliwell, R. A. (1967). A theory of discrete VLF emissions from the magnetosphere. *Journal of Geophysical Research*, **72**, 4773–90.

Helliwell, R. A. (1969). Low frequency waves in the magnetosphere. *Reviews of Geophysics*, **7**, 281–303.

Helliwell, R. A. (1988). VLF wave stimulation experiments in the magnetosphere from Siple Station, Antarctica. *Reviews of Geophysics*, **26**,

551–78.

Helliwell, R. A. & Carpenter, D. L. (1961).
Whistlers-West IGY-IGC Synoptic Program.
Final Report National Science Foundation
Grants IGY 6. 10/20 and G-8839, Radioscience
Laboratory. Stanford, CA, Stanford University.

Helliwell, R. A. & Carpenter, D. L. (1962).
Whistlers-West results from the IGY/IGC-59
synoptic program. Transactions of the
American Geophysical Union, 43(1), 125–133.

Helliwell, R. A., Crary, J. H., Katsufrakis, J. P. &
Trimpi, M. L. (1961). The Stanford University
Real-time Spectrum Analyzer. Tech. rept. no.
10, Air Force Contract no. AF 18(603)-126.
Radioscience Laboratory. Stanford, CA,
Stanford University.

Helliwell, R. A., Crary, J. H., Pope, J. H. & Smith,
R. L. (1956). The 'nose' whistler – a new
high-latitude phenomenon. Journal of
Geophysical Research, 61(1), 139–42.

Helliwell, R. A. & Gehrels, E. (1958). Observations
of magneto-ionic duct propagation using
man-made signals of very low frequency.
Proceedings of the Institute of Electrical and
Electronic Engineers, 46(4), 185–7.

Helliwell, R. A., Jean, A. G. & Taylor, W. L. (1958).
Some properties of lightning impulses which
produce whistlers. Proceedings I. R. E., 46(10),
1760–72.

Helliwell, R. A. & Katsufrakis, J. P. (1974). VLF
wave injection into the magnetosphere from
Siple Station, Antarctica. Journal of Geophysical
Research, 79, 2511–8.

Helliwell, R. A., Katsufrakis, J. P. & Trimpi, M.
(1973). Whistler-induced amplitude
perturbation in VLF propagation. Journal of
Geophysical Research, 78, 4679–88.

Helliwell, R. A., Mende, S. B. & Doolittle, J. H.,
Armstrong, W. C. & Carpenter, D. L. (1980).
Correlations between Å 4278 optical emissions
and VLF wave events observed at L ∼ 4 in the
Antarctic. Journal of Geophysical Research, 85,
3376–86.

Higel, B. & Wu L. (1984). Electron density and
plasmapause characteristics at 6. 6 RE : a
statistical study of the GEOS 2 relaxation
sounder data. Journal of Geophysical Research,
89, 1583–601.

Ho, D. & Carpenter, D. L. (1974). Outlying
plasmasphere structure detected by whistlers.
Planetary and Space Science, 24, 987–94.

Hoch, R. J. & Lemaire, J. (1975). Stable auroral red
arcs and their importance for the physics of the
plasmapause region. Annales de Géophysique,
31, 105–10.

Hoch, R. J. & Smith, L. L. (1971). Location in the
magnetosphere of field lines leading to SAR
arcs. Journal of Geophysical Research, 76,
3079–86.

Hoegy, W. R. & Grebowsky, J. M. (1994).
Comparison of ion composition from the
Goddard Comprehensive Ionosphere Database
with the International Reference Ionosphere
Model. Advances in Space Research, 14(12),
121–4 and 171.

Hoffman, J. H. & Dodson, W. H. (1980). Light ion
concentrations and fluxes in the polar regions
during magnetically quiet times. Journal of
Geophysical Research, 85, 626–32.

Hoffman, J. H., Dodson, W. H., Lippincott, C. R. &
Hammack, H. D. (1974). Initial ion
composition results from the ISIS 2 satellite.
Journal of Geophysical Research, 79, 4246–51.

Holzer, R. E., Farley, T. A., Burton, R. K. &
Chapman, M. C. (1974). A correlated study of
ELF waves and electron precipitation on OGO
6. Journal of Geophysical Research, 79, 1007–13.

Holzer, T. E., Fedder, J. A. & Banks, P. M. (1971).
A comparison of kinetic and hydrodynamic
models of an expanding ion-exosphere. Journal
of Geophysical Research, 76, 2453–68.

Horwitz, J. L. (1981). ISEE 1 observations of O^{++}
in the magnetosphere. Journal of Geophysical
Research, 86, 9225–9.

Horwitz, J. L. (1983). Plasmapause diffusion.
Journal of Geophysical Research, 88, 4950–2.

Horwitz, J. L. (1995). The ionosphere's wild ride in
outer space. Reviews of Geophysics, 33, Suppl.,
703–8.

Horwitz, J. L., Brace, L. H., Comfort, R. H. &
Chappell, C. R. (1986a). Dual spacecraft
measurements of plasmasphere-ionosphere
coupling. Journal of Geophysical Research, 91,
11 203–16.

Horwitz, J. L. & Chappell, C. R. (1979).
Observations of warm plasma in the dayside
plasma trough at geosynchronous orbit.
Journal of Geophysical Research, 84, 7075–90.

Horwitz, J. L., Comfort, R. H. & Chappell, C. R.
(1984). Thermal ion composition measurements
of the formation of the new outer plasmasphere
and double plasmapause during storm recovery
phase. Geophysical Research Letters, 11, 701–4.

Horwitz, J. L., Comfort, R. H. & Chappell, C. R.
(1986b). Plasmasphere and plasmapause region

characteristics as measured by DE-1. *Advances in Space Research*, **6**(3), 21–9.

Horwitz, J. L., Comfort, R. H. & Chappell, C. R. (1990a). A statistical characterization of plasmasphere density structure and boundary locations. *Journal of Geophysical Research*, **95**, 7937–47.

Horwitz, J. L., Comfort, R. H., Richards, P. G., Chandler, M. O., Chappell, C. R., Anderson, P., Hanson, W. B. & Brace, C. R. (1990b). Plasmasphere-Ionosphere coupling 2. Ion composition measurements at plasmaspheric and ionospheric altitudes and comparison with modeling results. *Journal of Geophysical Research*, **95**, 7949–59.

Horwitz, J. L., Ho, C. W., Scarbo, H. D., Wilson, G. R. & Moore, T. E. (1994). Centrifugal acceleration of the polar wind. *Journal of Geophysical Research*, **99**, 15 051–64.

Horwitz, J. L., Menteer, S., Turnley, J., Burch, J. L., Winningham, J. D., Chappell, C. R., Craven, J. D., Frank, L. A. & Slater, D. W. (1986c). Plasma boundaries in the inner magnetosphere. *Journal of Geophysical Research*, **91**, 8861–82.

Horwitz, J. L. & Singh, N. (1991). Refilling of the Earth's plasmasphere. *EOS*, **72**(37), 399–402.

Horwitz, J. L., Wilson, G. R., Lin, J., Brown, D. G. & Ho, C. W. (1993). Plasma transport in the ionosphere-magnetosphere system using semi-kinetic models. In *Rarefied Gas Dynamics Conference*, RGD-18, UBC, Vancouver, July 1992.

Huang, C. Y., Goertz, C. K. & Anderson, R. R. (1983). A theoretical study of plasmaspheric hiss generation. *Journal of Geophysical Research*, **88**, 7927–40.

Huang, T. S., Wolf, R. A. & Hill, T. W. (1990). Interchange instability of the Earth's plasmapause. *Journal of Geophysical Research*, **95**, 17 187–8.

Hultqvist, B. (1971). On the production of a magnetic field aligned electric field by interaction between the hot magnetospheric plasma and the cold ionosphere. *Planetary Space Science*, **19**, 749–59.

Hurren, P. J., Smith, A. J., Carpenter, D. L. & Inan, U. S. (1986). Burst precipitation induced perturbations on multiple VLF propagation paths in Antarctica. *Annales Geophysicae*, **4**, 311–18.

Imhof, W. L., Voss, H. D., Mobilia, J., Walt, M., Inan, U. S. & Carpenter, D. L. (1989). Characteristics of short-duration electron precipitation bursts and their relationship with VLF wave activity. *Journal of Geophysical Research*, **94**, 10 079–93.

Inan, U. S., Bell, T. F. & Anderson, R. R. (1977). Cold plasma diagnostics using satellite measurements of VLF signals from ground transmitters. *Journal of Geophysical Research*, **82**, 1167–76.

Inan, U. S., Knifsend, F. A. & Oh, J. (1990). Subionospheric VLF 'imaging' of lightning-induced electron precipitation from the magnetosphere. *Journal of Geophysical Research*, **95**, 17 217–31.

Iversen, I. B., Block, L. P., Brönstad, K., Grard, R., Haerendel, G., Junginger, H., Korth, A., Kremser, G., Madsen, M., Niskanen, J., Riedler, W., Tanskanen, P., Torkar, K. M. & Ullaland, S. (1984). Simultaneous observations of a pulsation event from the ground, with balloons and with a geostationary satellite on August 12, 1978. *Journal of Geophysical Research*, **89**, 6775–85.

Jacobson, A. B. & Erickson, W. C. (1993). Observations of electron-density irregularities in the plasmasphere using the VLA radio-interferometer. *Annales de Géophysique*, **11**, 869–88.

Jaggi, R. K. & Wolf, R. A. (1973). Self-consistent calculation of the motion of a sheet of ions in the magnetosphere. *Journal of Geophysical Research*, **78**, 2852–66.

James, H. G. (1972). Refraction of whistler-mode waves by large-scale gradients in the middle-latitude ionosphere. *Annales de Géophysique*, **28**, 301–39.

James, H. G. & Bell, T. F. (1987). Spin modulation of spectrally broadened VLF signals. *Journal of Geophysical Research*, **92**, 7560–8.

Jasna, D., Inan, U. S. & Bell, T. F. (1992). Precipitation of suprathermal (100 eV) electrons by oblique whistler waves. *Geophysical Research Letters*, **19**, 1639–42.

Jiricek, F., Smilauer, J., Triska, P., Triskova, L. & Kudela, K. (1996). Dynamics of the plasmasphere during magnetic storm as measured in the project ACTIVE. *Advances in Space Research*, **17**,(10), 129–34.

Johnstone, A. D. (1994). Pitch angle diffusion of low energy electrons and positive ions in the inner magnetosphere: A review of observations and theory (COSPAR abstract D2.2-005, Hamburg, 1994).

Jones, D. (1976). Source of terrestrial non-thermal

radiation. *Nature*, **260**, 686–9.

Jones, D. (1982). Terrestrial myriametric radiation from the earth's plasmapause. *Planetary and Space Science*, **30**, 399–410.

Jorjio, N. V., Kovrazhkin, R. A., Mogilevsky, M. M., Bosqued, J.-M., Reme, H., Sauvaud, J. A., Beghin C. & Rauch, J. L. (1985). Detection of suprathermal ionospheric O$^+$ ions inside the plasmasphere. *Advances Space Research*, **5**(4), 141–4.

Kamide, Y., Craven, J. D., Frank, L. A., Ahn, B.-H. & Akasofu, S.-I. (1986). Modeling substorm current systems using conductivity distributions inferred from DE auroral images. *Journal of Geophysical Research*, **91**, 11 235–56.

Karlson, E. T. (1963). Streaming of a plasma through a magnetic dipole field. *The Physics of Fluids*, **6**, 708–22.

Karlson, E. T. (1970). On the equilibrium of the magnetopause. *Journal of Geophysical Research*, **75**, 2438–48.

Karlson, E. T. (1971). Plasma flow in the magnetosphere. I. A two-dimensional model of stationary flow. *Cosmic Electrodynamics*, **1**, 474–95.

Kasahara, Y., Sawada, A., Yamamoto, M., Kimura, I., Kokubun & Karlson, E. T. (1971). Plasma flow in the magnetosphere. *Cosmic Electrodynamics*, **1**, 474–95.

Kavanagh, L. D., Jr., Freeman, J. W., Jr. & Chen, A. J. (1968). Plasma flow in the magnetosphere. *Journal of Geophysical Research*, **73**, 5511–9.

Kaye, S. M. & Kivelson, M. G. (1979). Time dependent convection electric fields and plasma injection. *Journal of Geophysical Research*, **84**, 4183–8.

Kennel, C. F. & Engelmann, F. (1966). Velocity space diffusion from weak plasma turbulence in a magnetic field. *Physics of Fluids*, **9**, 2377–88.

Kennel, C. F. & Petschek, H. E. (1966). Limit on stable trapped particle fluxes. *Journal of Geophysical Research*, **71**, 1–28.

Kennel, C. F., Scarf, F. L., Fredricks, R. W., McGehee, J. H. & Coroniti, F. V. (1970). VLF electric-field observations in the magnetosphere. *Journal of Geophysical Research*, **75**, 6136–52.

Kersley, L., Hajeb-Hosseiniem, H. & Edwards, K. J. (1978). Post-geomagnetic storm protonospheric replenishment. *Nature*, **271**, 429–30.

Khazanov, G. V., Gombosi, T. I., Nagy, A. F. &

Koen, M. A. (1992). Analysis of the Ionosphere - Plasmasphere transport of superthermal electrons, 1. Transport in the Plasmasphere. *Journal of Geophysical Research*, **97**, 16 887–95.

Khazanov, G. V., Koyen, M. A., Konikov, Yu. V. & Sidorov, I. M. (1984). Simulation of ionosphere-plasmasphere coupling taking into account ion inertia and temperature anisotropy. *Planetary and Space Sciences*, **32**, 585–98.

Khazanov, G. V. & Liemohn, M. W. (1995). Non-steady-state ionosphere–plasmasphere coupling of superthermal electrons. *Journal of Geophysical Research*, **100**, 9669–81.

Khazanov, G. V., Liemohn, M. W., Gombosi, T. I., & Nagy, A. F. (1993). Non-steady-state transport of superthermal electrons in the plasmasphere. *Geophysical Research Letters*, **20**, 2821–4.

Khazanov, G. V., Neubert, T. & Gefan, G. D. (1994). A unified theory of ionosphere-plasmasphere transport of suprathermal electrons. *IEEE Transactions on Plasma Science*, **22**, 187–97.

Kimura, I. (1966). Effects of ions on whistler-mode ray tracing. *Radio Science*, **1**, 269–83.

Kimura, I. (1989). Ray paths of electromagnetic and electrostatic waves in the Earth and planetary magnetospheres. In *Plasma Waves and Instabilities at Comets and in Magnetospheres*, eds B. T. Tsurutani and H. Oya, pp. 161–177. Washington, DC, American Geophysical Union.

Kintner, P. M. & Gurnett, D. A. (1978). Evidence of drift waves at the plasmapause. *Journal of Geophysical Research*, **83**, 39–44.

Kivelson, M. G. (1976). Magnetospheric electric fields and their variations with geomagnetic activity. *Reviews of Geophysics and Space Physics*, **14**, 189–97.

Kivelson, M. G., Kaye, S. M. & Southwood, D. J. (1979). The physics of plasma injection events, in *Dynamics of the magnetosphere*, ed. S.-I. Akasofu, D. Reidel, Dordrecht, Holland, pp. 385–405.

Kivelson, M. G. & Russell, C. T. (1973). Active experiments, magnetospheric modification, and a naturally occurring analogue. *Radio Science*, **8**, 1035–48.

Klumpar, D. M. (1979). Transversely accelerated ions: an ionospheric source of hot magnetospheric ions. *Journal of Geophysical Research*, **84**, 4229–37.

Knight, S. (1973). Parallel electric field. *Planetary Space Science*, **21**, 741–50.

Kockarts, G. & Nicolet, M. (1962). Le problème aéronomique de l'hélium et de l'hydrogène neutres. *Annales de Géophysique*, **18**, 269–90.

Köhnlein, W. (1986). A model of the electron and ion temperatures in the ionosphere. *Planetary and Space Science*, **34**, 609–30.

Köhnlein, W. & Raitt, W. J. (1977). Position of the mid-latitude trough in the topside ionosphere as deduced from ESRO 4 observations. *Planetary and Space Science*, **25**, 600–2.

Konikov, Yu., V., Gorbachev, O. A., Khazanov, G. V. & Chernov, A. A. (1989). Hydrodynamical equations for thermal electrons taking into account their scattering on ion-cyclotron waves in the outer plasmasphere of the Earth. *Planetary and Space Science*, **37**, 1157–68.

Koons, H. C. (1989). Observations of large-amplitude, whistler mode wave ducts in the outer plasmasphere. *Journal of Geophysical Research*, **94**, 15 393–7.

Kopal, Z. (1989). *The Roche Problem and its Significance for Double-star Astronomy*, Kluwer Academic Press, Dordrecht, ISBN 0-7923-0129-3, 263 pp.

Kowalkowski, L. & Lemaire, J. (1979). Contribution à l'étude des éléments de plasma détachés dans la magnétosphère. *Bulletin de la Classe des Sciences, Académie Royale de Belgique*, **65**, 159–73.

Kozyra, J. U., Brace, L. H., Cravens, T. E. & Nagy, A. F. (1986). A statistical study of the subauroral electron temperature enhancement using Dynamics Explorer 2 Langmuir probe observations. *Journal of Geophysical Research*, **91**, 11 270–80.

Kozyra, J. U., Shelley, E. G., Comfort, R. H., Brace, L. H., Cravens, T. E. & Nagy, A. F. (1987). The role of ring current O^+ in the formation of stable auroral red arcs. *Journal of Geophysical Research*, **92**, 7487–502.

Kozyra, J. U., Valladares, C. E., Carlson, H. C., Buonsanto, M. J. & Slater, D. W. (1990). A theoretical study of the seasonal and solar cycle variations at stable auroral red arcs. *Journal of Geophysical Research*, **95**, 12 219–34.

Krehbiel, J. P., Brace, L. G., Theis, R. F., Pinkus, W. H. & Kaplan, R. B. (1981). The dynamics explorer Langmuir probe instrument. *Space Science Instrumentation*, **5**, 493–502.

Krinberg, I. A. & Tashchilin, A. V. (1980). The influence of the ionosphere-plasmasphere coupling upon the latitude variations of ionospheric parameters. *Annales de Géophysique*, **36**, 537–48.

Krinberg, I. A. & Tashchilin, A. V. (1982). Refilling of geomagnetic force tubes with a thermal plasma after magnetic disturbance. *Annales de Géophysique*, **38**, 25–32.

Krinberg, I. A. & Tashchilin, A. V. (1984). *The Ionosphere and the Plasmasphere*, Nauka, Moscow (in Russian).

Kurita, K. & Hayakawa, M. (1985). Evaluation of the effectiveness of theoretical model calculation in determining the plasmapause structure. *Journal of Geophysics*, **7**, 130–5.

Kurt, V. G. & Moroz, V. I. (1961). The potential of a metal sphere in interplanetary space. *Iskusstvennye Sputniki Zemli*, **7**, 77–88.

Kurth, W. S. & Gurnett, D. A. (1991). Plasma waves in planetary magnetospheres. *Journal of Geophysical Research*, **96**, 18 977–91.

Kurth, W. S., Gurnett, D. A. & Anderson, R. R. (1981). Escaping nonthermal continuum radiation. *Journal of Geophysical Research*, **86**, 5519–31.

Laaspere, T., Morgan, M. G. & Johnson, W. C. (1963). Some results of five years of whistler observation from Labrador to Antarctica. *Proceedings of the Institute of Electrical and Electronic Engineers*, **51**, 554–68.

LaBelle, J., Treumann, R. A., Baumjohann, W., Haerendel, G., Sckopke, N., Paschmann, G. & Luhr, H. (1988). The duskside plasmapause/ring current interface: convection and plasma wave observations. *Journal of Geophysical Research*, **93**, 2573–90.

Lakhina, G. S., Mond, M. & Hameiri, E. (1990). Ballooning mode instability at the plasmapause. *Journal of Geophysical Research*, **95**, 4007–16.

Lanzerotti, L. J. & Fukunishi, H. (1975). Relationships of the characteristics of magnetohydrodynamic waves to plasma density gradients in the vicinity of the plasmapause. *Journal of Geophysical Research*, **80**, 4627–34.

Lanzerotti, L. J., Mellen, D. B. & Fukunishi, H. (1975). Excitation of plasma density gradients in the magnetosphere at ultralow frequencies. *Journal of Geophysical Research*, **80**, 3131–40.

Lanzerotti, L. J. & Southwood, D. J. (1979). Hydromagnetic waves. In *Solar System Plasma Physics*, Volume III, eds L. J. Lanzerotti, C. F. Kennel and E. N. Parker, pp. 109–135, North

Holland, Amsterdam.

LeDocq, M. J., Gurnett, D. A. & Anderson R. R. (1994). Electron number density fluctuations near the plasmapause observed by CRRES spacecraft. *Journal of Geophysical Research*, **99**, 23661–71.

Lemaire, J. (1974). The 'Roche-limit' of ionospheric plasma and the formation of the plasmapause. *Planetary and Space Science*, **22**, 757–66.

Lemaire, J. (1975). The mechanisms of formation of the plasmapause. *Annales de Géophysique*, **31**, 175–89.

Lemaire, J. (1976a). Rotating ion-exospheres. *Planetary and Space Science*, **24**, 975–85.

Lemaire, J. F. (1976b). Steady state plasmapause positions deduced from McIlwain's electric field models. *Journal of Atmospheric and Terrestrial Physics*, **38**, 1041–6.

Lemaire, J. (1978). Kinetic versus hydrodynamic solar wind models, in 'Pleins feux sur la physique solaire', Editions, CNRS, Paris, pp. 341–58.

Lemaire, J. (1983). Plasmapause formation and deformation, (VHS-Video animation), UCL–CAV, *Louvain-Neuve, Belgium*.

Lemaire, J. (1985). *Frontiers of the Plasmasphere (Theoretical Aspects)*. Université Catholique de Louvain, Faculté des Sciences, Editions Cabay, Louvain-la-Neuve, ISBN 2-87077-310-2; Aeronomica Acta A 298.

Lemaire, J. (1986). Plasma transport in the plasmasphere. *Advances in Space Research*, **6**(3), 157–75.

Lemaire, J. (1987). The plasmapause formation. *Physica Scripta*, **T18**, 111–8.

Lemaire, J. (1989). Plasma distribution models in a rotating magnetic dipole and refilling plasmaspheric flux tubes. *Physics of Fluids*, **32**, 1519–27.

Lemaire, J. (1994). Formation and deformations of the plasmapause during magnetospheric substorms, (abstract) presented at Taos, Aug. 14–19, 1994, *Workshop on the Earth's trapped particle environment*.

Lemaire, J. (1997). Convective instability of the outer plasmasphere (in preparation).

Lemaire, J., Barakat, A., Lesceux, J. M. & Shizgal, B. (1991). A method for solving Poisson's equation in geophysical and astrophysical plasmas, *Aeronomica Acta A 357*, 1990; in Rarefied Gas Dynamics, ed. A. E. Beylich, VCH Verlagsgesellschaft mbH, pp. 417–24.

Lemaire, J. & Burlaga, L. F. (1976). Diamagnetic boundary layers: A kinetic theory. *Astrophysics and Space Sciences*, **45**, 303–25.

Lemaire, J. & Kowalkowski, L. (1981). The role of plasma interchange motion for the formation of a plasmapause. *Planetary and Space Science*, **29**, 469–78.

Lemaire, J. & Roth, M. (1991). Non steady-state solar wind – magnetosphere interaction. *Space Science Reviews*, **57**, 59–108.

Lemaire, J., Roth, M. & De Keyser, J. (1996). An EMF driver of subauroral ion drifts (in preparation).

Lemaire, J. & Scherer, M. (1969). Le champ electrique de polarisation dans l'exosphere ionique polaire. *Comptes Rendus de Académie des Sciences*, Paris, **269B**, 666–9.

Lemaire, J. & Scherer, M. (1970). Models of the polar ion-exosphere. *Planetary and Space Science*, **18**, 103–20.

Lemaire, J. & Scherer, M. (1973). Kinetic models of the solar and polar winds. *Reviews of Geophysics and Space Physics*, **11**, 427–68.

Lemaire, J. & Scherer, M. (1974a). Exospheric models of the topside ionosphere. *Space Science Reviews*, **15**, 591–640.

Lemaire, J. & Scherer, M. (1974b). Ionosphere-plasmasheet field-aligned currents and parallel electric fields. *Planetary and Space Science*, **22**, 1485–90.

Lemaire, J. & Scherer, M. (1978). Field aligned distribution of plasma mantle and ionospheric plasmas. *Journal of Atmospheric and Terrestrial Physics*, **40**, 337–42.

Lemaire, J. & Scherer, M. (1983). Field-aligned current density versus electric potential characteristics for magnetospheric flux tubes. *Annales Geophysicae*, **1**, 91–5.

Lemaire, J. & Schunk, R. W. (1992). Plasmaspheric wind. *Journal of Atmospheric and Terrestrial Physics*, **54**, 467–77.

Lemaire, J. & Schunk, R. W. (1994). Plasmaspheric convection with non-closed streamlines. *Journal of Atmospheric and Terrestrial Physics*, **56**, 1629–33.

Lennartson, W. & Reasoner, D. C. (1978). Low-energy plasma observations at synchronous orbit. *Journal of Geophysical Research*, **83**, 2145–56.

Lester, M. & Smith, A. J. (1980). Whistler duct structure and formation. *Planetary and Space Science*, **28**, 645–54.

Li, W., Sojka, J. J. & Raitt, W. J. (1983). A study of plasmaspheric density distributions for diffusive equilibrium conditions. *Planetary and Space Science*, **31**, 1315–27.

Lie-Svendsen, . & Rees, M. H. (1996). An improved kinetic model for the polar outflow of minor ion. *Journal of Geophysical Research*, **101**, 2415–33.

Lin, J., Horwitz, J. L., Wilson, G. R., Ho, C. W. & Brown, D. G. (1992). A semikinetic model for early stage plasmasphere refilling : 2. Effects off wave-particle interactions. *Journal of Geophysical Research*, **97**, 1121–34.

Lin, J., Horwitz, J. L., Wilson, G. R. & Brown, D. G. (1994). Equatorial heating and hemispheric decoupling effects on inner magnetospheric core plasma evolution. *Journal of Geophysical Research*, **99**, 5727–44.

Lockwood, M., Waite, J. H. Jr., Moore, T. E., Johnson, J. F. E. & Chappell, C. R. (1985). A new source of suprathermal O^+ ions near the dayside polar cap boundary. *Journal of Geophysical Research*, **90**, 4099–116.

Longmire, C. L. (1963). *Elementary Plasma Physics*, Wiley Interscience, New York, 296 pp.

Lyons, L. R. & Williams, D. J. (1984). *Quantitative Aspects of Magnetospheric Physics*, D. Reidel, Dordrecht, 231 pp.

Maeda, H. (1964). Electric fields in the magnetosphere associated with daily geomagnetic variations and their effects on trapped particles. *Journal of Atmospheric and Terrestrial Physics*, **26**, 1133–8.

Mange, P. (1960). The distribution of minor ions in electrostatic equilibrium in the high atmosphere. *Journal of Geophysical Research*, **65**, 3833–4.

Marklund, G. T., Blomberg, L. G., Hardy, D. A. & Rich, F. J. (1987a). Instantaneous pictures of the high-latitude electrodynamics using Viking and DMSP/F7 observations, in *Proc. 8th ESA Symp.*, Sunne, Sweden, May 1987, ESA SP-270, 45.

Marklund, G. T., Blomberg, L. G., Potemra, T. A., Murphree, J. S., Rich, F. J. & Stasiewicz, K. (1987b). A new method to derive 'instantaneous' high-latitude potential distributions from satellite measurements including auroral imager data. *Geophysical Research Letters*, **14**, 439–42.

Marklund, G. T., Blomberg, L. G., Stasiewicz, K., Murphree, J. S., Pottelette, R., Potemra, T. A., Hardy, D. A. & Rich, F. J. (1988). Snapshots of high-latitude electrodynamics using Viking and DMSP/F7 observations. Report TRITA-EPP-88-01, The Royal Inst. of Technol., Stockholm, Sweden.

Marubashi, K. (1970). Escape of the polar-ionospheric plasma into the magnetospheric tail. *Rep. Ionosph. Space Research*, Japan, **24**, 322–46.

Mauk, B. H. & Co-I's (1995). The magnetospheric imaging mission: a proposal to NASA.

Mauk, B. H. & McIlwain, C. E. (1974). Correlation of K_p with substorm-injected plasma boundary. *Journal of Geophysical Research*, **79**, 3193–6.

Mauk, B. H. & Meng, C.-I. (1983). Characterization of geostationary particle signatures based on the 'injection boundary' model. *Journal of Geophysical Research*, **88**, 3055–71.

Maynard, N. C., Aggson, T. L. & Heppner, J. P. (1980). Magnetospheric observation of large sub-auroral electric fields. *Geophysical Research Letters*, **7**, 881–4.

Maynard, N. C. Aggson, T. L. & Heppner, J. R. (1983). The plasmaspheric electric field as measured by ISEE-1. *Journal of Geophysical Research*, **88**, 3991–4003.

Maynard, N. C. & Chen, A. J. (1975). Isolated cold plasma regions : Observations and their relation to possible production mechanisms. *Journal of Geophysical Research*, **80**, 1009–13.

Maynard, N. C. & Grebowsky, J. M. (1977). The plasmapause revisited. *Journal of Geophysical Research*, **82**, 1591–600.

Maynard, N. C., Heppner, J. P. & Egeland, A. (1982). Intense, variable electric fields at ionospheric altitudes in the high latitude regions as observed by DE-2. *Geophysical Research Letters*, **9**, 981–4.

McComas, D. J., Bame, S. J., Barraclough, B. L., Donart, J. R., Elphic, R. C., Gosling, J. T., Moldwin, M. B., Moore, K. R. and Thomsen, M. F. (1993). Magnetospheric plasma analyser: initial three-spacecraft observations from geosynchronous orbit. *Journal of Geophysical Research*, **98**, 13 453–65.

McIlwain, C. E. (1972). Plasma convection in the vicinity of the geosynchronous orbit. In *Earth Magnetospheric Processes*, ed. B. M. McCormac, pp. 268–279, D. Reidel, Dordrecht, Holland.

McIlwain, C. E. (1974). Substorm injection boundaries, in *Magnetospheric Physics*, ed. B. M. McCormac, D. Reidel, Dordrecht, Holland, pp. 143–54.

McIlwain, C. E. (1986). A Kp dependent equatorial electric field model. *Advances in Space Research*, **6(3)**, 187–97.

McNeill, F. A. (1967). Frequency shifts on whistler mode signals from a stabilized VLF transmitter. *Radio Science*, **2**, 589–94.

Mead, G. D. & Fairfield, D. H. (1975). A quantitative magnetospheric model derived from spacecraft magnetometer data. *Journal of Geophysical Research*, **80**, 523–34.

Melrose, D. B. (1967). Rotational effects on the distribution of thermal plasma in the magnetosphere of Jupiter. *Planetary and Space Science*, **15**, 381–93.

Mielke, T. A. & Helliwell, R. A. (1992). An experiment on the threshold effect in the coherent wave instability. *Geophysical Research Letters*, **19**, 2075–8.

Miller, N. J. (1970). The main electron trough during the rising solar cycle. *Journal of Geophysical Research*, **75**, 7175–81.

Miller, N. J. (1974). The dayside mid-latitude plasma trough. *Journal of Geophysical Research*, **79**, 3795–801.

Miller, R. H. & Combi, M. R. (1994). A Coulomb collision algorithm for weighted particle simulations. *Geophysical Research Letters*, **21**, 1735–8.

Miller, R. H. & Khazanov, G. V. (1993). Self-consistent electrostatic potential due to trapped plasma in the magnetosphere. *Geophysical Research Letters*, **20**, 1331–4.

Miller, R. H., Rasmussen, C. E., Gombosi, T. I., Khazanov, G. V. & Winske, D. (1993). Kinetic simulation of plasma flows in the inner magnetosphere. *Journal of Geophysical Research*, **98**, 19301–13.

Millward, G. H., Moffett, R. J., Quegan, S. & Fuller-Rowell, T. J. (1996). A coupled thermosphere-ionosphere-plasmasphere model (CTIP). *Solar Terrestrial Energy Program: Handbook of Ionospheric Models*, ed. R. W. Schunk, CASS-USU Logan, UT 84322-4405, pp. 239–80.

Mitchell, H. G. Jr., Ganguli, S. B. & Palmadesso, P. J. (1992). Diodelike response of high-latitude plasma in Magnetosphere-Ionosphere Coupling in the presence of field-aligned currents. *Journal of Geophysical Research*, **97**, 12045–56.

Mitchell, H. G. Jr. & Palmadesso, P. J. (1983) A

dynamic model for the auroral field line plasma in the presence of field-aligned current. *Journal of Geophysicl Research*, **88**, 2131–9.

Mitchell, H. G. Jr. & Palmadesso, P. J. (1984). O⁺ acceleration due to resistive momentum transfer in the auroral field line plasma. *Journal Geophysical Research*, **89**, 7573–6.

Miyazaki, S. (1979). Ion transition height distribution obtained with the satellite Taiyo. *Journal Geomagnetism and Geoelectricity*, **31**, Suppl., S113–S124.

Moldwin, M. B., Thomsen, M. F., Bame, S. J., McComas, D. J. & Moore, K. R. (1994). An examination of the structure and dynamics of the outer plasmasphere using multiple geosynchronous satellites. *Journal of Geophysical Research*, **99**, 11475–81.

Moldwin, M. B., Thomsen, M. F., Bame, S. J., McComas, D. J. & Reeves, G. D. (1995). The fine-scale structure of the outer plasmasphere. *Journal of Geophysical Research*, **100**, 8021–9. Correction: Ibid., **100**, 9649.

Moore, T. E. (1991). Origins of magnetospheric plasma. *Reviews of Geophysics*, **29**, Suppl., 1039–48.

Moore, T. E., Gallagher, D. L., Horwitz, J. L. & Comfort, R. H. (1987). MHD wave breaking in the outer plasmasphere. *Geophysical Research Letters*, **14**, 1007–10.

Morfill, G. E. (1978). A review of selected topics in magnetospheric physics. *Reports on Progress in Physics*, **41**, 303–94.

Morgan, M. G. & Allcock, G. M. (1956). Observations of whistling atmospherics at geomagnetically conjugate points. *Nature*, **177**, No 4497, 30–1.

Morgan, M. G., Brown, P. E., Johnson, W. C. & Taylor, H. A., Jr. (1977). Light ion and electron troughs observed in the midlatitude topside ionosphere on two passes of OGO 6 compared to coincident equatorial electron density deduced from whistlers. *Journal of Geophysical Research*, **82**, 2797–800.

Morgan, D. D. & Gurnett, D. A. (1991). The source location and beaming of terrestrial continuum radiation. *Journal of Geophysical Research*, **96**, 9595–613.

Mosier, S. R., Kaiser, M. L. & Brown, L. W. (1973). Observations of noise bands associated with the upper hybrid resonance by the IMP 6 radio astronomy experiment. *Journal of Geophysical Research*, **78**, 1673–80.

Mozer, F. S. & Serlin, R. (1969). Magnetospheric

electric field measurements with balloons. *Journal of Geophysical Research*, **74**, 4739–54.

Muldrew, D. B. (1965). F-layer ionization troughs deduced from Alouette data. *Journal of Geophysical Research*, **70**, 2635–50.

Muldrew, D. B. (1967). Medium frequency conjugate echoes observed in topside-sounder data. *Canadian Journal of Physics*, **45**, 3935–44.

Muldrew, D. B. & Hagg, E. L. (1969). Properties of high-latitude ionospheric ducts deduced from Alouette II two-hop echoes. *Proceedings of the Institute of Electrical and Electronic Engineers*, **57(6)**, 1128–34.

Muzzio J. L. R. & Angerami, J. J. (1972). OGO–4 observations of extremely low frequency hiss. *Journal of Geophysical Research*, **77**, 1157–73.

Nagai, T., Horwitz, J. L., Anderson, R. R. & Chappell, C. R. (1985). Structure of the plasmapause from ISEE 1 low-energy ion and plasma wave observations. *Journal of Geophysical Research*, **90**, 6622–6.

Nagai, T., Johnson, J. F. E. & Chappell, C. R. (1983). Low-energy (< 100 eV) ion pitch angle distributions in the magnetosphere by ISEE 1. *Journal of Geophysical Research*, **88**, 6944–60.

Nakada, M. P., Dungey, J. W. & Hess, W. N. (1965). On the origin of outer-belt protons. *Journal of Geophysical Research*, **70**, 3529–32.

Newberry, I. T., Comfort, R. H., Richards, P. G. & Chappell, C. R. (1989). Thermal He$^+$ in the plasmasphere : comparison of observations with numerical calculations. *Journal of Geophysical Research*, **94**, 15 265–76.

Newcomb, W. A. (1961). Convective instability induced by gravity in a plasma with a frozen-in magnetic field. *Physics of Fluids*, **4**, 391–6.

Newell, P. T. & Meng, C. I. (1993). Reply to comment by M. Lockwood and M. F. Smith on our paper entitled, 'Mapping the dayside ionosphere to the magnetosphere according to particle precipitation characteristics'. *Geophysical Research Letters*, **20**, 1741–2.

Nicolet, M. (1961). Helium, an important constituent in the lower-exosphere. *Journal of Geophysical Research*, **66**, 2263–4.

Nishida, A. (1966). Formation of plasmapause, or magnetospheric plasma knee, by the combined action of magnetospheric convection and plasma escape from the tail. *Journal of Geophysical Research*, **71**, 5669–79.

Norris, A. J., Johnson, J. F. E., Sojka, J. J. & Wrenn, G. L., Cornilleau-Wehrlin, N., Perraut, S. & Roux, A. (1983). Experimental evidence for the acceleration of thermal electrons by ion cyclotron waves in the magnetosphere. *Journal of Geophysical Research*, **88**, 889–98.

Northrop, T. G., & Teller, E. (1960). Stability of the adiabatic motion of charged particles in the Earth's field. *Physical Review*, **117**, 215–25.

Odera, T. J. (1986). Solar wind controlled pulsations: a review. *Reviews of Geophysics*, **24**, 55–74.

Oguti, T., Meek, J. H. & Hayashi, K (1984). Multiple correlation between auroral and magnetic pulsations. *Journal of Geophysical Research*, **89**, 2295–303.

Okada, T., H. Hayakawa, K. Tsuruda, A. Nishida & A. Matsuoka (1993). EXOS-D observations of enhanced electric fields during the giant magnetic storm in March 1989. *Journal of Geophysical Research*, **98**, 15417–24.

Oliver, W. L., Holt, J. M., Wand, R. H. and Evans, J. V. (1983). Millstone Hill incoherent scatter observations of auroral convection over $60° < \Lambda < 75°$, 3. Average patterns versus K_p. *Journal of Geophysical Research*, **88**, 5505–16.

Olsen, R. C. (1981). Equatorially trapped plasma populations. *Journal of Geophysical Research*, **86**, 11 235–45.

Olsen, R. C. (1982). The hidden ion population of the magnetosphere. *Journal of Geophysical Research*, **87**, 3481–8.

Olsen, R. C. (1992). The density minimum at the Earth's magnetic equator. *Journal of Geophysical Research*, **97**, 1135–50.

Olsen, R. C., Chappell, C. R., Gallagher, D. L., Green, J. L. & Gurnett, D. A. (1995). The hidden ion population: Revisited. *Journal of Geophysical Research*, **92**, 12 121–32.

Olsen, R. C., Scott, L. J. & Boardsen, S. (1994). Comparison between Liouville's theorem and observed latitudinal distribution of trapped ions in the plasmapause region. *Journal of Geophysical Research*, **99**, 2191–9.

Olsen, R. C., Shawhan, S. D., Gallager, D. L., Green, J. L., Chappell, C. R. & Anderson, R. R. (1987). Plasma observations at the magnetic equator. *Journal of Geophysical Research*, **92**, 2385–407.

Olson, W. P. & Pfitzer, K. A. (1974). A quantitative model of the magnetospheric magnetic field. *Journal of Geophysical Research*, **79**, 3739–48.

Olson, W. P. & Pfitzer, K. A. (1977). Magnetospheric magnetic field modeling.

Annual Scientific report, AFOSR contract No F 44620-75-C-0033, McDonnell Douglas Astronautics Company, Huntington Beach, CA.

Ondoh, T., Nakamura, Y., Watanabe, S. & Aikyo, K. (1989). Impulsive plasma waves observed by the DE 1 in nightside magnetosphere. *Journal of Geophysical Research*, **94**, 3779–84.

Ossakow, S. L. & Chaturvedi, P. K. (1978). Morphological studies of rising equatorial spread F bubbles. *Journal of Geophysical Research*, **83**, 2085–90.

Ott, E. (1978). Theory of Rayleigh-Taylor bubbles in the equatorial ionosphere. *Journal of Geophysical Research*, **83**, 2066–70.

Oya, H. (1991). Studies on plasma and plasma waves in the plasmasphere and auroral particle acceleration region, by PWS on board the EXOS-D (Akebono) satellite. *Journal of Geomagnetism and Geoelectricity*, **43**, 369–93.

Oya, H., Iizima, M. & Morioka, A. (1991). Plasma turbulence disc circulating in the equatorial region of the plasmasphere identified by the plasma wave detector (PWS) onboard the Akebono (EXOS-D) satellite. *Geophysical Research Letters*, **18**, 329–32.

Oya, H. & Ono, T (1987). Stimulation of plasma waves in the magnetosphere using satellite JIKIKEN (EXOS-B) Part II: Plasma density across the plasmapause. *Journal of Geomagnetism and Geoelectricity*, **39**, 591–607.

Paetzold, H. K. (1962). Corpuscular heating of the upper atmosphere. *Journal of Geophysical Research*, **67**, 2741–4.

Palmadesso, P. J., Ganguli, S. B. & Mitchell, H. G. Jr. (1988). Multifluid simulations of transport processes in the auroral zones in *Magnetosphere and ionosphere plasma models*, *Geophys. Monograph Series*, eds T. E. Moore & J. H. Waite, Jr., pp. 133–43, AGU, Washington DC.

Pannekoek, A. (1922). Ionization in stellar atmospheres. *Bulletin Astronomical Institute Mathematics*, **1**, 107–18.

Park, C. G. (1970). Whistler observations of the interchange of ionization between the ionosphere and the protonosphere. *Journal of Geophysical Research*, **75**, 4249–60.

Park, C. G. (1971). Westward electric fields as the cause of night-time enhancements in electron concentrations in the midlatitude F region. *Journal of Geophysical Research*, **76**, 4560–8.

Park, C. G. (1972). Methods of Determining

Electron Concentration in the Magnetosphere from Nose Whistlers. Radio Science Laboratory, Technical Report 3454-1, Stanford, CA, Stanford University.

Park, C. G. (1973). Whistler observations of the depletion of the plasmasphere during a magnetospheric substorm. *Journal of Geophysical Research*, **78**, 672–83.

Park, C. G. (1974a). Some features of plasma distribution in the plasmasphere deduced from Antarctic whistlers. *Journal of Geophysical Research*, **79**, 169–73.

Park, C. G. (1974b). A morphological study of substorm-associated disturbances in the ionosphere. *Journal of Geophysical Research*, **79**, 2821–7.

Park, C. G. (1978). Whistler observations of substorm electric fields in the nightside plasmasphere. *Journal of Geophysical Research*, **83**, 5773–7.

Park, C. G. & Banks, P. M. (1974). Influence of thermal plasma flow on the mid-latitude night-time F2 layer: Effects of electric fields and neutral winds inside the plasmasphere. *Journal of Geophysical Research*, **79**, 4661–8.

Park, C. G. & Carpenter, D. L. (1970). Whistler evidence of large-scale electron-density irregularities in the plasmasphere. *Journal of Geophysical Research*, **75**, 3825–35.

Park, C. G. & Carpenter, D. L. (1978). Very low frequency radio waves in the magnetosphere. In *Upper Atmosphere Research in Antarctica*, eds L. J. Lanzerotti and C. G Park, pp. 72–99, Washington DC, American Geophysical Union.

Park C. G., Carpenter, D. L. & Wiggin, D. B. (1978). Electron density in the plasmasphere: whistler data on the solar cycle, annual, and diurnal variations. *Journal of Geophysical Research*, **83**, 3137–44.

Park, C. G. & Helliwell, R. A. (1971). The formation by electric fields of field-aligned irregularities in the magnetosphere. *Radio Science*, **6**, 299–304.

Park, C. G. & Meng, C. I. (1971). Vertical motions of the midlatitude F2 layer during magnetospheric substorms. *Journal of Geophysical Research*, **76**, 8326–32.

Park, C. G. & Meng, C. I. (1973). Distortions of the nightside ionosphere during magnetospheric substorms. *Journal of Geophysical Research*, **78**, 3828–40.

Parker, E. N. (1958). Dynamics of the

interplanetary gas and magnetic field. *Astrophysical Journal*, **128**, 664–76.

Paschal, E. W. (1988). Phase Measurements of Very Low Frequency Signals from the Magnetosphere. PhD thesis, Stanford, CA, Stanford University.

Paschal, E. W. & Helliwell, R. A. (1983). Phase measurements of whistler mode signals from the Siple VLF transmitter. *Journal of Geophysical Research*, **89**, 1667–74.

Pedersen, A., Cattell, C. A., Fälthammar, C.-G., Knott, K., Lindqvist, P.-A., Manka, R. H. & Mozer, F. S. (1985). Electric fields in the plasma sheet and plasma sheet boundary layer. *Journal of Geophysical Research*, **90**, 1231–42.

Persson, H. (1966). Electric field parallel to the magnetic field in a low density plasma. *Physics of Fluids*, **9**, 1090–8.

Peterson, W. K., Collin, H. L., Doherty, M. F. & Bjorklund, C. M. (1992). O^+ and He^+ restricted and extended (bi-modal) ion conic distributions. *Geophysical Research Letters*, **19**, 1439–42.

Peymirat, C. & Fontaine, D. (1994). Numerical simulation of magnetospheric convection including the effect of field-aligned currents and electron precipitation. *Journal of Geophysical Research*, **99**, 11155–76.

Pfitzer, K. A., Olson, W. P. & Mogstad, T. (1988). A time dependent, source driven magnetospheric magnetic field model. *EOS*, **69**, 426.

Pierrard, V. (1996). New model of magnetospheric current-voltage relationship. *Journal of Geophysical Research*, **101**, 2669–75.

Pierrard, V. (1997). Fonction de distribution des vitesses dans la région de transition du vent polaire, PhD thesis, Université Catholique de Louvain, Louvain-Ld-Neuve, Belgium.

Pierrard, V. & Lemaire, J. F. (1996). New exospheric models based on Lorentzian velocity distributions. *Journal of Geophysical Research*, **101**, 7923–34.

Piggott, W. R., Lawden, M. D., Smith, R. W., Kendall, M. L. & Colbourne, M. B. (1970). The use of satellite data for prediction purposes. In *Proceedings of the AGARD EPC Symposium on Ionospheric Forcasting*, Ottawa, 3–6 September 1969, p. 70.

Poletti-Liuzzi, D. A., Yeh, K. C. & Liu, C. H. (1977). Radio beacon studies of the plasmasphere. *Journal of Geophysical Research*, **82**, 1106–14.

Potemra, T. A., Luhr, H., Zanetti, L. J., Takahashi, K., Erlandson, R. E., Marklund, G. T., Block, L. P., Blomberg, L. G. & Lepping, R. P. (1989). Multisatellite and ground-based observations of transient ULF waves. *Journal of Geophysical Research*, **94**, 2543–54.

Potemra, T. A. & Rosenberg, T. J. (1973). VLF propagation disturbances and electron precipitation at mid-latitudes. *Journal of Geophysical Research*, **78**, 1572–80.

Quegan S., (1989). The influence of convection on the structure of the high-latitude ionosphere. *Philosophical Transactions of the Royal Society (London) A*, **328**, 119–37.

Quegan, S., Bailey, G. J., Moffett, R. J., Heelis, R. A., Fuller-Rowell, T. J., Rees, D. & Spiro, R. W. (1982). A Theoretical study of the distribution of ionization in the high - latitude ionosphere and the plasmasphere: first results on the mid-latitude trough and the light ion trough. *Journal of Atmospheric and Terrestrial Physics*, **44**, 619–40.

Raitt, W. J. (1974). The temporal and spatial development of mid-latitude thermospheric electron temperature enhancements during a geomagnetic storm. *Journal of Geophysical Research*, **79**, 4703–8.

Raitt, W. J. & Dorling, E. B. (1976). The global morphology of light ions measured by the ESRO-4 satellite. *Journal of Atmospheric and Terrestrial Physics*, **38**, 1077–83.

Raitt, W. J. & Schunk, R. W. (1983). Composition and characteristics of the polar wind. In *Energetic Ion Composition*, ed. R. Johnson, Terra Scientific, Tokyo, pp. 99–141.

Raitt, W. J., Schunk, R. W. & Banks, P. M. (1978). Quantitative calculations of helium ion escape fluxes from the polar ionospheres. *Journal of Geophysical Research*, **83**, 5617–23.

Rasmussen, C. E., Guiter, S. M. & Thomas, S. G. (1993). A two-dimensional model of the plasmasphere: refilling time constants. *Planetary and Space Science*, **41**, 35–43.

Rasmussen, C. E. & Schunk, R. W. (1988). Multistream hydrodynamic modeling of interhemispheric plasma flow. *Journal of Geophysical Research*, **93**, 14 557–65.

Rasmussen, C. E. & Schunk, R. W. (1990). A three-dimensional time-dependent model of the plasmasphere. *Journal of Geophysical Research*, **95**, 6133–44 and 6125.

Rastani, K., Inan, U. S. & Helliwell, R. A. (1985). DE-1 observations of Siple transmitter signals and associated sidebands. *Journal of Geophysical Research*, **90**, 4128–40.

Ratcliffe, J. A. (1960). *Physics of the Upper Atmosphere*, London, Academic Press.

Rawer, K. & Bilitza, D. (1985). Study of ionospheric and tropospheric models. Final Report ESA contract 5375/83/D and *Journal of Atmospheric Terrestrial Physics*, **51**, 781–90.

Rawer, K. & Bilitza, D. (1989). Electron density profile description in the international reference ionosphere. *Journal of Atmospheric Terrestrial Physics*, **51**, 781–90.

Rawer, K., Bilitza, D. & Ramakrishnan, S. (1978). Goals and status of the international reference ionosphere. *Reviews of Geophysics and Space Physics*, **16**, 177–81.

Rees, D. & Fuller-Rowell, T. J. (1992). Modeling the response of the thermosphere-ionosphere system to time dependent forcing. *Advances in Space Research*, **12**(6), 69–87.

Remizov, A. P., Gringauz, K. I. & Bassolo, V. S. (1990). Specific tasks of Regatta-E measurements correlated with cluster/Regatta A in the ISTP and Regatta programmes, in *Proc. Int. Workshop on Space Plasma Phys.*, ESA SP-306, 137–40.

Retterer, J. M., Chang, T., Crew, G. B., Jaspere, J. R. & Winningham, J. D. (1987). Monte Carlo modeling of oxygen ion conic acceleration by cyclotron resonance. *Physical Review Letters*, **59**, 148–51.

Rich, F. J., Burke, W. J., Kelley, M. C. & Smiddy, M. (1980). Observations of field-aligned currents in association with strong convection electric fields at subauroral latitudes. *Journal of Geophysical Research*, **85**, 2335–40.

Richards, P. G., Schunk, R. W. & Sojka, J. J. (1983). Large-scale counterstreaming of H^+ and He^+ along plasmaspheric flux tubes. *Journal of Geophysical Research*, **88**, 7879–86.

Richards, P. G. & Torr, D. G. (1985). Seasonal, diurnal and solar cyclical variations of the limiting H^+ flux in the Earth's topside ionosphere. *Journal of Geophysical Research*, **90**, 5261–8.

Richards, P. G. & Torr, D. G. (1988). Hydrodynamic models of plasmasphere. In *Modelling of magnetospheric plasma*, Geophys. Meteor. Seq., V. 44, eds T. E. Moore and J. H. Waite, AGU, Washington, DC.

Richards, P. G., Torr, D. G., Reinisch, B. E., Gamache, R. R. & Wilkinson, P. J. (1994). F2 peak electron density at Millstone Hill and

Hobart: comparison of theory and measurement at solar maximum. *Journal of Geophysical Research*, **99**, 15 005–16.

Richmond, A. D. (1973). Self-induced motions of thermal plasma in the magnetosphere and the stability of the plasmapause. *Radio Science*, **8**, 1019–27.

Richmond, A. D. (1976). Electric field in the ionosphere and plasmasphere on quiet days. *Journal of Geophysical Research*, **81**, 1447–50.

Richmond, A. D., Blanc, M., Emery, B. A., Wand, R. H., Fejer, B. G., Woodman, R. F., Ganguly, S., Amayenc, P., Behnke, R. A., Calderon, C. & Evans, J. V. (1980). An empirical model of quiet–day ionospheric electric fields in middle- and low-latitude. *Journal of Geophysical Research*, **85**, 4658–64.

Richmond, A. D. & Kamide, Y. (1988). Mapping electrodynamic features of the high-latitude ionosphere from localized observations: Technique. *Journal of Geophysical Research*, **93**, 5741–59.

Richmond, A. D., Kamide, Y., Akasofu, S.-I., Alcaydé, D., Blanc, M., De la Beaujardière, O., Evans, D. S., Foster, J. C., Friis-Christensen, E., Holt, J. M., Pellinen, R. J., Senior, C. & Zaytsev, A. N. (1990). Global measures of ionospheric electrodynamic activity inferred from combined incoherent scatter radar and ground magnetometer observations. *Journal of Geophysical Research*, **95**, 1061–71.

Rishbeth, H. & Garriott, O. K. (1969). *Introduction to Ionospheric Physics*, Academic Press, New York, 331 pp.

Roberts, W. T., Jr., Horwitz, J. L., Comfort, R. H., Chappell, C. R., Waite, J. H. Jr. & Green, J. L. (1987). Heavy ion density enhancements in the outer plasmasphere. *Journal of Geophysical Research*, **92**, 13 499–512.

Roble, R. G. (1996). The NCAR thermosphere-ionosphere-mesosphere-electrodynamics general circulation model (TIME-GCM). *Solar Terrestrial Energy Program: Handbook of Ionospheric Models*, ed. R. W. Schunk, CASS-USU Logan, UT 84322-4405, pp. 281–8.

Roble, R. G., Ridley, E. C., Richmond, A. D. & Dickinson, R. E. (1988). A coupled thermosphere/ionosphere general circulation model. *Geophysical Research Letters*, **15**, 1325–8.

Rodger, A. S., Moffett, R. J. & Quegan, S. (1992). The role of ion drift in the formation of ionisation troughs in the mid- and high-latitude ionosphere – A review. *Journal of Atmospheric and Terrestrial Physics*, **54**, 1–30.

Rodger, A. S. & Pinnock, M. (1982). Movements of the mid-latitude ionospheric trough. *Journal of Atmospheric and Terrestrial Physics*, **44**, 985–92.

Rosenberg, T. J., Helliwell, R. A. & Katsufrakis, J. P. (1971). Electron precipitation associated with discrete very-low-frequency emissions. *Journal of Geophysical Research*, **76**, 8445–52.

Rosenberg, T. J., Siren, J. C., Matthews, D. L., Marthinsen, K., Holtet, J. A., Egeland, A., Carpenter, D. L. & Helliwell, R. A. (1981). Conjugacy of electron microbursts and VLF chorus. *Journal of Geophysical Research*, **86**, 5819–32.

Rosseland, S. (1924). Electrical state of a star. *Monthly Notices Royal Astronomical Society*, **84**, 720–8.

Roth, M. (1975a). The effects of different field-aligned ionization models on the electron densities and total flux tube contents deduced from whistlers. *Annales de Géophysique*, **31**, 69–75.

Roth, M. (1975b). The plasmapause as a plasma sheath: a minimum thickness. *Journal of Atmospheric and Terrestrial Physics*, **38**, 1065–70.

Russell, C. T., Holzer, R. E. & Smith, E. J. (1969). OGO 3 observations of ELF noise in the magnetosphere. 1. Spatial extent and frequency of occcurrence. *Journal of Geophysical Research*, **74**, 755–77.

Russell, C. T., Holzer, R. E. & Smith, E. J. (1970). OGO 3 observations of ELF noise in the magnetosphere. 2. The nature of the equatorial noise. *Journal of Geophysical Research*, **75**, 755–68.

Russell, C. T., McPherron, R. L. & Coleman, P. J. (1972). Fluctuating magnetic fields in the magnetosphere. 1. ELF and VLF fluctuations. *Space Science Reviews*, **12**, 810–56.

Rycroft, M. J. (1973). Enhanced energetic electron intensities at 100 km altitude and a whistler propagating through the plasmasphere. *Planetary and Space Science*, **21**, 239–51.

Rycroft, M. J. (1974a). Magnetospheric plasma flow and electric fields derived from whistler observations. In *Correlated Interplanetary and Magnetospheric Observations*, ed. D. E. Page, D. Reidel Publishing Company, Dordrecht, Holland, pp. 317–35.

Rycroft M. J. (1974b). Whistlers and discrete ELF/VLF emissions. In *ELF-VLF Radio Wave Propagation*, ed. J. A. Holtet, pp. 317–34. Dordrecht, Holland, D. Reidel.

Rycroft, M. J. (1975). A review of in situ observations of the plasmapause. *Annales de Géophysique*, **31**, 1–16.

Rycroft, M. J. (1987). Some aspect of geomagnetically conjugate phenomena. *Annales Geophysicae*, **5**, 463–77.

Rycroft, M. J. & Alexander, P. D. (1969). Model of hydrogen and helium ion concentrations in the plasmasphere. Paper A. 3. 11, COSPAR Meeting, Prague, Czechoslovakia.

Rycoft, M. J. & Burnell, S. J. (1970). Statistical analysis of movements of the ionospheric trough and plasmapause. *Journal of Geophysical Research*, **75**, 5600–4.

Rycroft, M. J. & Jones, I. R. (1985). Modelling the plasmasphere for the International Reference Ionosphere. *Advances in Space Research*, **5**(10), 21–7.

Rycroft, M. J. & Jones, I. R. (1987). A suggested model for the IRI plasmaspheric distribution. *Advances in Space Research*, **7**(6), 13–22.

Rycroft, M. J. & Thomas, J. O. (1970). The magnetospheric plasmapause and the electron density trough at the ALOUETTE I orbit. *Planetary and Space Science*, **18**, 65–80.

Sagalyn, R. C. & Smiddy, M. (1964). Electrical processes in the night-time exosphere. *Journal of Geophysical Research*, **69**, 1809–23.

Sagawa, E., Marubashi, K., Iwamoto, I., Watanabe, S., Mori, H., Whalen, B. A. & Yau, A. W. (1988). A model for long-term response to the magnetospheric activity and related measurement plans by EXOS-D. *Advances in Space Research*, **8**(8), 35–44.

Sagawa, E., Yau, A. W., Whalen, B. A. & Peterson, W. K. (1987). Pitch angle distributions of low-energy ions in the near-Earth magnetosphere. *Journal of Geophysical Research*, **92**, 12 241–54.

Sagredo, J. L. & Bullough, K. (1973). VLF goniometer observations at Halley Bay, Antarctica, 2, magnetospheric structure deduced from whistler observations. *Planetary and Space Science*, **21**, 913–23.

Saito, T. & Matsushita, S. (1968). Solar cycle effects on geomagnetic Pi2 pulsations. *Journal of Geophysical Research*, **73**, 267–86.

Sanatani, S. & Hanson, W. B. (1970). Plasma temperatures in the magnetosphere. *Journal of*

Geophysical Research, **75**, 769–75.

Sato, N. & Fukunishi, H. (1980). Interaction between ELF-VLF emissions and magnetic pulsations: classification of quasi-periodic ELF-VLF emissions based on frequency-time spectra. *Journal of Geophysical Research*, **86**, 19–29.

Sauvaud, J.-A., Crasnier, J. Galperin, Yu. I. & Feldstein, Ya. I. (1983). A statistical study of the dynamics of the equatorward boundary of the diffuse aurora in the pre-midnight sector. *Geophysical Research Letters*, **10**, 749–52.

Saxton J. M. & Smith, A. J. (1989). Quiet time plasmaspheric electric fields and plasmasphere-ionosphere coupling fluxes at L = 2. 5. *Planetary and Space Science*, **37**, 283-93.

Sazhin, S. S. & Hayakawa, M. (1992). Magnetospheric chorus emissions: a review. *Planetary and Space Science*, **40**, 681-97.

Sazhin, S. S., Hayakawa, M. & Bullough, K. (1992). Whistler diagnostics of magnetospheric parameters : A review. *Annales Geophysicae*, **10**, 293-308.

Sazhin, S. S., Smith, A. J. & Sazhina, E. M. (1990). Can magnetospheric electron temperature be inferred from whistler dispersion measurements? *Annales Geophysicae*, **8**, 273–85.

Scarf, F. L. & Chappell, C. R. (1973). An association of magnetospheric whistler dispersion characteristics with changes in local plasma density. *Journal of Geophysical Research*, **78**, 1597–602.

Schieldge, J. P. (1974). Quiet-time Currents and Electric Fields Produced by the Ionospheric Dynamo. PhD thesis, Los Angeles, University of California.

Schmerling, E. R. & Langille, R. C. (1969). Introduction, 1969 special issue on topside sounding. *Proceedings of the Institute of Electrical and Electronic Engineers*, **57**, 859–60.

Schmidt, G. (1966). *Physics of High Temperature Plasmas*, Academic Press, New York, pp. 211.

Schulz, M. & Koons, H. C. (1972). Thermalization of colliding ion streams beyond the plasmapause. *Journal of Geophysical Research*, **77**, 248–54.

Schunk, R. W. (1975). Transport equations for Aeronomy. *Planetary and Space Science*, **23**, 437–85.

Schunk, R. W. (1977). Mathematical structure of transport equations for multi-species flows.

Reviews of Geophysics and Space Physics, **15**, 429–45.

Schunk, R. W. (1988a). A mathematical model of the middle and high latitude ionosphere. *Pure and Applied Geophysics*, **127**, 255–303.

Schunk, R. W. (1988b). The polar wind, in *Modelling Magnetospheric Plasma*, eds T. E. Moore and J. J. Waiter, Jr., pp. 219–28, AGU, Washington, DC.

Schunk, R. W. & Nagy, A. F. (1980). Ionospheres of the terrestrial planets. *Reviews of Geophysics and Space Physics*, **18**, 813–52.

Schunk, R. W. & Sojka, J. J. (1989). A three-dimensional time-dependent model of the polar wind. *Journal of Geophysical Research*, **94**, 8973–91.

Schunk, R. W. & Sojka, J. J. (1992). Approaches to ionospheric modelling, simulation and prediction. *Advances in Space Research*, **12(6)**, 317–26.

Schunk, R. W. & Sojka, J. J. (1996a). USU model of the global ionosphere. *Solar Terrestrial Energy Program: Handbook of Ionospheric Models*, ed. R. W. Schunk, CASS-USU Logan, UT 84322-4405, pp. 153–73.

Schunk, R. W. & Sojka, J. J. (1996b). Ionospheric models. To appear in *Modern Ionospheric Science*, in press.

Schunk, R. W. & Watkins, D. S. (1979). Comparison of solutions to the thirteen-moment and standard transport equations for low speed thermal proton flows. *Planetary Space Science*, **27**, 433–44.

Schunk, R. W. & Watkins, D. S. (1981). Electron temperature anisotropy in the polar wind. *Journal of Geophysical Research*, **86**, 91–102.

Scudder, J. D. (1992a). On the causes of temperature change in inhomogeneous low-density astrophysical plasmas. *Astrophysical Journal*, **398**, 299–318.

Scudder J. D. (1992b). Why all stars should possess circumstellar temperature inversions. *Astrophysical Journal*, **398**, 319–49.

Sellek, R., Bailey, G. J., Moffett, R. J., Heelis, R. A. & Anderson, P. C. (1991). Effects of large zonal plasma drifts on the subauroral ionosphere. *Journal of Atmospheric and Terrestrial Physics*, **53**, 557–65.

Senior, C. & Blanc, M. (1984). On the control of magnetospheric convection by the spatial distribution of ionospheric conductivities. *Journal of Geophysical Research*, **89**, 261–84.

Senior, C., Fontaine, D., Caudal, G., Alcaydé, D. & Fontanari, J. (1989). Convection electric fields and electrostatic potential over $61° < \Lambda < 72°$ invariant latitude observed with the European incoherent scatter facility. 2. Statistical results. *Annales Geophysicae*, **8**, 257–72.

Senior, C., Sharber, J. R., de la Beaujardiere, O., Heelis, R. A., Evans, D. S., Winningham, J. D., Sugiura, M. & Hoegy, W. R. (1987). E- and F-region study of the evening sector auroral oval: A Chatanika/Dynamics Explorer 2/NOAA 6 comparison. *Journal of Geophysical Research*, **92**, 2477–94.

Serbu, G. P. & Maier, E. J. R. (1966). Low energy electrons measured on IMP 2. *Journal of Geophysical Research*. 71, 3755–66.

Serbu, G. P. & Maier, E. J. R. (1967). Thermal plasma measurements within the magnetosphere. In *Space Research VII*, eds R. L. Smith-Rose, S. A. Bowhill and J. W. King, Vol. 1, pp. 527–34, Amsterdam, North-Holland.

Serbu, G. P. & Maier, E. J. R. (1970). Observations from OGO 5 of the thermal ion density and temperature within the magnetosphere. *Journal of Geophysical Research*, **78**, 6102–13.

Sharma, R. P. & Muldrew, D. B. (1975). Seasonal and longitudinal variations in the occurrence frequency of magnetospheric ionization ducts. *Journal of Geophysical Research*, **80**, 977–84.

Sharp, G. W. (1966). Mid-latitude trough in the night ionosphere. *Journal of Geophysical Research*, **71**, 1345–56.

Shaw, R. R. & Gurnett, D. A. (1975). Electrostatic noise bands associated with the electron gyrofrequency and plasma frequency in the outer magnetosphere. *Journal of Geophysical Research*, **80**, 4259–71.

Shawhan, S. D., Gurnett, D. A., Odem, D. L., Helliwell, R. A. & Park, C. G. (1981). The plasma wave and quasi-static electric field instrument (PWI) for DE-A. *Space Science Instruments*, **5**, 535–54.

Shizgal, B. & Blackmore, R. (1984). A discrete ordinate method of solution of linear and boundary value problems. *Journal of Computational Physics*, **55**, 313–27.

Shizgal, B., Weinert, U. & Lemaire, J. (1986). Collisional kinetic theory of the escape of light ions from the polar wind. In *Proceedings of the 15th International Symposium on rarefied gas dynamics*, Vol. II, edited by Boffi, V. & Cercignani, C., Teubner, Stuttgart, pp. 374–83.

Singh, N. (1988). Refilling of a plasmaspheric flux tube: microscopie plasma processes. In *Modelling Plasma*, Geophys. Monogr. Ser., Vol. 44, eds T. E. Moore and J. H. Waite, Jr., p. 87, AGU, Washington DC.

Singh, N. (1990). Comment on 'Multistream hydrodynamic modeling of interhemispheric plasma flow' by C. E. Rasmussen & R. W. Schunk. *Journal of Geophysical Research*, **95**, 17273–9.

Singh, N. & Chan, C. B. (1992). Effects of equatorally trapped ions on refilling of the plasmasphere. *Journal of Geophysical Research*, **97**, 1167–79.

Singh, N. & Horwitz, J. L. (1992). Plasmaspheric refilling : recent observations and modelling. *Journal of Geophysical Research*, **97**, 1049–79.

Singh, N., Schunk, R. W. & Thiemann, H. (1986). Temporal features of the refilling of a plasmaspheric flux tube. *Journal of Geophysical Research*, **91**, 13433–54.

Singh, N. & Torr, D. G. (1990). Effects of ion temperature anisotropy on the interhemispheric plasma transport during plasmaspheric refilling. *Geophysical Research Letters*, **17**, 925–8.

Singh, N., Wilson, G. R. & Horwitz, J. L. (1994). Comparison of hydrodynamic and semi-kinetic treatments for a plasma flow along closed field lines. *Journal of Geophysical Research*, **99**, 11495–506.

Sivtseva, L. D., Filippov, V. M., Khalipov, V. L., Galperin, Yu. I., Ershova, V. A., Nikolayenko, L. M., Ponomarev, Yu. N. & Sinitsyn, V. M. (1984). Coordinated investigations of processes in the subauroral upper ionosphere and the light-ion concentration trough. *Kosmicheskie Issledovaniya*, **22**, 720–41.

Sivtseva, L. D., Filippov, V. M., Khalipov, V. L., Galperin, Yu. I., Ershova, V. A., Nikolaenko, L. M., Ponomarev, Yu. N. & Sinitsyn, V. M. (1983). Coordinated experiments to study midltitude trough by ground-based geophysical observations and measurements from the AUREOL-1 and AUREOL-2 satellites. *Kosmicheskie Issledovaniya*, **21**, 584–608.

Smiddy, M., Kelley, M. C., Burke, W., Rich, F., Sagalyn, R., Shuman, B., Hays, R. & Lai, S. (1977). Intense poleward directed electric fields near the ionospheric projection of the plasmapause. *Geophysical Research Letters*, **4**, 543–6.

Smilauer, J., Truhlik, V., Boskova, J., Triska, P. & Shultchishin, J. (1996). Observations of thermal

O^{++} ions in the outer ionosphere. *Advances in Space Research*, **17**, (10)135–40.

Smith, A. J. & Carpenter, D. L. (1982). Echoing mixed-path whistlers near the dawn plasmapause, observed by direction-finding receivers at two Antarctic stations. *Journal of Atmospheric and Terrestrial Physics*, **44**, 973–84.

Smith, A. J., Carpenter, D. L. & Inan, U. S. (1985). Whistler-triggered VLF noise bursts observed on the DE–1 satellite and simultaneously at Antarctic ground stations. *Annales Geophysicae*, **3**, 81–8.

Smith, A. J., Carpenter, D. L. & Lester, M. (1981). Longitudinal variations of plasmapause radius and the propagation of VLF noise within small ($\Delta L \sim 0.5$) extensions of the plasmasphere. *Geophysical Research Letters*, **8**, 980–3.

Smith, A. J., Rodger, A. S. & Thomas, D. W. P. (1987). Simultaneous ground-based observations of the plasmapause and the F-region mid-latitude trough. *Journal of Atmospheric and Terrestrial Physics*, **49**, 43–7.

Smith, R. L. (1960). The Use of Nose Whistlers in the Study of the Outer Ionosphere. Ph D thesis, Stanford, CA, Stanford University.

Smith, R. L. (1961a). Propagation characteristics of whistlers trapped in field-aligned columns of enhanced ionization. *Journal of Geophysical Research*, **66**, 3699–707.

Smith, R. L. (1961b). Properties of the outer ionosphere deduced from nose whistlers. *Journal of Geophysical Research*, **66**, 3709–16.

Smith, R. L. & Angerami, J. J. (1968). Magnetospheric properties deduced from OGO-1 observations of ducted and nonducted whistlers. *Journal of Geophysical Research*, **73**, 1–20.

Smith, R. L. & Carpenter, D. L. (1961). Extension of nose whistler analysis. *Journal of Geophysical Research*, **66**, 2582–86.

Smith, R. L. & Helliwell, R. A. (1960). Electron densities to 5 Earth radii deduced from nose whistlers. *Journal of Geophysical Research*, **65**, 2583.

Smith, R. L. & Angerami, J. J. (1968). Magnetospheric properties deduced from OGO 1 observations of ducted and non-ducted whistlers. *Journal of Geophysical Research*, **73**, 1–20.

Soicher, H. (1976). Response of electrons in ionosphere and plasmasphere to magnetic storms. *Nature*, **259**, 33–5.

Sojka, J. J. (1989). Global scale, physical models of the F-region ionosphere. *Reviews of Geophysics*, **27**, 371–403.

Sojka, J. J., Bowline, M., Schunk, R. W., Craven, J. D., Frank, L. A., Sharber, J. R., Winningham, J. D. & Brace, L. H. (1992). Ionospheric simulation compared with Dynamics Explorer observations for November 22, 1981. *Journal of Geophysical Research*, **97**, 1245–56.

Sojka, J. J. & Schunk, R. W. (1985). A theoretical study of the global F-region for June solstice, solar maximum and low magnetic activity. *Journal of Geophysical Research*, **90**, 5285–98.

Sojka, J. J., Schunk, R. W., Johnson, J. F. E., Waite, J. H. & Chappell, C. R. (1983). Characteristics of thermal and suprathermal ions associated with the dayside plasma trough as measured by the Dynamics Explorer retarding ion mass spectrometer. *Journal of Geophysical Research*, **88**, 7895–911.

Sojka, J. J., Rasmussen, C. E. & Schunk, R. W. (1986). An interplanetary magnetic field dependent model of the ionospheric convection electric field. *Journal of Geophyscial Research*, **91**, 11 281–90.

Sojka, J. J. & Wrenn, G. L. (1985). Refilling of geosynchronous flux tubes as observed at the equator by GEOS–2. *Journal of Geophysical Research*, **90**, 6379–85.

Solomon, J., Cornilleau-Wehrlin, N., Korth, A. & Kremser, G. (1988). An experimental study of ELF/VLF hiss generation in the Earth's magnetosphere. *Journal of Geophysical Research*, **93**, 1839–47.

Soloviev, V. S., Galperin, Yu. I., Zinin, L. V., Sivtseva, L. D., Filippov, V. M. & Khalipov, V. L. (1989). The diffuse auroral zone, IX. Equatorial border of diffuse precipitation of the plasmasheet electrons as the boundary of the large-scale convection in the magnetosphere. *Kosmicheskie Issledovaniya*, **27**, 232–47.

Song, X. T., Caudal, G. & Gendrin, R. (1988a). Refilling of the plasmasphere at the geostationary orbit : A Kp-dependent model deduced from GEOS-1 measurements of the cold plasma density. *Advances in Space Research*, **8**(8), 45–8.

Song, X. T., Gendrin, R. & Blanc, M. (1988b). Determination of the Volland convection electric field parameters and computation of the associated field-aligned current distribution. *Planetary and Space Science*, **36**, 631–9.

Sonnerup, B. U. O. & Laird, M. J. (1963). On magnetospheric interchange instability. *Journal of Geophysical Research*, **68**, 131–9.

Sonwalkar, V. S. (1994). Magnetospheric LF, VLF, and ELF waves. In *Handbook of Atmospheric Electrodynamics*, Vol. II, ed. H. Volland, CRC Press, Boca Raton, FL.

Sonwalkar, V. S., Bell, T. F., Helliwell, R. A. & Inan, U. S. (1984). Direct multiple path magnetospheric propagation: a fundamental property of nonducted VLF waves. *Journal of Geophysical Research*, **89**, 2823–30.

Sonwalkar, V. S., Helliwell, R. A. & Inan, U. S. (1990). Wideband VLF electromagnetic bursts observed on the DE 1 satellite. *Geophysical Research Letters*, **17**, 1861–4.

Sonwalkar, V. S. & Inan, U. S. (1988). Wave normal direction and spectral properties of whistler mode hiss observed on the DE 1 satellite. *Journal of Geophysical Research*, **93**, 7493–514.

Sonwalkar, V. S. & Inan, U. S. (1989). Lightning as an embryonic source of VLF hiss. *Journal of Geophysical Research*, **94**, 6986–94.

Southwood, D. J. (1977). The role of hot plasma in magnetospheric convection. *Journal of Geophysical Research*, **82**, 5512–20.

Southwood, D. J. & Hughes, W. J. (1983). Theory of hydromagnetic waves in the magnetosphere. *Space Science Reviews*, **35**, 301–66.

Southwood, D. J. & Kivelson, M. G. (1987). Magnetospheric interchange instability. *Journal of Geophysical Research*, **92**, 109–16.

Southwood, D. J. & Kivelson, M. G. (1989). Magnetospheric interchange motions. *Journal of Geophysical Research*, **94**, 299–308.

Spiro, R. W., Harel, M., Wolf, R. A. & Reiff, P. H. (1981). Quantitative simulation of a magnetospheric substorm. 3. Plasmaspheric electric fields and evolution of the plasmapause, *Journal of Geophysical Research*, **86**, 2261–72.

Spiro, R. W., Heelis, R. A. & Hanson, W. B. (1978). Ion convection and the formation of the mid-latitude F region ionization trough. *Journal of Geophysical Research*, **83**, 4255–64.

Spiro, R. W., Heelis, R. A. & Hanson, W. B. (1979). Rapid subauroral ion drifts observed by Atmospheric Explorer C. *Geophysical Research Letters*, **6**, 657–60.

Spitzer, L., Jr. (1956). *Physics of Fully Ionized Gases*. Interscience, New York, 105 pp.

Stern, D. P. (1977). Large-scale electric fields in the earth's magnetosphere. *Reviews of Geophysics and Space Physics*, **15**, 156–94.

Stern, D. P. & Tsyganenko, N. A. (1992). Uses and limitations of the Tsyganenko magnetic field model. *EOS*, **73**, 489–94.

Storey, L. R. O. (1951). An Investigation of Whistling Atmospherics. PhD thesis, Cambridge, Cambridge University.

Storey, L. R. O. (1953). An investigation of whistling atmospherics. *Philosophical Transactions of the Royal Society (London)* A, **246**, 113–41.

Storey, L. R. O. (1956). A method to detect the presence of ionised hydrogen in the outer atmosphere. *Canadian Journal of Physics*, **34**, 1153–9.

Storey, L. R. O. (1958). Protons outside the earth's atmosphere. *Annales de Géophysique*, **14**, 144–53.

Storey, L. R. O. & Lefeuvre, F. (1979). The analysis of 6-component measurements of a random electromagnetic wave field in a magnetoplasma, I, The direct problem. *Geophysical Journal of Royal Astronomical Society*, **56**, 255–69.

Storey, L. R. O. & Lefeuvre, F. (1980). The analysis of 6-component measurements of a random electromagnetic wave field in a magnetoplasma, II, The integration kernels. *Geophysical Journal of Royal Astronomical Society*, **62**, 173–94.

Storey, L. R. O., Lefeuvre, F., Parrot, M., Cairo, L. & Anderson, R. R. (1991). Initial survey of the wave distribution functions for plasmaspheric hiss observed by ISEE 1. *Journal of Geophysical Research*, **96**, 19 469–89.

Strangeways, H. J. (1991). The upper cut-off frequency of nose whistlers and implications for duct structure. *Journal of Atmospheric and Terrestrial Physics*, **53**, 151–69.

Strangeways, H. J. & Rycroft, M. J. (1980). Trapping of whistler-waves through the side of ducts. *Journal of Atmospheric and Terrestrial Physics*, **42**, 983–94.

Szuszczewicz, E. P., Fejer, B., Roelof, E., Schunk, R. W., Wolf, R., Abdu, M., Bateman, T., Blanchard, P., Emery, B. A., Feldstein, A., Hanbaba, R., Joselyn, J., Kikuchi, T., Leitinger, R., Lester, M., Sobral, J., Reddy, B. M., Richmond, A. D., Sica, R., Walker, G. O. & Wilkinson, P. (1992). Modeling and measurements of global-scale ionospheric behavior under solar minimum, equinoctial conditions. *Advances in Space Research*, **12**(6), 105–15.

Takahashi, K. & McPherron, R. L. (1982). Harmonic structure of Pc 3–4 pulsations.

Journal of Geophysical Research, **87**, 1504–16.

Tam, S. W. Y., Yasseen, F. & Chang, T. (1995). Self-consistent kinetic photoelectron effect on the polar wind. *Geophysical Research Letters*, **22**, 2107–10.

Tanaka, Y. & Hayakawa, M. (1985). On the propagation of daytime whistlers at low latitudes. *Journal of Geophysical Research*, **90**, 3457–64.

Tanskanen, P., Kangas, J., Block, L., Kremser, G., Korth, A., Woch, J., Iversen, I. B., Torkar, K. M., Riedler, W., Ullaland, S., Stadnes, J. and Glassmeier, K.-H. (1987). Different phases of a magnetospheric substorm on June 23, 1979. *Journal of Geophysical Research*, **92**, 7443–57.

Tarcsai, Gy. (1975). Routine whistler analysis by means of accurate curve fitting. *Journal of Atmospheric and Terrestrial Physics*, **37**, 1447–57.

Tarcsai, Gy. (1985). Ionosphere–plasmasphere electron fluxes at middle latitudes obtained from whistlers. *Advances in Space Research*, **5**(4), 155–8.

Tarcsai, Gy., Szemerédy, P. & Hegymegi, L. (1988). Average electron density profiles in the plasmasphere between L = 1.4 and 3.2 deduced from whistlers. *Journal of Atmospheric and Terrestrial Physics*, **50**, 607–11.

Taylor, H. A., Jr. (1971). Evidence of solar geomagnetic seasonal control of the topside ionosphere. *Planetary and Space Science*, **19**, 77–93.

Taylor, H. A., Jr. (1972a). Observed solar geomagnetic control of the ionosphere: implications for the reference ionospheres. *Space Research*, **12**, 1275–90.

Taylor, H. A., Jr. (1972b). The light ion trough. *Planetary and Space Science*, **20**, 1593–1605.

Taylor, H. A, Jr., Brinton, H. C., Carpenter, D. L., Bonner, F. M. & Heyborne, R. L. (1969). Ion depletion in the high-latitude exosphere. Simultaneous OGO 2 observations of the light ion trough and the VLF cutoff. *Journal of Geophysical Research*, **74**, 3517–28.

Taylor, H. A. Jr., Brinton, H. C. & Pharo, M. W., III (1968a). Contraction of the plasmasphere during geomagnetically disturbed periods. *Journal of Geophysical Research*, **73**, 961–8.

Taylor, H. A. Jr., Brinton, H. C., Pharo, M. W., III & Rahman, N. K. (1968b). Thermal ions in the exosphere; evidence of solar and geomagnetic control. *Journal of Geophysical Research*, **73**, 5521–33.

Taylor, H. A., Brinton, H. C. & Smith, C. R. (1965a). Preliminary results from measurements of hydrogen and helium ions below 50 000 km. Paper presented at AGU meeting, Washington DC, April 19–22.

Taylor, H. A., Brinton, H. C. & Smith, C. R. (1965b). Positive ion composition in the magnetosphere obtained from OGO-A satellite. *Journal of Geophysical Research*, **70**, 5769–81.

Taylor, H. A., Jr. & Cordier, G. R. (1974). In-situ observations of irregular ionospheric structure associated with the plasmapause. *Planetary and Space Science*, **22**, 1289–96.

Taylor, H. A., Jr., Mayr, H. G. & Brinton, H. C. (1970). Observations of hydrogen and helium ions during a period of rising solar activity. *Space Research*, **10**, 663–78.

Taylor, H. A., Jr. & Walsh, W. J. (1972). The light ion trough, the main trough, and the plasmapause. *Journal of Geophysical Research*, **77**, 6716–23.

Temerin, M., Cerny, K., Lotko, W. & Mozer, F. S. (1982). Observations of double layers and solitary waves in the auroral plasma. *Physical Review Letters*, **48**, 1175.

Thomas, J. O., & Sader, A. Y. (1964). The electron density at the ALOUETTE orbit. *Journal of Geophysical Research*, **69**, 4561–81.

Thomson, R. J. (1978). The formation and lifetime of whistler ducts. *Planetary and Space Science*, **26**, 423–30

Thomson, R. J. & Dowden, R. L. (1977). Simultaneous ground and satellite reception of whistlers. 1. Ducted whistlers. *Journal of Atmospheric and Terrestrial Physics*, **39**, 869–77.

Thorne, R. M., Church, S. R. & Gorney, D. J. (1979). On the origin of plasmaspheric hiss: The importance of wave propagation and the plasmapause. *Journal of Geophysical Research*, **84**, 5241–7.

Thorne, R. M., Smith, E. J., Burton, R. K. & Holzer, R. E. (1973), Plasmaspheric hiss. *Journal of Geophysical Research*, **78**, 1581–96.

Titheridge, J. E. (1976a). Ion transition heights from topside electron density profiles. *Planetary and Space Science*, **24**, 229–45.

Titheridge, J. E. (1976b). Plasmapause effects in the topside ionosphere. *Journal of Geophysical Research*, **81**, 3227–33.

Titova, E. E., Di, V. I., Yurov, V. E., Raspopov, O. M., Trakhtengerts, V. Yu., Jiricek, F. & Triska, P. (1984). Interaction between VLF waves and the turbulent ionosphere. *Geophysical Research Letters*, **11**, 323–6.

Torr, M. R. & Torr, D. G. (1982). The role of metastable species in the thermosphere. *Reviews on Geophysics and Space Physics*, **20**, 91–144.

Tserkovnikov, Y. A. (1960). The question of convectional instability of a plasma. *Soviet Physics Doklady* (Engl. Translation), **5**, 87–90.

Tsyganenko, N. A. (1987). Global quantitative models of the geomagnetic field in the cislunar magnetosphere for different disturbance levels. *Planetary and Space Science*, **35**, 1347–58.

Tsyganenko, N. A. (1989). A magnetospheric magnetic field model with a warped tail current sheet. *Planetary and Space Science*, **37**, 5–20.

Valchuk, T. E., Galperin, Yu. I., Nikolayenko, L. M., Feldstein, Ya. I., Bosqued, J.-M., Sauvaud, J. A. & Crasnier, J. (1986). The diffuse auroral zone. VIII. Equatorial boundary of the zone of diffuse precipitation of auroral electrons in the morning sector. *Kosmicheskie Issledovaniya*, **24**, 875–83.

Van Allen, J. A. (1959). The geomagnetically trapped corpuscular radiation. *Journal of Geophysical Research*, **64**, 1683–89.

Van Allen, J. A. (1960). The first public lecture on the discovery of the geomagnetically trapped radiation, State University of Iowa Rept., p. 60.

Vasyliunas, V. M. (1970). Mathematical models of magnetospheric convection and its coupling to the ionosphere, in *Particles and Fields in the Magnetosphere*, ed B. M. McCormac, D. Reidel, Dordrecht, Holland, pp. 60–71.

Vasyliunas, V. M. (1972). The interrelationship of magnetospheric processes. In *Earth's Magnetospheric Processes*, ed. B. M. McCormac, D. Reidel, pp. 29–38.

Vernov, S. N. & Chudakov, A. E. (1960). Terrestrial corpuscular radiation and cosmic rays. *Space Research*, **1**, 751–96.

Vickrey, J. F., Swartz, W. E. & Farley, D. T. (1979). Post-sunset observations of ionospheric-protonospheric coupling at Arecibo. *Journal of Geophysical Research*, **84**, 1310–4.

Vinas, A. F. & Madden, T. R. (1986). Shear flow-ballooning instability as a possible mechanism for hydromagnetic fluctuations. *Journal of Geophysical Research*, **91**, 1519–28.

Volland, H. A. (1973). A semi-empirical model of large-scale magnetospheric electric fields.

Journal of Geophysical Research, **78**, 171–80.

Volland, H. A. (1975). Models of global electric fields within the magnetosphere. *Annales de Géophysique*, **31**, 159–73.

Volland, H. A. (1978). A model of the magnetospheric electric convection field. *Journal of Geophysical Research*, **83**, 2695–9.

Voss, H. D., Imhof, W. L., Walt, M., Mobilia, J., Gaines, E. E., J. B., Inan, U. S., Helliwell, R. A., Carpenter, D. L., Katsufrakis, J. P. & Chang, H. C. (1984). Lightning-induced electron precipitation. *Nature*, **312**, 740–2.

Waite, J. H., Jr., Nagai, T., Johnson, J. F. E., Chappell, C. R., Burch, J. L., Killeen, T. L., Hays, P. B., Carignan, G. R., Peterson, W. K. & Shelley, E. G. (1985). Escape of suprathermal O^+ ions in the polar cap. *Journal of Geophysical Research*, **90**, 1619–30.

Watanabe, S., Oyama, K.-I & Abe, T. (1989). Electron temperature structure around mid-latitude ionospheric trough. *Planetary and Space Science*, **37**, 1453–60.

Watanabe, S., Whalen, B. A. & Yau, A. W. (1992). Thermal ion observations of depletion and refilling in the plasmaspheric trough. *Journal of Geophysical Research*, **97**, 1081–96.

Weiss, L. A., Lambour, R. L., Elphic, R. C. & Thomsen M. F. (1997). Study of plasmaspheric evolution using geosynchronous observations and global modeling. *Geophysical Research Letter*, **24**, 599–602.

Whipple, E. C. (1981). Potential of surfaces in space. *Reports on Progress in Physics*, **44**, 1197.

Williams, D. J., Roelof, E. C. & Mitchell, D. G. (1992). Global magnetospheric imaging. *Reviews of Geophysics*, **30**, 183–208.

Wilson, G. R. (1992). Semikinetic modeling of the outflow of ionospheric plasma through the topside collisional to collisionless transition region. *Journal of Geophysical Research*, **97**, 10 551–65.

Wilson, G. R., Horwitz, J. L. & Lin, J. (1992). A semi kinetic model for early stage plasmasphere refilling : 1. Effects of Coulomb collisions. *Journal of Geophysical Research*, **97**, 1109–19.

Wilson, G. R., Horwitz, J. L. & Lin, J. (1993). A semi-kinetic modelling of plasma flow on outer plasmaspheric field lines. *Advances in Space Research*, **13**(4), 107–16.

Wolf, R. A. (1970). Effects of ionospheric conductivity on convective flow of plasma in the magnetosphere. *Journal of Geophysical*

Research, **75**, 4677–98.

Wolf, R. A. (1983). The quasi-static (slow-flow) region of the magnetosphere. In *Solar-Terrestrial Physics, Principles*, eds R. L. Carovillano and J. M. Forbes, Reidel, Dordrecht, Holland, pp. 303–68.

Wolf, R. A. (1995). Magnetospheric configuration, in M. G. Kivelson, and C. T. Russell, (eds.), *Introduction to Space Physics*, Cambridge University Press, Cambridge, pp. 288–329.

Wrenn, G. L., Johnson, J. F. E. & Sojka, J. J. (1979). Stable 'pancake' distributions of low energy electrons in the plasma trough. *Nature*, **279**, 512–4.

Wrenn, G. L., Sojka, J. J. & Johnson, J. F. E. (1984). Thermal protons in the morning magnetosphere: filling and heating near the equatorial plasmapause. *Planetary and Space Science*, **32**, 351–63.

Yabroff, I. (1961). Computation of whistler ray paths. *Journal of Research*, National Bureau of Standards, **65D**, 485–505.

Yang, Y. S., Wolf, R. A., Spiro, R. A. & Dessler, A. J. (1992). Numerical simulation of plasma transport driven by the Io torus. *Journal of Geophysical Research*, **19**, 957–60.

Yasseen, F., Retterer, J. M., Chang, T. & Winningham, J. D. (1989). Monte-Carlo modeling of polar wind photoelectron distributions with anomalous heat flux. *Geophysical Research Letters*, **16**, 1023–6.

Yeh, H.-C. & Flaherty, B. J. (1966). Ionospheric electron content at temperate latitudes during the declining phase of the sunspot cycle. *Journal of Geophysical Research*, **71**, 4557–70.

Yeh, H.-C. & Foster, J. C.. (1990). Storm-time heavy ion outflow at mid-latitude. *Journal of Geophysical Research*, **95**, 7881–91.

Yeh, H.-C., Foster, J. C., Rich, F. J. & Swider, W. (1991). Storm time electric field penetration observed at mid-latitude. *Journal of Geophysical Research*, **96**, 5707–21.

Yeoman, T. K. & Orr, D. (1989). Phase and spectral power of mid-latitude Pi2 pulsations: evidence for a plasmaspheric cavity resonance. *Planetary and Space Science*, **37**, 1367–83.

Yoshida, S. & Hatanaka, T. (1962). The disturbances of exosphere as seen from the VLF emissions. *Journal of the Physical Society of Japan*, **17**, (supplement A-11), 78–83.

Young, D. T., Geiss, J., Balsiger, H., Eberhardt, P., Ghielmetti, A. & Rosenbauer H. (1977).

Discovery of He^{++} and O^{++} ions of terrestrial origin in the outer magnetosphere. *Geophysical Research Letters*, **4**, 561–4.

Young, E. R., Torr, D. G., Richards, P. & Nagy, A. F. (1980). A computer simulation of the midlatitude plasmasphere and ionosphere. *Planetary Space Science*, **28**, 881–93.

Zinin, L. V., Galperin, Yu. I., Latyshev, K. S. & Grigoriev, S. A. (1985). Non-stationary field-aligned fluxes of thermal ions O^+ and H^+ outside the plasmapause: a refinement of the polar wind theory, Results of the ARCAD 3 PROJECT and of the recent programmes in magnetospheric and ionospheric physics, Trans. of International Symposium, Toulouse, May, 1984, ed. by CNES, CEPADUES-EDITIONS, Toulouse, 391–408.

Index

Printed in the United States
By Bookmasters